Undergraduate Texts in Mathematics

Editors
J.H. Ewing
F.W. Gehring
P.R. Halmos

Undergraduate Texts in Mathematics

Apostol: Introduction to Analytic Number Theory. Second edition.
Armstrong: Groups and Symmetry.
Armstrong: Basic Topology.
Bak/Newman: Complex Analysis.
Banchoff/Wermer: Linear Algebra Through Geometry. Second edition.
Brémaud: An Introduction to Probabilistic Modeling.
Bressoud: Factorization and Primality Testing.
Bressoud: Second Year Calculus.
Readings in Mathematics.
Brickman: Mathematical Introduction to Linear Programming and Game Theory.
Cederberg: A Course in Modern Geometries.
Childs: A Concrete Introduction to Higher Algebra.
Chung: Elementary Probability Theory with Stochastic Processes. Third edition.
Cox/Little/O'Shea: Ideals, Varieties, and Algorithms.
Curtis: Linear Algebra: An Introductory Approach. Fourth edition.
Devlin: The Joy of Sets: Fundamentals of Contemporary Set Theory. Second edition.
Dixmier: General Topology.
Driver: Why Math?
Ebbinghaus/Flum/Thomas: Mathematical Logic.
Edgar: Measure, Topology, and Fractal Geometry.
Fischer: Intermediate Real Analysis.
Flanigan/Kazdan: Calculus Two: Linear and Nonlinear Functions. Second edition.
Fleming: Functions of Several Variables. Second edition.
Foulds: Optimization Techniques: An Introduction.
Foulds: Combinatorial Optimization for Undergraduates.
Franklin: Methods of Mathematical Economics.
Halmos: Finite-Dimensional Vector Spaces. Second edition.
Halmos: Naive Set Theory.
Hämmerlin/Hoffmann: Numerical Mathematics.
Readings in Mathematics.
Iooss/Joseph: Elementary Stability and Bifurcation Theory. Second edition.
James: Topological and Uniform Spaces.
Jänich: Topology.
Klambauer: Aspects of Calculus.
Kinsey: Topology of Surfaces
Lang: A First Course in Calculus. Fifth edition.
Lang: Calculus of Several Variables. Third edition.
Lang: Introduction to Linear Algebra. Second edition.
Lang: Linear Algebra. Third edition.
Lang: Undergraduate Algebra. Second edition.
Lang: Undergraduate Analysis.
Lax/Burstein/Lax: Calculus with Applications and Computing. Volume 1.
LeCuyer: College Mathematics with APL.
Lidl/Pilz: Applied Abstract Algebra.
Macki-Strauss: Introduction to Optimal Control Theory.

(continued after index)

Thomas Banchoff
John Wermer

Linear Algebra
Through Geometry

Second Edition

1992

With 92 Illustrations

Springer-Verlag
New York Berlin Heidelberg London Paris
Tokyo Hong Kong Barcelona Budapest

Thomas Banchoff
John Wermer
Department of Mathematics
Brown University
Providence, RI 02912
USA

QA
184
B36
1992

Library of Congress Cataloging-in-Publication Data
Banchoff, Thomas.
   Linear algebra through geometry, 2nd ed./Thomas Banchoff, John Wermer.
      p.   cm.—(Undergraduate texts in mathematics)
   Includes bibliographical references and index.
   ISBN 0-387-97586-1
   1. Algebras, Linear.   I. Wermer, John.   II. Title.   III. Series.
QA184.B36   1991
512′.5—dc20   91-18083

Printed on acid-free paper.

Typeset by Asco Trade Typesetting Ltd., Hong Kong.
Printed and bound by Braun-Brumfield, Ann Arbor, MI.
Printed in the United States of America.

9 8 7 6 5 4 3 2 (Corrected second printing, 1993)

ISBN 0-387-97586-1 Springer-Verlag New York Berlin Heidelberg
ISBN 3-540-97586-1 Springer-Verlag Berlin Heidelberg New York

To our wives Lynore and Kerstin

Preface to the Second Edition

In this book we lead the student to an understanding of elementary linear algebra by emphasizing the geometric significance of the subject.

Our experience in teaching undergraduates over the years has convinced us that students learn the new ideas of linear algebra best when these ideas are grounded in the familiar geometry of two and three dimensions. Many important notions of linear algebra already occur in these dimensions in a non-trivial way, and a student with a confident grasp of the ideas will encounter little difficulty in extending them to higher dimensions and to more abstract algebraic systems. Moreover, we feel that this geometric approach provides a solid basis for the linear algebra needed in engineering, physics, biology, and chemistry, as well as in economics and statistics.

The great advantage of beginning with a thorough study of the linear algebra of the plane is that students are introduced quickly to the most important new concepts while they are still on the familiar ground of two-dimensional geometry. In short order, the student sees and uses the notions of dot product, linear transformations, determinants, eigenvalues, and quadratic forms. This is done in Chapters 2.0–2.7.

Then, the very same outline is used in Chapters 3.0–3.7 to present the linear algebra of three-dimensional space, so that the former ideas are reinforced while new concepts are being introduced.

In Chapters 4.0–4.2, we deal with geometry in $\mathbb{R}^n$ for $n \geq 4$. We introduce linear transformations and matrices in $\mathbb{R}^4$, and we point out that the step from $\mathbb{R}^4$ to $\mathbb{R}^n$ with $n > 4$ is now almost immediate. In Chapters 4.3 and 4.4, we treat systems of linear equations in n variables.

In the present edition, we have added Chapter 5 on vector spaces, Chapter 6 on inner products on a vector space, and Chapter 7 on symmetric $n \times n$ matrices and quadratic forms in n variables. Finally, in Chapter 8 we deal with three applications:

(1) differential systems, that is, systems of linear first-order differential equations;
(2) least-squares method in data analysis; and
(3) curvature of surfaces in $\mathbb{R}^3$, which are given as graphs of functions of two variables.

Except for Chapter 8, the student need only know basic high-school algebra and geometry and introductory trigonometry in order to read this book. In fact, we believe that high-school seniors who are interested in mathematics could read much of this book on their own. To read Chapter 8, students should have a knowledge of elementary calculus.

Acknowledgments

We would like to thank the many students in our classes, whose interest and suggestions have helped in the development of this book. We particularly thank Curtis Hendrickson and Davide Cervone, who produced the computer-generated illustrations in this new edition. Our special thanks go to Dale Cavanaugh, Natalie Johnson, and Carol Oliveira of the Brown University Mathematics Department for their assistance.

Contents

CHAPTER 1.0

Vectors in the Line

Analytic geometry begins with the line. Every point on the line has a real number as its coordinate and every real number is the coordinate of exactly one point. A *vector in the line* is a directed line segment from the origin to a point with coordinate x. We denote this vector by a single capital letter $\mathbf{X}$. The collection of all vectors in the line is denoted by $\mathbb{R}^1$.

We add two vectors by adding their coordinates, so if $\mathbf{U}$ has coordinate u, then $\mathbf{X} + \mathbf{U}$ has coordinate $x + u$. To multiply a vector $\mathbf{X}$ by a real number r, we multiply the coordinate by r, so the coordinate of $r\mathbf{X}$ is rx. The vector with coordinate zero is denoted by $\mathbf{0}$. (See Fig. 1.1.)

The familiar properties of real numbers then lead to corresponding properties for vectors in 1-space. For any vectors $\mathbf{X}, \mathbf{U}, \mathbf{W}$ and any real numbers r and s we have:

$\mathbf{X} + \mathbf{U} = \mathbf{U} + \mathbf{X}$.
$(\mathbf{X} + \mathbf{U}) + \mathbf{W} = \mathbf{X} + (\mathbf{U} + \mathbf{W})$.
For all $\mathbf{X}$, $\mathbf{0} + \mathbf{X} = \mathbf{X} = \mathbf{X} + \mathbf{0}$.
For any $\mathbf{X}$, there is a vector $-\mathbf{X}$ such that $\mathbf{X} + (-\mathbf{X}) = \mathbf{0}$.
$r(\mathbf{X} + \mathbf{U}) = r\mathbf{X} + r\mathbf{U}$
$(r + s)\mathbf{X} = r\mathbf{X} + s\mathbf{X}$
$r(s\mathbf{X}) = (rs)\mathbf{X}$
$1\mathbf{X} = \mathbf{X}$

We can define the length of a vector $\mathbf{X}$ with coordinate x as the absolute value of x, i.e., the distance from the point labelled x to the origin. We denote this length by $|\mathbf{X}|$ and we may write $|\mathbf{X}| = \sqrt{x^2}$. (We always understand this symbol to stand for the non-negative square root.) Then $\mathbf{0}$ is the

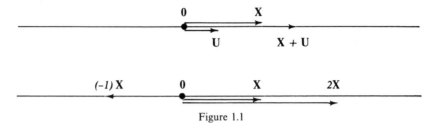

Figure 1.1

unique vector of the length 0 and there are just two vectors with length 1, with coordinates 1 and -1.

CHAPTER 2.0

The Geometry of Vectors in the Plane

Many of the familiar theorems of plane geometry appear in a new light when we rephrase them in the language of *vectors*. This is particularly true for theorems which are usually expressed in the language of analytic or coordinate geometry, because vector notation enables us to use a single symbol to refer to a pair of numbers which gives the coordinates of a point. Not only does this give us convenient notations for expressing important results, but it also allows us to concentrate on algebraic properties of vectors, and these enable us to apply the techniques used in plane geometry to study problems in space, in higher dimensions, and also in situations from calculus and differential equations which at first have little resemblance to plane geometry. Thus, we begin our study of linear algebra with the study of the geometry of vectors in the plane.

§1. The Algebra of Vectors

In vector geometry we define a *vector* in the plane as a pair of numbers $\begin{pmatrix} x \\ y \end{pmatrix}$ written in column form, with the *first coordinate* x written above the *second coordinate* y. We designate this vector by a single capital letter $\mathbf{X}$, i.e., we write $\mathbf{X} = \begin{pmatrix} x \\ y \end{pmatrix}$. We can picture the vector $\mathbf{X}$ as an arrow, or directed line segment, starting at the origin in the coordinate plane and ending at the point with coordinates x and y. We illustrate the vectors $\mathbf{A} = \begin{pmatrix} 3 \\ 1 \end{pmatrix}$, $\mathbf{B} = \begin{pmatrix} 1 \\ 2 \end{pmatrix}$, $\mathbf{C} = \begin{pmatrix} 4 \\ 3 \end{pmatrix}$, and $\mathbf{D} = \begin{pmatrix} 2 \\ 4 \end{pmatrix}$ in Figure 2.1.

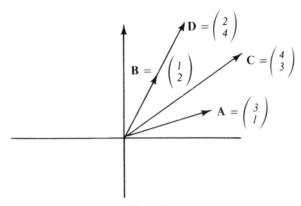

Figure 2.1

We *add* two vectors by adding their components, so if $\mathbf{X} = \begin{pmatrix} x \\ y \end{pmatrix}$ and $\mathbf{U} = \begin{pmatrix} u \\ v \end{pmatrix}$, we have

$$\mathbf{X} + \mathbf{U} = \begin{pmatrix} x + u \\ y + v \end{pmatrix}. \tag{1}$$

Thus, in the above diagram, we have $\mathbf{A} + \mathbf{B} = \mathbf{C}$, since

$$\mathbf{A} + \mathbf{B} = \begin{pmatrix} 3 \\ 1 \end{pmatrix} + \begin{pmatrix} 1 \\ 2 \end{pmatrix} = \begin{pmatrix} 3 + 1 \\ 1 + 2 \end{pmatrix} = \begin{pmatrix} 4 \\ 3 \end{pmatrix} = \mathbf{C}.$$

We *multiply* a vector $\mathbf{X}$ by a number r by multiplying each coordinate of $\mathbf{X}$ by r, i.e.,

$$r\mathbf{X} = r\begin{pmatrix} x \\ y \end{pmatrix} = \begin{pmatrix} rx \\ ry \end{pmatrix}. \tag{2}$$

In Fig. 2.1, $\mathbf{D} = \begin{pmatrix} 2 \\ 4 \end{pmatrix} = 2\begin{pmatrix} 1 \\ 2 \end{pmatrix} = 2\mathbf{B}$, and we also have $\mathbf{B} = \frac{1}{2}\mathbf{D}$.

Multiplying by a number r *scales* the vector $\mathbf{X}$ giving a longer vector $r\mathbf{X}$ if $r > 1$ and a shorter vector $r\mathbf{X}$ if $0 < r < 1$. Such multiplication of a vector by a number is called *scalar multiplication*, and the number r is called a *scalar*. If $r = 1$, then the result is the vector itself, so $1\mathbf{X} = \mathbf{X}$. If $r = 0$, then multiplication of any vector by $r = 0$ yields the *zero* vector $\begin{pmatrix} 0 \\ 0 \end{pmatrix}$, denoted by $\mathbf{0} = \begin{pmatrix} 0 \\ 0 \end{pmatrix}$. If $\mathbf{X}$ is not the zero vector, then the scalar multiples of $\mathbf{X}$ all lie on a line through the origin and the point at the endpoint of the arrow representing $\mathbf{X}$. We call this line the *line along* $\mathbf{X}$. If $r > 0$, we get the points on the ray from $\begin{pmatrix} 0 \\ 0 \end{pmatrix}$ through $\begin{pmatrix} x \\ y \end{pmatrix}$, while if $r < 0$, we get the points on the opposite ray. In particular, if $r = -1$, we get the vector $(-1)\mathbf{X} = (-1)\begin{pmatrix} x \\ y \end{pmatrix} = \begin{pmatrix} -x \\ -y \end{pmatrix}$ which has the same length as $\mathbf{X}$ but the opposite

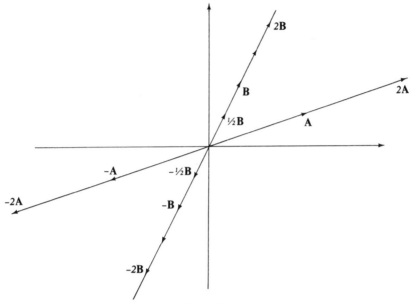

Figure 2.2

direction. We denote this vector by $-\mathbf{X} = \begin{pmatrix} -x \\ -y \end{pmatrix}$ and we note that

$$\mathbf{X} + (-\mathbf{X}) = \begin{pmatrix} x \\ y \end{pmatrix} + \begin{pmatrix} -x \\ -y \end{pmatrix} = \begin{pmatrix} x + (-x) \\ y + (-y) \end{pmatrix} = \begin{pmatrix} 0 \\ 0 \end{pmatrix} = \mathbf{0}. \qquad (3)$$

We say that the vector $-\mathbf{X}$ is the *negative* of $\mathbf{X}$ or the *additive inverse* of $\mathbf{X}$.

In Figure 2.2, we indicate some scalar multiples of the vectors $\mathbf{A} = \begin{pmatrix} 3 \\ 1 \end{pmatrix}$ and $\mathbf{B} = \begin{pmatrix} 1 \\ 2 \end{pmatrix}$.

Two particularly important vectors are $\mathbf{E}_1 = \begin{pmatrix} 1 \\ 0 \end{pmatrix}$ and $\mathbf{E}_2 = \begin{pmatrix} 0 \\ 1 \end{pmatrix}$, which we call the *basis vectors* of the plane. The collection of all scalar multiples $r\mathbf{E}_1 = r\begin{pmatrix} 1 \\ 0 \end{pmatrix} = \begin{pmatrix} r \\ 0 \end{pmatrix}$ of $\mathbf{E}_1$ then gives the *first coordinate axis*, and the *second coordinate axis* is given similarly by $s\mathbf{E}_2 = s\begin{pmatrix} 0 \\ 1 \end{pmatrix} = \begin{pmatrix} 0 \\ s \end{pmatrix}$. Since $\mathbf{X} = \begin{pmatrix} x \\ y \end{pmatrix}$
$= \begin{pmatrix} x \\ 0 \end{pmatrix} + \begin{pmatrix} 0 \\ y \end{pmatrix} = x\begin{pmatrix} 1 \\ 0 \end{pmatrix} + y\begin{pmatrix} 0 \\ 1 \end{pmatrix} = x\mathbf{E}_1 + y\mathbf{E}_2$, we may express any vector $\mathbf{X}$ uniquely as a sum of one vector from the first coordinate axis and one vector from the second coordinate axis. Thus,

$$\mathbf{A} = \begin{pmatrix} 3 \\ 1 \end{pmatrix} = \begin{pmatrix} 3 \\ 0 \end{pmatrix} + \begin{pmatrix} 0 \\ 1 \end{pmatrix} = 3\begin{pmatrix} 1 \\ 0 \end{pmatrix} + 1\begin{pmatrix} 0 \\ 1 \end{pmatrix} = 3\mathbf{E}_1 + \mathbf{E}_2,$$

and, similarly, $\mathbf{D} = \begin{pmatrix} 2 \\ 4 \end{pmatrix} = 2\mathbf{E}_i + 4\mathbf{E}_2$.

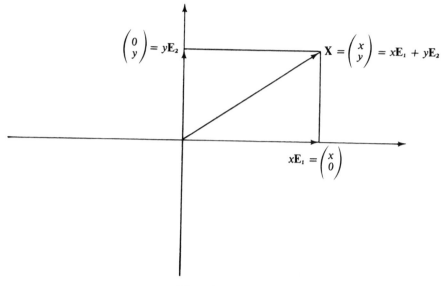

Figure 2.3

Writing a vector in this way expresses the point $\begin{pmatrix} x \\ y \end{pmatrix}$ as the fourth vertex of a rectangle whose other three coordinates are $\begin{pmatrix} x \\ 0 \end{pmatrix}$, $\begin{pmatrix} 0 \\ 0 \end{pmatrix}$, and $\begin{pmatrix} 0 \\ y \end{pmatrix}$. (See Fig. 2.3.)

More generally, we may obtain a geometric interpretation of vector addition as follows. If we start with the triangle with vertices $\begin{pmatrix} 0 \\ 0 \end{pmatrix}$, $\begin{pmatrix} x \\ 0 \end{pmatrix}$, $\begin{pmatrix} x \\ y \end{pmatrix}$

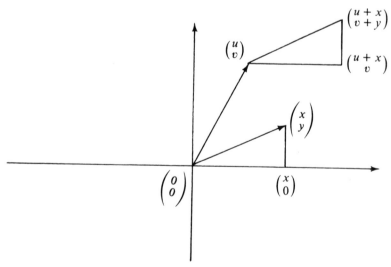

Figure 2.4

and move it parallel to itself so that its first vertex lies on $\left(\begin{smallmatrix} u \\ v \end{smallmatrix}\right)$, then the other two vertices lie on $\left(\begin{smallmatrix} u + x \\ v \end{smallmatrix}\right)$ and $\left(\begin{smallmatrix} u + x \\ v + y \end{smallmatrix}\right)$, respectively. (See Fig. 2.4.) Thus, the sum of the vectors $\mathbf{X} = \left(\begin{smallmatrix} x \\ y \end{smallmatrix}\right)$ and $\mathbf{U} = \left(\begin{smallmatrix} u \\ v \end{smallmatrix}\right)$ can be obtained by translating the directed segment from $\mathbf{0}$ to $\mathbf{X}$ parallel to itself until its beginning point lies at $\mathbf{U}$. The new endpoint will represent $\mathbf{U} + \mathbf{X}$, and this will be the fourth coordinate of a parallelogram with $\mathbf{U}$, $\mathbf{0}$, and $\mathbf{X}$ as the other three vertices.

In our diagrams we have pictured addition of a vector $\mathbf{X}$ with positive coordinates, but a similar argument shows that the parallelogram interpretation is still valid if one or both coordinates are negative or zero.

By referring either to the coordinate description or the geometric description, we can establish the following algebraic properties of vector addition and scalar multiplication which are analogous to familiar facts about arithmetic of numbers:

(4)	$\mathbf{X} + \mathbf{U} = \mathbf{U} + \mathbf{X}$.	Commutative law for vectors
(5)	$(\mathbf{X} + \mathbf{U}) + \mathbf{A} = \mathbf{X} + (\mathbf{U} + \mathbf{A})$.	Associative law for vectors
(6)	There is a vector $\mathbf{0}$ such that $\mathbf{X} + \mathbf{0} = \mathbf{X} = \mathbf{0} + \mathbf{X}$ for all $\mathbf{X}$.	Additive identity
(7)	For any $\mathbf{X}$ there is a vector $-\mathbf{X}$ such that $\mathbf{X} + (-\mathbf{X}) = \mathbf{0}$.	Additive inverse
(8)	$r(\mathbf{X} + \mathbf{U}) = r\mathbf{X} + r\mathbf{U}$.	Distributive law for vectors
(9)	$(r + s)(\mathbf{X}) = r\mathbf{X} + s\mathbf{X}$.	Distributive law for scalars
(10)	$r(s\mathbf{X}) = (rs)\mathbf{X}$.	Associative law for scalars
(11)	$1 \cdot \mathbf{X} = \mathbf{X}$ for each $\mathbf{X}$.	

Note that it is possible for the parallelogram to collapse to a doubly covered line segment if we add two multiples of the same vector. In Fig. 2.5, we show the parallelograms for $\mathbf{B} + \mathbf{B}$, $\mathbf{A} + \mathbf{B}$, and $\mathbf{A} + (-\mathbf{A})$.

We can use the negative of a vector to help define the notion of *difference* $\mathbf{U} - \mathbf{X}$ of the vectors $\mathbf{X}$ and $\mathbf{U}$. (See Fig. 2.6.) We define

$$\mathbf{U} - \mathbf{X} = \mathbf{U} + (-\mathbf{X}),$$

so, in coordinates,

$$\left(\begin{smallmatrix} u \\ v \end{smallmatrix}\right) - \left(\begin{smallmatrix} x \\ y \end{smallmatrix}\right) = \mathbf{U} - \mathbf{X} = \mathbf{U} + (-\mathbf{X}) = \left(\begin{smallmatrix} u \\ v \end{smallmatrix}\right) + \left(\begin{smallmatrix} -x \\ -y \end{smallmatrix}\right) = \left(\begin{smallmatrix} u - x \\ v - y \end{smallmatrix}\right).$$

Since $(\mathbf{U} - \mathbf{X}) + \mathbf{X} = \mathbf{U} + ((-\mathbf{X}) + \mathbf{X}) = \mathbf{U} + \mathbf{0} = \mathbf{U}$, we see that $\mathbf{U} - \mathbf{X}$ is the vector we add to $\mathbf{X}$ to get $\mathbf{U}$. Thus, if we move $\mathbf{U} - \mathbf{X}$ parallel to itself until its beginning point lies on $\mathbf{X}$, we get the directed line segment from $\mathbf{X}$ to $\mathbf{U}$. Thus,

$$\mathbf{A} - \mathbf{B} = \left(\begin{smallmatrix} 3 \\ 1 \end{smallmatrix}\right) - \left(\begin{smallmatrix} 1 \\ 2 \end{smallmatrix}\right) = \left(\begin{smallmatrix} 2 \\ -1 \end{smallmatrix}\right) \quad \text{and} \quad \mathbf{B} - \mathbf{A} = \left(\begin{smallmatrix} 1 \\ 2 \end{smallmatrix}\right) - \left(\begin{smallmatrix} 3 \\ 1 \end{smallmatrix}\right) = \left(\begin{smallmatrix} -2 \\ 1 \end{smallmatrix}\right).$$

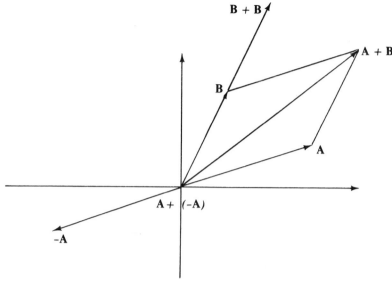

Figure 2.5

A pair of vectors $\mathbf{A}, \mathbf{B}$ is said to be *linearly dependent* if one of them is a multiple of the other. If $\mathbf{A} = \mathbf{0}$, then the pair $\mathbf{A}, \mathbf{B}$ is linearly dependent, since $\mathbf{0} = 0 \cdot \mathbf{B}$ no matter what $\mathbf{B}$ is. If $\mathbf{A} \neq \mathbf{0}$ and the pair $\mathbf{A}, \mathbf{B}$ is linearly dependent, then $\mathbf{B} = t\mathbf{A}$ for some t. If $\mathbf{B} = \mathbf{0}$, then we use $t = 0$, but if $\mathbf{A}$ and $\mathbf{B}$ are both nonzero, we have $\mathbf{B} = t\mathbf{A}$ and $(1/t)\mathbf{B} = \mathbf{A}$, so each of the vectors is a multiple of the other.

If $\mathbf{A}, \mathbf{B}$ is a linearly dependent pair of vectors and both $\mathbf{A}$ and $\mathbf{B}$ are nonzero, then the vectors $r\mathbf{A}$ for different values of r all lie on a line through the origin. The fact that $\mathbf{A}, \mathbf{B}$ is a linearly dependent pair means that $\mathbf{B}$ lies on the line determined by $\mathbf{A}$.

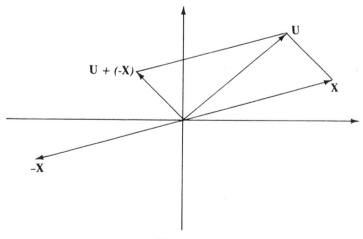

Figure 2.6

Exercise 1. For which choice of x will the following pairs be linearly dependent?

(a) $\left(\binom{x}{1}, \binom{4}{2} \right)$,   (b) $\left(\binom{x}{x^2}, \binom{-3}{9} \right)$,

(c) $\left(\binom{x}{1}, \binom{9}{x} \right)$,   (d) $\left(\binom{x}{x^2}, \binom{1}{x} \right)$.

Exercise 2. True of false? If A is a scalar multiple of B, then B is a multiple of A.

Exercise 3. True or false? If A is a nonzero scalar multiple of B, then B is a nonzero scalar multiple of A.

Just as the multiples tX of a nonzero vector X give a description of a line through the origin, we may describe a line through a point U parallel to the vector X by taking the sum of U and all multiples of X. The line is then given by $U + tX$ for all real t. (See Fig. 2.7.)

For example, the line through $B = \binom{2}{1}$ parallel to the vector $A = \binom{3}{1}$ is given by $X = B + tA = \binom{2}{1} + t\binom{3}{1} = \binom{2}{1} + \binom{3t}{t} = \binom{2 + 3t}{1 + t}$. This is called the *parametric representation* of a line in the plane, since the coordinates $x = 2 + 3t$ and $y = 1 + t$ are given linear functions of the *parameter* t. Similarly, the line given by the parametric equation in coordinates $\binom{x}{y} = \binom{3 + 4t}{1 + 2t}$ can be written in vector form as $X = \binom{3}{1} + \binom{4t}{2t} = \binom{3}{1} + t\binom{4}{2} = A + tD$.

Exercise 4. Write an equation of the line through $\binom{1}{4}$ parallel to the vector $\binom{2}{2}$.

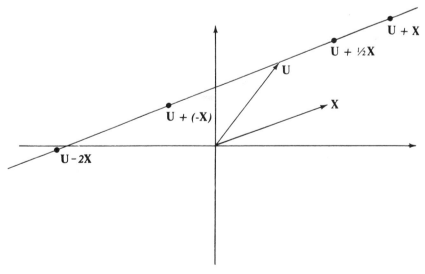

Figure 2.7

Exercise 5. Write an equation for the line through $\mathbf{A} = \begin{pmatrix} 3 \\ 1 \end{pmatrix}$ and $\mathbf{B} = \begin{pmatrix} 1 \\ 2 \end{pmatrix}$. *Hint:* This line will go through $\mathbf{B}$ and be parallel to the vector $\mathbf{B} - \mathbf{A}$.

Exercise 6. Show that the parametric equation $\mathbf{X} = \mathbf{V} + t(\mathbf{U} - \mathbf{V})$ represents the line through $\mathbf{U}$ and $\mathbf{V}$ if $\mathbf{U}$ and $\mathbf{V}$ are any two vectors which are not equal.

By the Pythagorean Theorem, the distance from a point $\begin{pmatrix} x \\ y \end{pmatrix}$ to the origin $\begin{pmatrix} 0 \\ 0 \end{pmatrix}$ is $\sqrt{x^2 + y^2}$, and we define this number to be the *length* of the vector $\mathbf{X} = \begin{pmatrix} x \\ y \end{pmatrix}$, written $|\mathbf{X}|$. For example, if $\mathbf{X} = \begin{pmatrix} 3 \\ 4 \end{pmatrix}$, then $|\mathbf{X}| = \sqrt{3^2 + 4^2}$ $= 5$, while $|\mathbf{E}_1| = \left| \begin{pmatrix} 1 \\ 0 \end{pmatrix} \right| = 1$ and $|\mathbf{0}| = \sqrt{0^2 + 0^2} = 0$. Since the square root is always considered to be positive or zero, the length of a vector is never negative, and in fact $|\mathbf{X}|$ is positive unless $\mathbf{X} = \begin{pmatrix} 0 \\ 0 \end{pmatrix}$.

EXAMPLE 1. $|\mathbf{X} - \mathbf{U}| = \left| \begin{pmatrix} x \\ y \end{pmatrix} - \begin{pmatrix} u \\ v \end{pmatrix} \right| = \left| \begin{pmatrix} x - u \\ y - v \end{pmatrix} \right| = \sqrt{(x - u)^2 + (y - v)^2}$.

For any scalar r, we have

$$|r\mathbf{X}| = \left| \begin{pmatrix} rx \\ ry \end{pmatrix} \right| = \sqrt{(rx)^2 + (ry)^2} = \sqrt{r^2 x^2 + r^2 y^2} = |r|\sqrt{x^2 + y^2} = |r||\mathbf{X}|.$$

Thus, the length of a scalar multiple of a vector is the length of the vector multiplied by the absolute value of the scalar. For example, $|-5\mathbf{X}| = |-5||\mathbf{X}| = 5|\mathbf{X}|$.

Exercise 7. Show that the midpoint of the segment joining points $\mathbf{X}$ and $\mathbf{U}$ is $\frac{1}{2}(\mathbf{X} + \mathbf{U})$, i.e. show that this point lies on the line through $\mathbf{X}$ and $\mathbf{U}$ and is equidistant from $\mathbf{X}$ and $\mathbf{U}$.

If $\mathbf{X} \neq \begin{pmatrix} 0 \\ 0 \end{pmatrix}$, then $\mathbf{X} \neq \mathbf{0}$, so we may scale by the reciprocal $(1/|\mathbf{X}|)$ to get a vector $(1/|\mathbf{X}|)\mathbf{X}$. This vector lies along the ray from $\mathbf{0}$ to $\mathbf{X}$ and it has length equal to 1 since

$$\left| \left(\frac{1}{|\mathbf{X}|} \right) \mathbf{X} \right| = \left| \frac{1}{|\mathbf{X}|} \right| |\mathbf{X}| = \frac{1}{|\mathbf{X}|} |\mathbf{X}| = 1.$$

The vectors of length 1 are called *unit vectors*, and they are represented by the points on the unit circle in the coordinate plane. The vector $(1/|\mathbf{X}|)\mathbf{X}$ is represented by the point where the ray from $\mathbf{0}$ to $\mathbf{X}$ intersects this unit circle. (See Fig. 2.8).

Any vector on the unit circle may be described by its angle θ from the ray along the positive x-axis to the ray along the unit vector. We call θ the *polar angle* of the vector. We may then write the unit vector using trigonometric functions as $\begin{pmatrix} \cos\theta \\ \sin\theta \end{pmatrix}$.

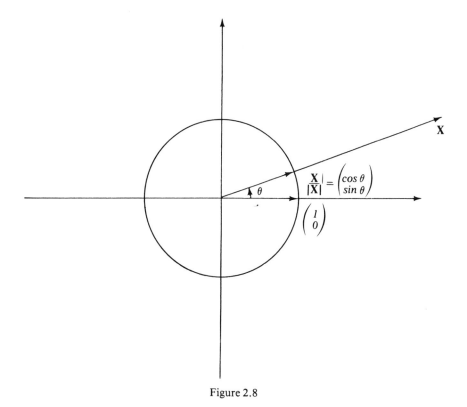

Figure 2.8

If **X** is any vector, we have

$$\mathbf{X} = |\mathbf{X}|\left(\frac{1}{|\mathbf{X}|}\mathbf{X}\right) = |\mathbf{X}|\begin{pmatrix} \cos\theta \\ \sin\theta \end{pmatrix} = \begin{pmatrix} |\mathbf{X}|\cos\theta \\ |\mathbf{X}|\sin\theta \end{pmatrix}$$

for some angle θ. This representation of X as a positive scalar multiple of a unit vector is called the *polar form* of the vector **X**, since we have written the coordinates of **X** in the form of polar coordinates.

EXAMPLE 2. If $\mathbf{X} = \begin{pmatrix} 3 \\ 0 \end{pmatrix}$, we have $\mathbf{X} = 3\mathbf{E}_1$, where $\mathbf{E}_1$ is the unit vector $\begin{pmatrix} 1 \\ 0 \end{pmatrix} = \begin{pmatrix} \cos(0) \\ \sin(0) \end{pmatrix}$. If $\mathbf{X} = \begin{pmatrix} 1 \\ 1 \end{pmatrix}$, then $\mathbf{X} = \sqrt{2}\begin{bmatrix} 1/\sqrt{2} \\ 1/\sqrt{2} \end{bmatrix} = \sqrt{2}\begin{pmatrix} \cos\theta \\ \sin\theta \end{pmatrix}$, where $\theta = 45° = \pi/4$.

§2. The Dot Product

An extremely useful notion in linear algebra is the *dot product* of two vectors **X** and **U** defined by

$$\mathbf{X} \cdot \mathbf{U} = \begin{pmatrix} x \\ y \end{pmatrix} \cdot \begin{pmatrix} u \\ v \end{pmatrix} = xu + yv. \tag{12}$$

The dot product of two vectors is then a number formed by adding the product of their first coordinates to the product of their second coordinates. For example, if $\mathbf{A} = \begin{pmatrix} 3 \\ 1 \end{pmatrix}$ and $\mathbf{B} = \begin{pmatrix} 1 \\ 2 \end{pmatrix}$ then $\mathbf{A} \cdot \mathbf{B} = 3 \cdot 1 + 1 \cdot 2 = 5$ and $\mathbf{A} \cdot \mathbf{A} = 3 \cdot 3 + 1 \cdot 1 = 10$. Note that $\mathbf{E}_1 \cdot \mathbf{E}_1 = 1 \cdot 1 + 0 \cdot 0 = 1$ while $\mathbf{E}_1 \cdot \mathbf{E}_2 = \begin{pmatrix} 1 \\ 0 \end{pmatrix} \cdot \begin{pmatrix} 0 \\ 1 \end{pmatrix} = 1 \cdot 0 + 0 \cdot 1 = 0$.

In general, $\mathbf{X} \cdot \mathbf{X} = \begin{pmatrix} x \\ y \end{pmatrix} \cdot \begin{pmatrix} x \\ y \end{pmatrix} = x^2 + y^2 = |\mathbf{X}|^2$, so the length of any vector is the square root of the dot product of the vector with itself. We therefore have $\mathbf{X} \cdot \mathbf{X} \geqslant 0$ for all $\mathbf{X}$, with equality if and only if $\mathbf{X} = \mathbf{0}$.

The dot product has certain algebraic properties that are similar to properties of ordinary multiplication of numbers:

(13) $\mathbf{X} \cdot \mathbf{U} = \mathbf{U} \cdot \mathbf{X}$ Commutative law for dot product

(14) $(r\mathbf{X}) \cdot \mathbf{U} = r(\mathbf{X} \cdot \mathbf{U})$ Associative law for scalar and dot product

(15) $\mathbf{A} \cdot (\mathbf{X} + \mathbf{U}) = \mathbf{A} \cdot \mathbf{X} + \mathbf{A} \cdot \mathbf{U}$ Distributive law for dot product

Each of these properties can be established easily by referring to the coordinate definition. For example, if $\mathbf{A} = \begin{pmatrix} a \\ b \end{pmatrix}$, $\mathbf{X} = \begin{pmatrix} x \\ y \end{pmatrix}$, and $\mathbf{U} = \begin{pmatrix} u \\ v \end{pmatrix}$, we have

$$\mathbf{A} \cdot (\mathbf{X} + \mathbf{U}) = \begin{pmatrix} a \\ b \end{pmatrix} \cdot \begin{pmatrix} x + u \\ y + v \end{pmatrix} = a(x + u) + b(y + v)$$

$$= (ax + by) + (au + bv) = \begin{pmatrix} a \\ b \end{pmatrix} \cdot \begin{pmatrix} x \\ y \end{pmatrix} + \begin{pmatrix} a \\ b \end{pmatrix} \cdot \begin{pmatrix} u \\ v \end{pmatrix}$$

$$= \mathbf{A} \cdot \mathbf{X} + \mathbf{A} \cdot \mathbf{U}.$$

In ordinary multiplication of real numbers, the product ax equals zero only if either $a = 0$ or $x = 0$. Note, however, that it is possible for the dot product of two vectors to be zero even if neither vector is equal to zero. For example, $\begin{pmatrix} 2 \\ 2 \end{pmatrix} \cdot \begin{pmatrix} -1 \\ 1 \end{pmatrix} = 2(-1) + 2 \cdot 1 = 0$. (See Fig. 2.9.)

Exercise 8. Show that if $r\mathbf{X} = \mathbf{0}$, then either $r = 0$ or $\mathbf{X} = \mathbf{0}$.

We may ask under which circumstances the dot product of two nonzero vectors $\mathbf{X}$ and $\mathbf{U}$ will be zero, i.e., when do we have $\mathbf{X} \cdot \mathbf{U} = \begin{pmatrix} x \\ y \end{pmatrix} \cdot \begin{pmatrix} u \\ v \end{pmatrix}$ $= xu + yv = 0$? One possibility is that one vector lies in the first coordinate axis and the other lies in the second coordinate axis, in which case the two vectors are perpendicular. If $\mathbf{X}$ does not lie in either coordinate axis, then $x \neq 0$ and $y \neq 0$. The slope of the line from the origin through $\begin{pmatrix} x \\ y \end{pmatrix}$ is y/x, and this is not equal to zero. Since $xu + yv = 0$, it follows that $yv = -ux$. If $u = 0$, then $v = 0$ as well. If $u \neq 0$, then $ux \neq 0$, so $-1 = yv/ux = (y/x)(v/u)$. Thus, either $\mathbf{U} = \mathbf{0}$ or the line from the origin to $\mathbf{U} = \begin{pmatrix} u \\ v \end{pmatrix}$ has

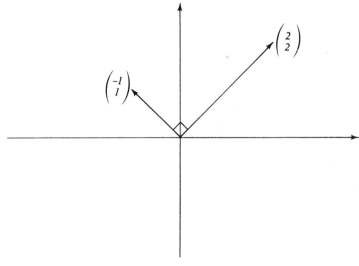

Figure 2.9

slope v/u equal to the negative reciprocal of y/x, the slope of the line from the origin to **X**. Therefore, these two lines must be perpendicular. It follows that in every case if the dot product of two nonzero vectors is zero, then the two vectors are perpendicular.

Retracing our steps, we easily see that, conversely, if **X** and **U** are any two perpendicular vectors, then $\mathbf{X} \cdot \mathbf{U} = 0$.

Exercise 9. Show that for any vector $\begin{pmatrix} x \\ y \end{pmatrix}$, we have $\begin{pmatrix} x \\ y \end{pmatrix}$ perpendicular to the vector $\begin{pmatrix} -y \\ x \end{pmatrix}$.

Note that the line with equation

$$ax + by = 0$$

may be described in two equivalent ways:

(i) The set of vectors $\mathbf{X} = \begin{pmatrix} x \\ y \end{pmatrix}$ which are perpendicular to the vector $\begin{pmatrix} a \\ b \end{pmatrix}$.

(ii) The line along the vector $\begin{pmatrix} -b \\ a \end{pmatrix}$ (by Exercise 9, the vector $\begin{pmatrix} -b \\ a \end{pmatrix}$ is perpendicular to $\begin{pmatrix} a \\ b \end{pmatrix}$).

Exercise 10. Find a vector **U** such that the line with equation $5x + 2y = 0$ lies along **U**.

Exercise 11. Find an equation of the form $ax + by = 0$ for the line along the vector $\begin{pmatrix} 3 \\ 1 \end{pmatrix}$.

Exercise 12. Find a vector **U** such that the line with equation $y = 2x$ lies along **U**.

For any vector $\mathbf{X} = \begin{pmatrix} x \\ y \end{pmatrix}$, we have $\mathbf{X} \cdot \mathbf{E}_1 = \begin{pmatrix} x \\ y \end{pmatrix} \cdot \begin{pmatrix} 1 \\ 0 \end{pmatrix} = (x \cdot 1) +$ $(y \cdot 0) = x$. Thus, the dot product of $\mathbf{X}$ with the unit vector $\mathbf{E}_1$ is the coordinate of the *projection of $\mathbf{X}$ to the first coordinate axis*. Similarly, the dot product $\mathbf{X} \cdot \mathbf{E}_2 = \begin{pmatrix} x \\ y \end{pmatrix} \cdot \begin{pmatrix} 0 \\ 1 \end{pmatrix} = y$ of the vector $\mathbf{X}$ with the unit vector $\mathbf{E}_2$ is the coordinate of the *projection of $\mathbf{X}$ to the second coordinate axis*.

More generally, if we have any unit vector $\mathbf{W} = \begin{pmatrix} \cos\phi \\ \sin\phi \end{pmatrix}$, we may use the polar form of the vector $\mathbf{X} = |\mathbf{X}| \begin{pmatrix} \cos\theta \\ \sin\theta \end{pmatrix}$ to get a geometric interpretation of the dot product of $\mathbf{X}$ and $\mathbf{W}$ (see Fig. 2.10). We have

$$\mathbf{X} \cdot \mathbf{W} = \left(|\mathbf{X}| \begin{pmatrix} \cos\theta \\ \sin\theta \end{pmatrix} \right) \cdot \begin{pmatrix} \cos\phi \\ \sin\phi \end{pmatrix} = |\mathbf{X}| \left(\begin{pmatrix} \cos\theta \\ \sin\theta \end{pmatrix} \cdot \begin{pmatrix} \cos\phi \\ \sin\phi \end{pmatrix} \right)$$

$$= |\mathbf{X}|(\cos\theta \cos\phi + \sin\theta \sin\phi).$$

There is a basic trigonometric identity that states that

$$\cos\theta \cos\phi + \sin\theta \sin\phi = \cos(\theta - \phi) = \cos(\phi - \theta), \qquad (16)$$

so we have $\mathbf{X} \cdot \mathbf{W} = |\mathbf{X}| \cdot \cos(\theta - \phi)$. Therefore, the dot product of a vector $\mathbf{X}$ with a unit vector $\mathbf{W}$ is the product of the length of $\mathbf{X}$ and the cosine of the angle between $\mathbf{X}$ and $\mathbf{W}$. If this angle $(\theta - \phi)$ is between 0 and $\pi/2$,

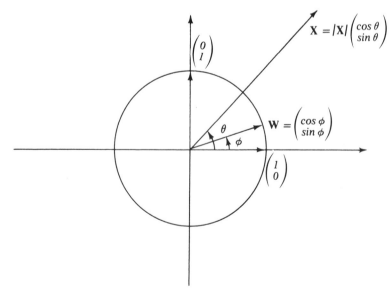

Figure 2.10

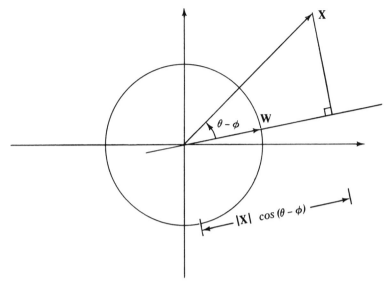

Figure 2.11

then this number $|\mathbf{X}|\cos(\theta - \phi)$ is the length of the adjacent side when the hypotenuse is $|\mathbf{X}|$. Thus, if $\mathbf{X}$ is a vector which makes an acute angle with the unit vector $\mathbf{W}$, then the dot product $\mathbf{X} \cdot \mathbf{W}$ is the *length of the projection of $\mathbf{X}$ to the line from the origin through $\mathbf{W}$* (see Fig. 2.11).

If the angle between $\mathbf{X}$ and $\mathbf{W}$ is greater than $\pi/2$, then $\cos(\theta - \phi)$ is negative and the dot product $\mathbf{X} \cdot \mathbf{W}$ is a negative number. The projection of $\mathbf{X}$ to the line from $\mathbf{0}$ through $\mathbf{W}$ will lie on the ray opposite the ray from $\mathbf{0}$ through $\mathbf{W}$, and the length of this projection is the absolute value of the dot product of $\mathbf{X}$ and $\mathbf{W}$. (See Fig. 2.12). In all cases, then, we can say that the dot product $\mathbf{X} \cdot \mathbf{W}$ represents the coordinate of the projection of the vector $\mathbf{X}$ to the directed line from the origin through the unit vector $\mathbf{W}$.

In general, if we take the dot product of two nonzero vectors in polar form $\mathbf{X} = |\mathbf{X}|\begin{pmatrix} \cos\theta \\ \sin\theta \end{pmatrix}$ and $\mathbf{U} = |\mathbf{U}|\begin{pmatrix} \cos\phi \\ \sin\phi \end{pmatrix}$, we get

$$\mathbf{X} \cdot \mathbf{U} = |\mathbf{X}|\begin{pmatrix} \cos\theta \\ \sin\theta \end{pmatrix} \cdot |\mathbf{U}|\begin{pmatrix} \cos\phi \\ \sin\phi \end{pmatrix} = |\mathbf{X}||\mathbf{U}|\begin{pmatrix} \cos\theta \\ \sin\theta \end{pmatrix} \cdot \begin{pmatrix} \cos\phi \\ \sin\phi \end{pmatrix} = |\mathbf{X}||\mathbf{U}|\cos(\theta - \phi).$$

(17)

Thus, the dot product of two nonzero vectors is the product of their lengths multiplied by the cosine of the angle between them.

We may use the dot product to calculate the angle between two nonzero vectors just by writing

$$\cos(\phi - \theta) = \frac{\mathbf{X} \cdot \mathbf{U}}{|\mathbf{X}||\mathbf{U}|} = \frac{xu + yv}{\sqrt{x^2 + y^2}\sqrt{u^2 + v^2}}.$$

(18)

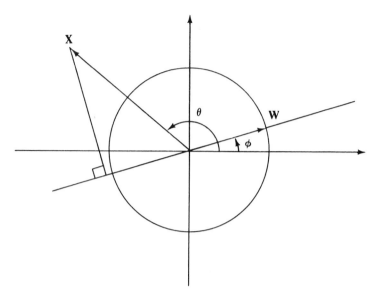

Figure 2.12

For example, if $\mathbf{A} = \begin{pmatrix} 3 \\ 1 \end{pmatrix}$, and $\mathbf{B} = \begin{pmatrix} 1 \\ 2 \end{pmatrix}$, then $|\mathbf{A}| = \sqrt{10}$, $|\mathbf{B}| = \sqrt{5}$, and $\mathbf{A} \cdot \mathbf{B} = 5$. Thus,

$$\cos(\phi - \theta) = \frac{5}{\sqrt{5} \cdot \sqrt{10}} = \frac{1}{\sqrt{2}}$$

and $\theta - \phi = \pi/4$.

Exercise 13. Find the angle between $\begin{pmatrix} 2 \\ 2 \end{pmatrix}$ and $\begin{pmatrix} 0 \\ 3 \end{pmatrix}$.

Exercise 14. Find the angle between $\begin{pmatrix} 3 \\ 1 \end{pmatrix}$ and $\begin{pmatrix} -1 \\ -2 \end{pmatrix}$.

Similarly, using the trigonometric relation

$$\sin(\phi - \theta) = \sin\phi \cos\theta - \cos\phi \sin\theta, \tag{19}$$

we obtain an expression for $\sin(\phi - \theta)$. Setting $\mathbf{X} = \begin{pmatrix} x \\ y \end{pmatrix}$, $\mathbf{U} = \begin{pmatrix} u \\ v \end{pmatrix}$, we have

$$\cos\theta = \frac{x}{\sqrt{x^2 + y^2}}, \qquad \sin\theta = \frac{y}{\sqrt{x^2 + y^2}}, \qquad \cos\phi = \frac{u}{\sqrt{u^2 + v^2}}$$

and

$$\sin\phi = \frac{v}{\sqrt{u^2 + v^2}},$$

and so

$$\sin(\phi - \theta) = \frac{vx - uy}{\sqrt{x^2 + y^2}\,\sqrt{u^2 + v^2}}. \tag{20}$$

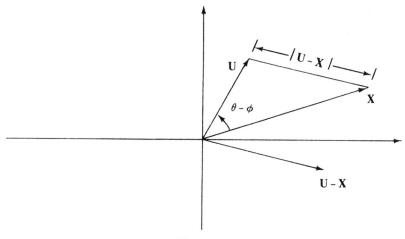

Figure 2.13

Note that the sine of the angle from $\begin{pmatrix} u \\ v \end{pmatrix}$ to $\begin{pmatrix} x \\ y \end{pmatrix}$ is the opposite of the sine of the angle from $\begin{pmatrix} x \\ y \end{pmatrix}$ to $\begin{pmatrix} u \\ v \end{pmatrix}$. We will return to this formula in Chapter 2.5.

If we apply this notion of dot product to the difference of two vectors, we obtain an important result from trigonometry. We calculate the square of the length of the segment from $\mathbf{X} = |\mathbf{X}|\begin{pmatrix} \cos\theta \\ \sin\theta \end{pmatrix}$ to $\mathbf{U} = |\mathbf{U}|\begin{pmatrix} \cos\phi \\ \sin\phi \end{pmatrix}$ (see Fig. 2.13):

$$|\mathbf{U} - \mathbf{X}|^2 = (\mathbf{U} - \mathbf{X}) \cdot (\mathbf{U} - \mathbf{X}) = \mathbf{U} \cdot \mathbf{U} - \mathbf{X} \cdot \mathbf{U} - \mathbf{U} \cdot \mathbf{X} + \mathbf{X} \cdot \mathbf{X}$$

$$= |\mathbf{U}|^2 - 2\mathbf{U} \cdot \mathbf{X} + |\mathbf{X}|^2 = |\mathbf{U}|^2 + |\mathbf{X}|^2 - 2|\mathbf{U}||\mathbf{X}|\cos(\theta - \phi). \quad (21)$$

Thus, the square of the length of one side of a triangle is the sum of the squares of the lengths of the other two sides minus twice the product of those lengths and the cosine of the angle between them. This result is known in trigonometry as the *law of cosines*.

In particular, if the vectors $\mathbf{U}$ and $\mathbf{X}$ are perpendicular, so that the angle between them is $\theta - \phi = \pi/2$, then $|\mathbf{U} - \mathbf{X}|^2 = |\mathbf{U}|^2 + |\mathbf{X}|^2$, so $\mathbf{X} \cdot \mathbf{U} = |\mathbf{X}||\mathbf{U}|\cos(\theta - \phi) = 0$. We thus have another proof of the result that two nonzero vectors are *perpendicular* if and only if $\mathbf{X} \cdot \mathbf{U} = 0$. In linear algebra, we use the convention that the zero vector is perpendicular to every vector, and we frequently use the synonym *orthogonal* instead of perpendicular. We may thus say that two vectors $\mathbf{X}$ and $\mathbf{U}$ are *orthogonal* if and only if $\mathbf{X} \cdot \mathbf{U} = 0$.

We use the notion of dot product to solve some geometric problems which will be crucial in our further development of linear algebra:

(i) To find the projection of a given vector to a given line through the origin.

(ii) To compute the distance from a given point to the line through the origin with equation $ax + by = 0$.

(iii) To calculate the area of a parallelogram with one vertex at the origin.

(iv) To give a geometric interpretation of a system of two linear equations in two unknowns (where both lines go through the origin).

(i) We already know that if $\mathbf{W}$ is a unit vector, then the dot product of $\mathbf{X}$ and $\mathbf{W}$ represents the coordinate of the projection of the point $\mathbf{X}$ to the line from the origin through $\mathbf{W}$. We set $P_\mathbf{W}(\mathbf{X})$ (read "P sub $\mathbf{W}$ of $\mathbf{X}$") equal to the *vector* on this line which is the projection of $\mathbf{X}$ to the line. Thus, $P_\mathbf{W}(\mathbf{X}) = (\mathbf{X} \cdot \mathbf{W})\mathbf{W}$.

If $\mathbf{U}$ is an arbitrary vector, then we can find a formula for the projection $P_\mathbf{U}(\mathbf{X})$ of $\mathbf{X}$ to the line from the origin along $\mathbf{U}$ by using the above formula to find the projection of $\mathbf{X}$ to the line from the origin through the unit vector $\mathbf{U}/|\mathbf{U}|$, i.e.,

$$P_\mathbf{U}(\mathbf{X}) = \left(\mathbf{X} \cdot \frac{\mathbf{U}}{|\mathbf{U}|}\right)\frac{\mathbf{U}}{|\mathbf{U}|} = \frac{(\mathbf{X} \cdot \mathbf{U})\mathbf{U}}{|\mathbf{U}|^2} = \frac{(\mathbf{X} \cdot \mathbf{U})\mathbf{U}}{(\mathbf{U} \cdot \mathbf{U})} . \qquad (22)$$

Hence the length of the projection of $\mathbf{X}$ to the line along $\mathbf{U}$ is given by

$$\left|\mathbf{X} \cdot \frac{\mathbf{U}}{|\mathbf{U}|}\right| = \frac{|\mathbf{X} \cdot \mathbf{U}|}{|\mathbf{U}|} \qquad (23)$$

Alternatively, we may try to find the projection of $\mathbf{X}$ to the line along the nonzero vector $\mathbf{U}$ by finding a scalar t such that $\mathbf{X} - t\mathbf{U}$ is orthogonal to $\mathbf{U}$. (See Fig. 2.14). In terms of the dot product, we obtain

$$0 = (\mathbf{X} - t\mathbf{U}) \cdot \mathbf{U} = (\mathbf{X} \cdot \mathbf{U}) - t(\mathbf{U} \cdot \mathbf{U}),$$

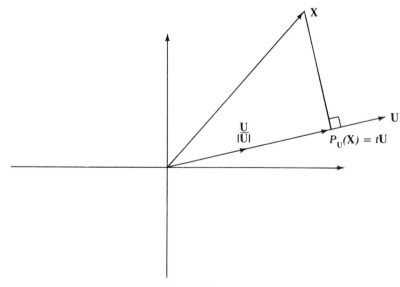

Figure 2.14

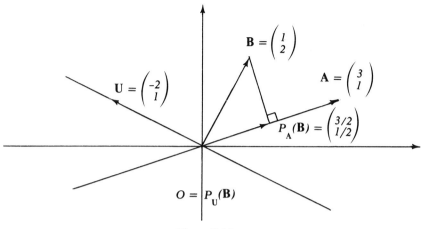

Figure 2.15

so $t = (\mathbf{X} \cdot \mathbf{U})/(\mathbf{U} \cdot \mathbf{U})$ and the projection of $\mathbf{X}$ to the line along $\mathbf{U}$ is

$$P_{\mathbf{U}}(\mathbf{X}) = t\mathbf{U} = \left(\frac{\mathbf{X} \cdot \mathbf{U}}{\mathbf{U} \cdot \mathbf{U}} \right)\mathbf{U}$$

which agrees with formula (22).

EXAMPLE 3. To find the projection of $\mathbf{B} = \begin{pmatrix} 1 \\ 2 \end{pmatrix}$ to the line along $\mathbf{A} = \begin{pmatrix} 3 \\ 1 \end{pmatrix}$, we have (see Fig. 2.15)

$$P_{\mathbf{A}}(\mathbf{B}) = \left(\frac{\mathbf{A} \cdot \mathbf{B}}{\mathbf{A} \cdot \mathbf{A}} \right)\mathbf{A} = \frac{5}{10}\mathbf{A} = \frac{1}{2}\mathbf{A} = \begin{bmatrix} \frac{3}{2} \\ \frac{1}{2} \end{bmatrix}.$$

To find the projection of $\mathbf{B} = \begin{pmatrix} 1 \\ 2 \end{pmatrix}$ to the line along $\mathbf{U} = \begin{pmatrix} -2 \\ 1 \end{pmatrix}$, we have

$$P_{\mathbf{U}}(\mathbf{B}) = \left(\frac{\mathbf{B} \cdot \mathbf{U}}{\mathbf{U} \cdot \mathbf{U}} \right)\mathbf{U} = \left(\frac{0}{2} \right)\mathbf{U} = 0.$$

(This fits with our intuition that if $\mathbf{X}$ is orthogonal to $\mathbf{U}$, the projection of $\mathbf{X}$ to the line along $\mathbf{U}$ will be the origin itself.)

Exercise 15. Find the projection of $\begin{pmatrix} -1 \\ 2 \end{pmatrix}$ to the line along $\begin{pmatrix} 3 \\ 1 \end{pmatrix}$.

Exercise 16. Find the projection of $\begin{pmatrix} 3 \\ 1 \end{pmatrix}$ to the line along $\begin{pmatrix} 1 \\ 2 \end{pmatrix}$.

Exercise 17. Find the projection of $\begin{pmatrix} 3 \\ 1 \end{pmatrix}$ to the line along $\begin{pmatrix} -2 \\ 6 \end{pmatrix}$.

Exercise 18. Find the distance from $\begin{pmatrix} 3 \\ 1 \end{pmatrix}$ to the line along $\begin{pmatrix} 1 \\ 2 \end{pmatrix}$.

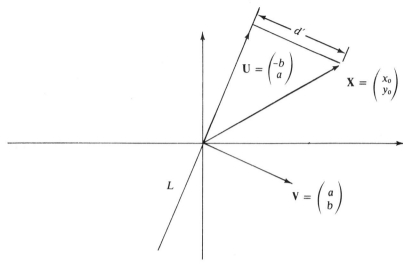

Figure 2.16

(ii) Now we want to find the distance d' from a point $\begin{pmatrix} x_0 \\ y_0 \end{pmatrix}$ to a line L with equation $ax + by = 0$, where a and b are not both 0 (see Fig. 2.16). The vector $\mathbf{U} = \begin{pmatrix} -b \\ a \end{pmatrix}$ is a nonzero vector on this line and the vector $\mathbf{V} = \begin{pmatrix} a \\ b \end{pmatrix}$ is a non-zero vector perpendicular to this line. The distance from $\mathbf{X} = \begin{pmatrix} x_0 \\ y_0 \end{pmatrix}$ to the line L is then given by the length of the projection of $\mathbf{X}$ to the line along $\mathbf{V}$. By (23) we obtain

$$d' = \frac{|\mathbf{X} \cdot \mathbf{V}|}{|\mathbf{V}|} = \frac{\left| \begin{pmatrix} x_0 \\ y_0 \end{pmatrix} \cdot \begin{pmatrix} a \\ b \end{pmatrix} \right|}{\left| \begin{pmatrix} a \\ b \end{pmatrix} \right|} = \frac{|ax_0 + by_0|}{\sqrt{a^2 + b^2}} \tag{24}$$

If $\begin{pmatrix} x_0 \\ y_0 \end{pmatrix}$ is a point on L, then the expression $ax_0 + by_0 = 0$, so by (24), $d' = 0$, as we expected.

Exercise 19. Find the distance from $\begin{pmatrix} 3 \\ 1 \end{pmatrix}$ to the line $y = 2x$.

Exercise 20. Verify that the sum of the squares of the distances from a point $\mathbf{X}$ to the perpendicular lines $ax + by = 0$ and $bx - ay = 0$ is equal to the square of the length of $\mathbf{X}$.

(iii) Once we have the formula for the distance from a point to the line along a given vector, it is an easy matter to find a formula for the area of a parallelogram with one vertex at the origin. If the other three vertices are $\mathbf{A} = \begin{pmatrix} a \\ c \end{pmatrix}$, $\mathbf{B} = \begin{pmatrix} b \\ d \end{pmatrix}$, and $\mathbf{A} + \mathbf{B} = \begin{pmatrix} a + b \\ c + d \end{pmatrix}$, then the distance from $\mathbf{A}$ to the

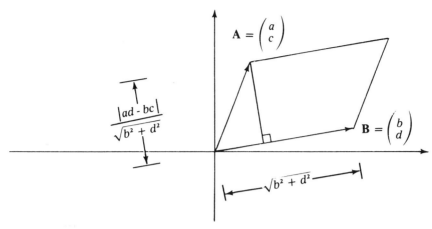

Figure 2.17

line along **B** is given by formula (24) by the expression

$$\frac{|ad - bc|}{\sqrt{b^2 + d^2}}.$$

Multiplying this distance by the length $\sqrt{b^2 + d^2}$ of the base **B**, we get the formula

$$|ad - bc| = \text{area of the parallelogram with sides } \begin{pmatrix} a \\ c \end{pmatrix} \text{ and } \begin{pmatrix} b \\ d \end{pmatrix}. \quad (25)$$

(See Figure 2.17.)

(iv) Let us try to solve the system of equations

$$\begin{align} 2x + 3y &= 0, \\ 4x - y &= 0 \end{align} \quad (26)$$

for the unknowns x and y.

Suppose x, y is a solution. We define the vector $\mathbf{X} = \begin{pmatrix} x \\ y \end{pmatrix}$. The first equation then says that $\mathbf{X} \cdot \begin{pmatrix} 2 \\ 3 \end{pmatrix} = 0$ and the second that $\mathbf{X} \cdot \begin{pmatrix} 4 \\ -1 \end{pmatrix} = 0$. Thus, the vector $\mathbf{X}$ is orthogonal to both the vectors $\begin{pmatrix} 2 \\ 3 \end{pmatrix}$ and $\begin{pmatrix} 4 \\ -1 \end{pmatrix}$. This is possible only if $\mathbf{X} = \mathbf{0}$.

Hence, $\begin{pmatrix} x \\ y \end{pmatrix} = \begin{pmatrix} 0 \\ 0 \end{pmatrix}$, so the only solution to (26) is $x = 0$, $y = 0$.

Now look at the general case of a system

$$\begin{align} ax + by &= 0, \\ cx + dy &= 0, \end{align} \quad (27)$$

where a, b, c, d are given constants such that not both a and b are zero and not both c and d are zero.

Of course, $x = 0$, $y = 0$ is a solution of (27). Are there others, and if so what are they?

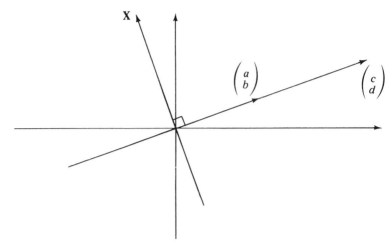

Figure 2.18

Let x, y be a solution other than 0, 0. Set $\mathbf{X} = \begin{pmatrix} x \\ y \end{pmatrix}$. Then $\mathbf{X} \neq \mathbf{0}$ and

$$\mathbf{X} \cdot \begin{pmatrix} a \\ b \end{pmatrix} = \mathbf{0} \quad \text{and} \quad \mathbf{X} \cdot \begin{pmatrix} c \\ d \end{pmatrix} = 0.$$

Thus, there is a nonzero vector orthogonal to both vectors $\begin{pmatrix} a \\ b \end{pmatrix}$ and $\begin{pmatrix} c \\ d \end{pmatrix}$. (See Fig. 2.18.) This can only happen where $\begin{pmatrix} a \\ b \end{pmatrix}$ and $\begin{pmatrix} c \\ d \end{pmatrix}$ lie on the same line through the origin. Since $\begin{pmatrix} c \\ d \end{pmatrix} \neq \mathbf{0}$, there is some scalar t with $\begin{pmatrix} a \\ b \end{pmatrix} = t\begin{pmatrix} c \\ d \end{pmatrix}$, and so $a = tc$ and $b = td$. Also $t \neq 0$. Every vector $\mathbf{X}$ on the line perpendicular to $\begin{pmatrix} a \\ b \end{pmatrix}$ is then orthogonal to both $\begin{pmatrix} a \\ b \end{pmatrix}$ and $\begin{pmatrix} c \\ d \end{pmatrix}$. Our result is this:

(27) has a solution other than $x = 0$, $y = 0$ only when there is some scalar t such that

$$a = tc, \qquad b = td.$$

In that case, there is a line consisting of solutions x, y, namely the line orthogonal to $\begin{pmatrix} a \\ b \end{pmatrix}$.

Exercise 21. Find all solutions to the system of equations
$$3x + 2y = 0,$$
$$4x - y = 0.$$

Exercise 22. Find all solutions to the system of equations
$$5x + y = 0,$$
$$-10x - 2y = 0.$$

CHAPTER 2.1

Transformations of the Plane

Recall the notion of a "function." A function is a rule which assigns to each number some number. This suggests the following definition: *A transformation of the plane* is a rule which assigns to each vector in the plane some vector in the plane.

We denote transformation by letters A, B, R, S, T, etc.

EXAMPLE 1. Let P be the transformation which assigns to each vector $\mathbf{X}$ the projection of $\mathbf{X}$ on the line along the vector $\mathbf{U} = \begin{pmatrix} 1 \\ 2 \end{pmatrix}$.

We write $P(\mathbf{X})$ for the vector which P assigns to $\mathbf{X}$ and we call $P(\mathbf{X})$ the *image* of $\mathbf{X}$ (see Fig. 2.19).

Let $\mathbf{X} = \begin{pmatrix} x \\ y \end{pmatrix}$ and let us calculate $P(\mathbf{X})$. By formula (22) of Chapter 2.0,

$$P(\mathbf{X}) = \left(\frac{\mathbf{X} \cdot \mathbf{U}}{\mathbf{U} \cdot \mathbf{U}} \right) \mathbf{U} = \frac{x + 2y}{1 + 4} \begin{bmatrix} 1 \\ 2 \end{bmatrix} = \begin{bmatrix} \dfrac{x + 2y}{5} \\ \dfrac{2}{5}(x + 2y) \end{bmatrix}. \tag{1}$$

EXAMPLE 2. Let S be the transformation which assigns to each vector $\mathbf{X}$ the reflection of $\mathbf{X}$ in the line along the vector $\begin{pmatrix} 1 \\ 2 \end{pmatrix}$.

Given $\mathbf{X} = \begin{pmatrix} x \\ y \end{pmatrix}$, we want to find the coordinates of the point $S(\mathbf{X})$ such that the midpoint of the segment from $\mathbf{X}$ to $S(\mathbf{X})$ is the projection of $\mathbf{X}$ to the line along $\mathbf{U} = \begin{pmatrix} 1 \\ 2 \end{pmatrix}$. Denote the coordinates of $S(\mathbf{X})$ by x', y'. Then

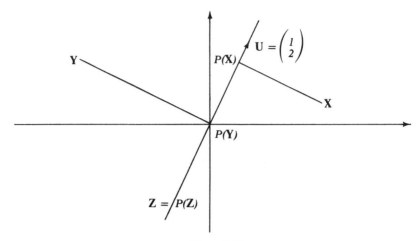

Figure 2.19

$\frac{1}{2}(X + S(X)) = P(X)$, where P was defined in the preceding example. So,

$$X + S(X) = 2P(X)$$

and

$$S(X) = 2P(X) - X.$$

From formula (1), we then obtain

$$S(X) = \begin{bmatrix} x' \\ y' \end{bmatrix} = 2 \begin{bmatrix} \dfrac{x + 2y}{5} \\ \dfrac{2(x + 2y)}{5} \end{bmatrix} - \begin{bmatrix} x \\ y \end{bmatrix}$$

$$= \begin{bmatrix} 2\dfrac{(x + 2y)}{5} - x \\ 4\dfrac{(x + 2y)}{5} - y \end{bmatrix}.$$

Thus

$$\begin{pmatrix} x' \\ y' \end{pmatrix} = \begin{pmatrix} -\frac{3}{5}x + \frac{4}{5}y \\ \frac{4}{5}x + \frac{3}{5}y \end{pmatrix}.$$

For example, if $X = \begin{pmatrix} 0 \\ 10 \end{pmatrix}$, then $S(X) = \begin{pmatrix} x' \\ y' \end{pmatrix} = \begin{pmatrix} 8 \\ 6 \end{pmatrix}$.

Exercise 1. In each of the following problems, U is a nonzero vector and P denotes the transformation which projects each vector X to the line along U. Let $X = \begin{pmatrix} x \\ y \end{pmatrix}$ and $P(X) = \begin{pmatrix} x' \\ y' \end{pmatrix}$, and calculate x' and y' in terms of x and y.

(a) $\mathbf{U} = \begin{pmatrix} 1 \\ 0 \end{pmatrix}$,

(b) $\mathbf{U} = \begin{pmatrix} 0 \\ 1 \end{pmatrix}$,

(c) $\mathbf{U} = \begin{pmatrix} 1 \\ -1 \end{pmatrix}$,

(d) $\mathbf{U} = \begin{pmatrix} 3 \\ 1 \end{pmatrix}$.

In each case draw a diagram and indicate several vectors and their images.

Exercise 2. Consider the line $5x - 2y = 0$ and let P denote projection on this line. If $\begin{pmatrix} a \\ b \end{pmatrix}$ is a given vector and $\begin{pmatrix} a' \\ b' \end{pmatrix} = P\begin{pmatrix} a \\ b \end{pmatrix}$, express a' and b' in terms of a and b.

Exercise 3. For each of the vectors $\mathbf{U}$ in Exercise 1, let $S(\mathbf{X}) = \begin{pmatrix} x' \\ y' \end{pmatrix}$ denote the reflection of $\begin{pmatrix} x \\ y \end{pmatrix}$ in the line along $\mathbf{U}$. Calculate the coordinates x' and y' in terms of x and y. In each case draw a diagram and indicate several vectors and their images.

Exercise 4. Let L be the line $5x - 2y = 0$, and let S denote reflection in L. If $\begin{pmatrix} a \\ b \end{pmatrix}$ is a given vector and $\begin{pmatrix} a' \\ b' \end{pmatrix} = S\begin{pmatrix} a \\ b \end{pmatrix}$, express a' and b' in terms of a and b.

EXAMPLE 3. Let D_2 be the transformation which sends each vector into twice itself:

$$D_2(\mathbf{X}) = 2\mathbf{X}.$$

If $x = \begin{pmatrix} x \\ y \end{pmatrix}$ and $D_2(\mathbf{X}) = \begin{pmatrix} x' \\ y' \end{pmatrix}$, let us calculate x' and y'.

$$\begin{pmatrix} x' \\ y' \end{pmatrix} = D_2(\mathbf{X}) = 2\mathbf{X} = 2\begin{pmatrix} x \\ y \end{pmatrix} = \begin{pmatrix} 2x \\ 2y \end{pmatrix}.$$

So

$$\begin{cases} x' = 2x, \\ y' = 2y. \end{cases} \tag{2}$$

An obvious generalization of this example consists in replacing the number 2 by the number r and defining the transformation D_r by $D_r(\mathbf{X}) = r\mathbf{X}$. For $\mathbf{X} = \begin{pmatrix} x \\ y \end{pmatrix}$, $D_r(\mathbf{X}) = \begin{pmatrix} x' \\ y' \end{pmatrix}$, then, we find that $x' = rx$, $y' = ry$. We call D_r the transformation of *stretching by r.*

Fix a scalar θ with $0 \leqslant \theta \leqslant 2\pi$. We define the transformation R_θ of *rotation by θ radians* as follows (see Fig. 2.20): Let $\mathbf{X}$ be a vector. Rotate the segment from $\mathbf{0}$ to $\mathbf{X}$ around $\mathbf{0}$ counterclockwise through an angle of θ radians. The endpoint of the new segment is $R_\theta(\mathbf{X})$.

EXAMPLE 4. Let $\mathbf{X} = \begin{pmatrix} x \\ y \end{pmatrix}$ and set $\begin{pmatrix} x' \\ y' \end{pmatrix} = R_{\pi/2}(\mathbf{X})$. Find x', y'. (See Fig. 2.21.) By rotating the triangle with vertices $\begin{pmatrix} 0 \\ 0 \end{pmatrix}$, $\begin{pmatrix} x \\ y \end{pmatrix}$, $\begin{pmatrix} x \\ 0 \end{pmatrix}$ through a right

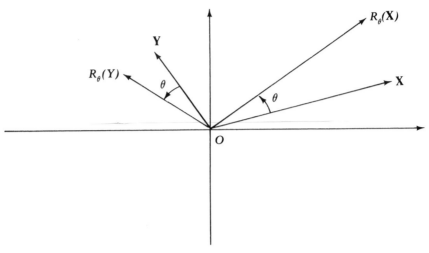

Figure 2.20

angle at the origin, we see that

$$x' = -y, \qquad y' = x.$$

Exercise 5. Let $\mathbf{X} = \begin{pmatrix} x \\ y \end{pmatrix}$. Calculate

(a) $R_{3\pi/2}(\mathbf{X})$,
(b) $R_{\pi}(\mathbf{X})$,
(c) $R_{2\pi}(\mathbf{X})$,
(d) $R_{\pi/4}(\mathbf{X})$.

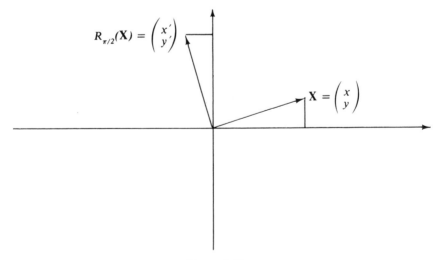

Figure 2.21

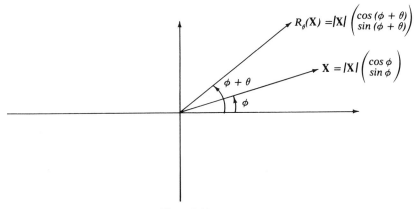

$$R_\theta(X) = |X| \begin{pmatrix} \cos(\phi + \theta) \\ \sin(\phi + \theta) \end{pmatrix}$$

$$X = |X| \begin{pmatrix} \cos\phi \\ \sin\phi \end{pmatrix}$$

$\phi + \theta$

ϕ

Figure 2.22

You may have found part (d) of Exercise 5 a bit difficult. Here is a method that lets us calculate $R_\theta(X)$ for any θ:

Set $X = \begin{pmatrix} x \\ y \end{pmatrix}$ and $\begin{pmatrix} x' \\ y' \end{pmatrix} = R_\theta(X)$. We can write X and $R_\theta(X)$ in the form

$$X = |X| \begin{pmatrix} \cos\phi \\ \sin\phi \end{pmatrix}, \qquad R_\theta(X) = |R_\theta(X)| \begin{pmatrix} \cos\phi' \\ \sin\phi' \end{pmatrix},$$

where ϕ is the polar angle of X, and ϕ' is the polar angle of $R_\theta(X)$. Then $\phi' = \phi + \theta$ and $|R_\theta(X)| = |X|$. (See Fig. 2.22.) So

$$R_\theta(X) = |X| \begin{pmatrix} \cos(\phi + \theta) \\ \sin(\phi + \theta) \end{pmatrix} = |X| \begin{pmatrix} \cos\phi\cos\theta - \sin\phi\sin\theta \\ \sin\phi\cos\theta + \cos\phi\sin\theta \end{pmatrix}$$

$$= \begin{pmatrix} |X|\cos\phi\cos\theta - |X|\sin\phi\sin\theta \\ |X|\sin\phi\cos\theta + |X|\cos\phi\sin\theta \end{pmatrix}.$$

Now

$$|X|\cos\phi = x, \qquad |X|\sin\phi = y.$$

So

$$\begin{pmatrix} x' \\ y' \end{pmatrix} = R_\theta(X) = \begin{pmatrix} x\cos\theta - y\sin\theta \\ y\cos\theta + x\sin\theta \end{pmatrix},$$

or

$$\begin{cases} x' = (\cos\theta)x - (\sin\theta)y, \\ y' = (\sin\theta)x + (\cos\theta)y. \end{cases} \tag{3}$$

Exercise 6. Interpret the results you obtained for Exercise 5 as corollaries of formula (3), by giving θ suitable values.

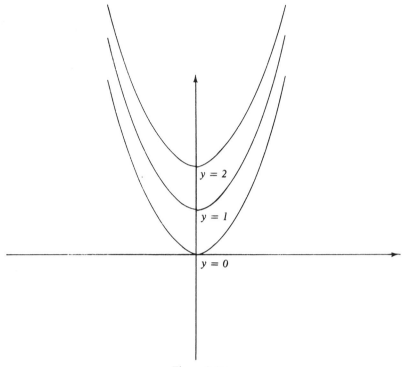

Figure 2.23

EXAMPLE 5. Let T_0 be the following transformation: T_0 sends every horizontal line $y = c$ into the parabola $y = x^2 + c$ by sending $\begin{pmatrix} x \\ y \end{pmatrix}$ into $\begin{pmatrix} x \\ x^2 + y \end{pmatrix}$. (See Fig. 2.23.) In other words, if $\mathbf{X} = \begin{pmatrix} x \\ y \end{pmatrix}$ and $\begin{pmatrix} x' \\ y' \end{pmatrix} = \mathbf{X}' = T_0(\mathbf{X})$, then

$$\begin{cases} x' = x, \\ y' = x^2 + y. \end{cases} \tag{4}$$

Let S, T be two transformations. When do we say that they are *equal*, i.e., $S = T$? Recall that two functions f, g were called equal if $f(x) = g(x)$ for every number x. In a similar spirit, we say $S = T$ provided

$$S(\mathbf{X}) = T(\mathbf{X}) \qquad \text{for every vector } \mathbf{X}.$$

EXAMPLE 6. The transformation $R_{-\pi/2}$, which rotates each vector clockwise by $\pi/2$ radians, and the transformation $R_{3\pi/2}$, which rotates each vector counterclockwise by $3\pi/2$ radians, are equal, i.e.,

$$R_{-\pi/2} = R_{3\pi/2}.$$

CHAPTER 2.2

Linear Transformations and Matrices

In Chapter 2.1, we looked at a number of transformations of the plane. Let us list the results we obtained for each transformation T, giving x', y' in terms of x, y, where $\begin{pmatrix} x' \\ y' \end{pmatrix} = T\begin{pmatrix} x \\ y \end{pmatrix}$.

(i) P = projection to the line along $\begin{pmatrix} 1 \\ 2 \end{pmatrix}$.

$$x' = \frac{x + 2y}{5},$$
$$y' = \frac{2(x + 2y)}{5}.$$

(ii) S is reflection about the line along $\begin{pmatrix} 1 \\ 2 \end{pmatrix}$.

$$x' = -\frac{3}{5}x + \frac{4}{5}y,$$
$$y' = \frac{4}{5}x + \frac{3}{5}y.$$

(iii) D_r is stretching by r.
$$x' = rx,$$
$$y' = ry.$$

(iv) R_θ is rotation by θ radians.
$$x' = (\cos\theta)x - (\sin\theta)y,$$
$$y' = (\sin\theta)x + (\cos\theta)y.$$

(v) T_0 is the transformation of Example 5.

$$x' = x,$$
$$y' = x^2 + y.$$

Can we describe some single general type of transformation, expressing x' and y' each in terms of x and y, which includes the above examples as special cases?

Let a, b, c, d be scalars. Denote by A the transformation which sends each vector $\mathbf{X} = \begin{pmatrix} x \\ y \end{pmatrix}$ into the vector $\mathbf{X}' = \begin{pmatrix} x' \\ y' \end{pmatrix}$, where

$$\begin{cases} x' = ax + by, \\ y' = cx + dy. \end{cases} \tag{1}$$

Setting $a = \frac{1}{5}$, $b = \frac{2}{5}$, $c = \frac{2}{5}$, $d = \frac{4}{5}$, we re-obtain example (i) above. Setting $a = -\frac{3}{5}$, $b = \frac{4}{5}$, $c = \frac{4}{5}$, $d = \frac{3}{5}$, we get (ii). Setting $a = r$, $b = 0$, $c = 0$, $d = r$, we get (iii). If we take $a = \cos\theta$, $b = -\sin\theta$, $c = \sin\theta$, $d = \cos\theta$, we obtain (iv). However, no choice of a, b, c, d will give us (v).

A transformation A given by a system (1) is called a *linear transformation of the plane* and the symbol

$$\begin{pmatrix} a & b \\ c & d \end{pmatrix}$$

is called the *matrix* of A, denoted $m(A)$. The plural of "matrix" is "matrices." Reflection through a line and projection to a line are linear transformations, provided the line goes through the origin. Stretchings D_r and rotations are also linear transformations. It is not possible to describe *all* linear transformations in simple geometrical terms. However, Equations (1) provide a simple algebraic description.

Let us list the matrices of the linear transformations (i)–(iv) considered above.

(i) $$m(P) = \begin{pmatrix} \frac{1}{5} & \frac{2}{5} \\ \frac{2}{5} & \frac{4}{5} \end{pmatrix},$$

(ii) $$m(S) = \begin{pmatrix} -\frac{3}{5} & \frac{4}{5} \\ \frac{4}{5} & \frac{3}{5} \end{pmatrix},$$

(iii) $$m(D_r) = \begin{pmatrix} r & 0 \\ 0 & r \end{pmatrix},$$

(iv) $$m(R_\theta) = \begin{pmatrix} \cos\theta & -\sin\theta \\ \sin\theta & \cos\theta \end{pmatrix},$$

(v) T_0 is *not* a linear transformation.

We need the linear transformation which is the analogue of the function $f(x) = x$. That function sends every number into itself. The *identity transformation*, denoted I, sends every vector into itself:

$$I(\mathbf{X}) = \mathbf{X}, \qquad \text{for every vector } \mathbf{X}.$$

Since I sends $\mathbf{X} = \begin{pmatrix} x \\ y \end{pmatrix}$ into $I(\mathbf{X}) = \begin{pmatrix} x \\ y \end{pmatrix}$, the system

$$x' = x,$$
$$y' = y$$

describes I. Thus the matrix of I is

(vi) $$m(I) = \begin{pmatrix} 1 & 0 \\ 0 & 1 \end{pmatrix}.$$

Finally, we need the linear transformation *zero*, denoted 0, which sends every vector into the zero vector:

$$0(\mathbf{X}) = \mathbf{0}, \qquad \text{for all } \mathbf{X}.$$

Evidently, the matrix of 0 is

(vii) $$m(0) = \begin{pmatrix} 0 & 0 \\ 0 & 0 \end{pmatrix}.$$

Next we introduce some useful notation. Let A be the linear transformation with matrix $\begin{pmatrix} p & q \\ r & s \end{pmatrix}$ and let $\mathbf{X} = \begin{pmatrix} x \\ y \end{pmatrix}$ be a vector. We shall write

$$\begin{pmatrix} p & q \\ r & s \end{pmatrix}\begin{pmatrix} x \\ y \end{pmatrix} = A\begin{pmatrix} x \\ y \end{pmatrix} = A(\mathbf{X}). \tag{2}$$

For instance, if D is stretching by 2, then

$$\begin{pmatrix} 2 & 0 \\ 0 & 2 \end{pmatrix}\begin{pmatrix} x \\ y \end{pmatrix} = D\begin{pmatrix} x \\ y \end{pmatrix} = \begin{pmatrix} 2x \\ 2y \end{pmatrix},$$

or if P is projection on the line along $\begin{pmatrix} 1 \\ 2 \end{pmatrix}$, then

$$\begin{pmatrix} \frac{1}{5} & \frac{2}{5} \\ \frac{2}{5} & \frac{4}{5} \end{pmatrix}\begin{pmatrix} x \\ y \end{pmatrix} = P\begin{pmatrix} x \\ y \end{pmatrix} = \begin{pmatrix} \frac{1}{5}x + \frac{2}{5}y \\ \frac{2}{5}x + \frac{4}{5}y \end{pmatrix}.$$

Let A have the matrix $\begin{pmatrix} p & q \\ r & s \end{pmatrix}$, $\mathbf{X} = \begin{pmatrix} x \\ y \end{pmatrix}$ and $A(\mathbf{X}) = \begin{pmatrix} x' \\ y' \end{pmatrix}$. Then

$$x' = px + qy,$$
$$y' = rx + sy.$$

By definition (2),

$$\begin{pmatrix} p & q \\ r & s \end{pmatrix}\begin{pmatrix} x \\ y \end{pmatrix} = A\begin{pmatrix} x \\ y \end{pmatrix} = \begin{pmatrix} x' \\ y' \end{pmatrix},$$

and so

$$\begin{pmatrix} p & q \\ r & s \end{pmatrix}\begin{pmatrix} x \\ y \end{pmatrix} = \begin{pmatrix} px + qy \\ rx + sy \end{pmatrix}. \tag{3}$$

Formula (3) is basic. We interpret (3) as saying that the matrix $\begin{pmatrix} p & q \\ r & s \end{pmatrix}$ *acts on the vector* $\begin{pmatrix} x \\ y \end{pmatrix}$ to yield the vector $\begin{pmatrix} px + qy \\ rx + sy \end{pmatrix}$.

EXAMPLE 1.

$$\begin{pmatrix} 5 & 6 \\ 7 & 8 \end{pmatrix}\begin{pmatrix} x \\ y \end{pmatrix} = \begin{pmatrix} 5x + 6y \\ 7x + 8y \end{pmatrix},$$

$$\begin{pmatrix} 5 & 6 \\ 7 & 9 \end{pmatrix}\begin{pmatrix} 1 \\ 1 \end{pmatrix} = \begin{pmatrix} 11 \\ 16 \end{pmatrix},$$

$$\begin{pmatrix} 0 & 0 \\ 2 & 2 \end{pmatrix}\begin{pmatrix} x \\ y \end{pmatrix} = \begin{pmatrix} 0 \\ 2x + 2y \end{pmatrix},$$

$$\begin{pmatrix} 1 & 0 \\ 0 & 1 \end{pmatrix}\begin{pmatrix} 5 \\ \pi \end{pmatrix} = \begin{pmatrix} 5 \\ \pi \end{pmatrix}.$$

Let A be an arbitrary linear transformation. We claim that A sends the origin into the origin, i.e.,

$$A(\mathbf{0}) = \mathbf{0},$$

for if $\begin{pmatrix} a & b \\ c & d \end{pmatrix}$ is the matrix of A, then

$$A(\mathbf{0}) = \begin{pmatrix} a & b \\ c & d \end{pmatrix}\begin{pmatrix} 0 \\ 0 \end{pmatrix} = \begin{pmatrix} 0 \\ 0 \end{pmatrix} = \mathbf{0}.$$

A basic reason why linear transformations are interesting is that a linear transformation acts in a simple way on the sum of two vectors. Let A be the linear transformation with matrix $\begin{pmatrix} a & b \\ c & d \end{pmatrix}$ and let $\mathbf{X} = \begin{pmatrix} x \\ y \end{pmatrix}$, $\overline{\mathbf{X}} = \begin{pmatrix} \bar{x} \\ \bar{y} \end{pmatrix}$ be two vectors.

$$A(\mathbf{X} + \overline{\mathbf{X}}) = A\begin{pmatrix} x + \bar{x} \\ y + \bar{y} \end{pmatrix} = \begin{pmatrix} a & b \\ c & d \end{pmatrix}\begin{pmatrix} x + \bar{x} \\ y + \bar{y} \end{pmatrix}$$

$$= \begin{pmatrix} a(x + \bar{x}) + b(y + \bar{y}) \\ c(x + \bar{x}) + d(y + \bar{y}) \end{pmatrix} = \begin{pmatrix} (ax + by) + (a\bar{x} + b\bar{y}) \\ (cx + dy) + (c\bar{x} + d\bar{y}) \end{pmatrix}$$

$$= \begin{pmatrix} a & b \\ c & d \end{pmatrix}\begin{pmatrix} x \\ y \end{pmatrix} + \begin{pmatrix} a & b \\ c & d \end{pmatrix}\begin{pmatrix} \bar{x} \\ \bar{y} \end{pmatrix} = A(\mathbf{X}) + A(\overline{\mathbf{X}}).$$

Thus, we have found

$$A(\mathbf{X} + \overline{\mathbf{X}}) = A(\mathbf{X}) + A(\overline{\mathbf{X}}) \tag{4}$$

for every pair of vectors $\mathbf{X}$, $\overline{\mathbf{X}}$.

A similar calculation shows

$$A(t\mathbf{X}) = tA(\mathbf{X}), \tag{5}$$

if $\mathbf{X}$ is a vector and t is a scalar.

Exercise 1. Verify that formula (5) is true.

Conversely, let B be a transformation of the plane. Let us not assume that B is linear, but instead let us suppose that (4) and (5) are valid for B, i.e., suppose

$$B(\mathbf{X} + \overline{\mathbf{X}}) = B(\mathbf{X}) + B(\overline{\mathbf{X}}), \qquad B(t\mathbf{X}) = tB(\mathbf{X}), \qquad (6)$$

whenever $\mathbf{X}$ and $\overline{\mathbf{X}}$ are vectors and t is a scalar. We claim that it follows that B is a linear transformation, i.e., B is given by a system (1) for suitable a, b, c, d.

To see this, set $\mathbf{E}_1 = \begin{pmatrix} 1 \\ 0 \end{pmatrix}$ and $\mathbf{E}_2 = \begin{pmatrix} 0 \\ 1 \end{pmatrix}$. Then an arbitrary vector $\mathbf{X} = \begin{pmatrix} x \\ y \end{pmatrix}$

can be expressed as

$$\mathbf{X} = x\mathbf{E}_1 + y\mathbf{E}_2 .$$

Set $B(\mathbf{X}) = \begin{pmatrix} x' \\ y' \end{pmatrix}$. By hypothesis,

$$B(\mathbf{X}) = B(x\mathbf{E}_1) + B(y\mathbf{E}_2) = xB(\mathbf{E}_1) + yB(\mathbf{E}_2).$$

$B(\mathbf{E}_1)$ can be written

$$B(\mathbf{E}_1) = \begin{pmatrix} u \\ v \end{pmatrix},$$

and similarly $B(\mathbf{E}_2) = \begin{pmatrix} w \\ z \end{pmatrix}$. Thus

$$\begin{pmatrix} x' \\ y' \end{pmatrix} = B(\mathbf{X}) = x\begin{pmatrix} u \\ v \end{pmatrix} + y\begin{pmatrix} w \\ z \end{pmatrix} = \begin{pmatrix} ux + wy \\ vx + zy \end{pmatrix}.$$

So

$$x' = ux + wy,$$
$$y' = vx + zy.$$

Thus, x', y' have the form of Eq. (1) of this chapter. Hence B is a linear transformation, by definition. The matrix of B is $\begin{pmatrix} u & w \\ v & z \end{pmatrix}$. Thus we have proved that if B is a transformation satisfying (6), then B is a linear transformation. Summing up, we have shown:

Theorem 2.1. *Let A be a transformation of the plane. Then A is a linear transformation if and only if for every pair of vectors $\mathbf{X}$ and $\overline{\mathbf{X}}$ and every scalar t:*

$$A(\mathbf{X} + \overline{\mathbf{X}}) = A(\mathbf{X}) + A(\overline{\mathbf{X}}), \qquad (7a)$$

and

$$A(t\mathbf{X}) = tA(\mathbf{X}). \qquad (7b)$$

Note: (7a) and (7b) together imply

$$A(t\mathbf{X} + s\overline{\mathbf{X}}) = tA(\mathbf{X}) + sA(\overline{\mathbf{X}}) \qquad (8)$$

for every pair of scalars t, s and every pair of vectors $\mathbf{X}$, $\overline{\mathbf{X}}$. This is so, since by (7a),

$$A(t\mathbf{X} + s\overline{\mathbf{X}}) = A(t\mathbf{X}) + A(s\overline{\mathbf{X}}),$$

while by (7b), $A(t\mathbf{X}) = tA(\mathbf{X})$ and $A(s\overline{\mathbf{X}}) = sA(\overline{\mathbf{X}})$. On the other hand, (8) clearly implies both (7a) and (7b). Thus, in Theorem 2.1, we may replace the two conditions (7a) and (7b) by the single condition (8). From now on, when presented with a transformation T, if we wish to show that T is a linear transformation, we can do either of the following: Show that T satisfies (7a) and (7b) (or, equivalently, (8)), or show that there is some matrix $\begin{pmatrix} a & b \\ c & d \end{pmatrix}$ such that for every vector $\mathbf{X} = \begin{pmatrix} x \\ y \end{pmatrix}$, the vector $T(\mathbf{X}) = \begin{pmatrix} x' \\ y' \end{pmatrix}$ is given by:

$$x' = ax + by,$$
$$y' = cx + dy.$$

If A is the linear transformation with matrix $\begin{pmatrix} a & b \\ c & d \end{pmatrix}$, then $A(\mathbf{E}_1)$

$= \begin{pmatrix} a & b \\ c & d \end{pmatrix}\begin{pmatrix} 1 \\ 0 \end{pmatrix} = \begin{pmatrix} a \\ c \end{pmatrix}$ and $A(\mathbf{E}_2) = \begin{pmatrix} a & b \\ c & d \end{pmatrix}\begin{pmatrix} 0 \\ 1 \end{pmatrix} = \begin{pmatrix} b \\ d \end{pmatrix}$. Thus we can describe the matrix of A by saying that its first column is the image of the first basis vector $\mathbf{E}_1$ and the second column is the image of $\mathbf{E}_2$.

EXAMPLE 2. Let P denote projection on the line along $\mathbf{U} = \begin{pmatrix} u \\ v \end{pmatrix}$, so $P(\mathbf{X}) = \left(\dfrac{\mathbf{X} \cdot \mathbf{U}}{\mathbf{U} \cdot \mathbf{U}} \right)\mathbf{U}$. Then

$$P(\mathbf{E}_1) = \left(\frac{u}{u^2 + v^2} \right)\begin{pmatrix} u \\ v \end{pmatrix} = \begin{bmatrix} \dfrac{u^2}{u^2 + v^2} \\[2ex] \dfrac{uv}{u^2 + v^2} \end{bmatrix},$$

$$P(\mathbf{E}_2) = \left(\frac{v}{u^2 + v^2} \right)\begin{pmatrix} u \\ v \end{pmatrix} = \begin{bmatrix} \dfrac{uv}{u^2 + v^2} \\[2ex] \dfrac{v^2}{u^2 + v^2} \end{bmatrix}.$$

Thus the matrix of P is given by

$$m(P) = \begin{bmatrix} \dfrac{u^2}{u^2 + v^2} & \dfrac{vu}{u^2 + v^2} \\[2ex] \dfrac{uv}{u^2 + v^2} & \dfrac{v^2}{u^2 + v^2} \end{bmatrix}. \tag{9}$$

For example, if $\mathbf{U}$ is the unit vector $\mathbf{U} = \begin{pmatrix} \cos\theta \\ \sin\theta \end{pmatrix}$, then $u^2 + v^2 = 1$, so

$$m(P) = \begin{pmatrix} \cos^2\theta & \sin\theta\cos\theta \\ \cos\theta\sin\theta & \sin^2\theta \end{pmatrix}. \tag{10}$$

Theorem 2.1 allows us to give a simple solution of the following geometric problem: Let A be a linear transformation and let L be a straight line. By the *image of L under A* we mean the collection of all vectors $A(\mathbf{X})$ when $\mathbf{X}$ is a vector whose endpoint lies on L. We denote this image by $A(L)$.

EXAMPLE 3. The image of the x-axis under the transformation $R_{\pi/2}$, which is rotation by $\pi/2$ radians, is the y-axis.

What kind of geometric object does the image of L under A turn out to be? The answer is given by:

Theorem 2.2. *Let A be a linear transformation and let L be a straight line. Then the image of L under A is either a straight line or a single point.*

PROOF. Remember from Chapter 2.0 that we can choose vectors $\mathbf{X}_0$ and $\mathbf{U}$ in such a way that L is described by

$$\mathbf{X} = \mathbf{X}_0 + t\mathbf{U}, \qquad t \text{ a real scalar.}$$

Thus, for each $\mathbf{X}$ on L,

$$\mathbf{X} = \mathbf{X}_0 + t\mathbf{U}.$$

Hence, by (7a) and (7b),

$$A(\mathbf{X}) = A(\mathbf{X}_0 + t\mathbf{U}) = A(\mathbf{X}_0) + tA(\mathbf{U}).$$

If $A(\mathbf{U}) \neq 0$, then as $\mathbf{X}$ runs through all vectors with endpoint on L, $A(\mathbf{X})$ runs through the collection of points

$$A(\mathbf{X}_0) + tA(\mathbf{U}), \qquad t \text{ real.}$$

This is a straight line, and so the image of L under A is this line. (See Fig. 2.24.) If $A(\mathbf{U}) = 0$, then $A(\mathbf{X}) = A(\mathbf{X}_0)$ for each $\mathbf{X}$ with endpoint on L. So the image of L under A is the single point $A(\mathbf{X}_0)$.

EXAMPLE 4. Let A denote reflection in the x-axis and let L be the line along $\begin{pmatrix} 1 \\ 2 \end{pmatrix}$. Find the image of L under A. The point $\begin{pmatrix} t \\ 2t \end{pmatrix}$ has image $\begin{pmatrix} t \\ -2t \end{pmatrix}$ so the line along $\begin{pmatrix} 1 \\ -2 \end{pmatrix}$ is the image of L under A. (See Fig. 2.25.)

EXAMPLE 5. Let P denote projection on the y-axis and let L be the x-axis. Find the image of L under P. (See Fig. 2.26.) If $\mathbf{X}$ lies on L, then $P(\mathbf{X}) = \mathbf{0}$. Hence, the image of L under P is a single point, the origin.

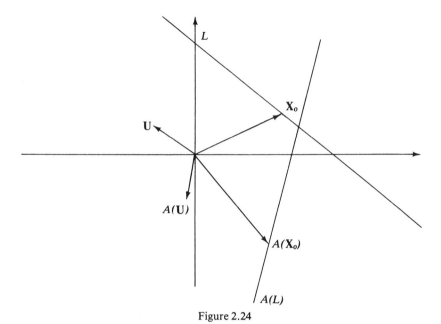

Figure 2.24

Exercise 2. For each of the following transformation T calculate the image of the x-axis.

(a) T is rotation by $45°$.
(b) T is reflection in the line $y = 2x$.
(c) T is projection on the line $y = x$.

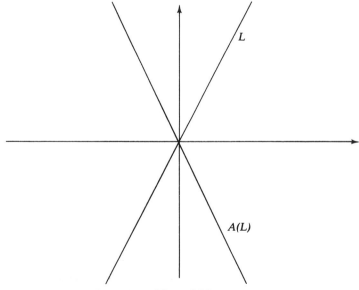

Figure 2.25

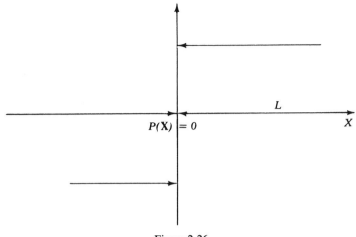

Figure 2.26

Exercise 3. A is the transformation with matrix $\begin{pmatrix} 2 & 3 \\ -1 & 0 \end{pmatrix}$.

(a) Find the image of the line along $\begin{pmatrix} 3 \\ 1 \end{pmatrix}$ under A.

(b) Find the image of the line along $\begin{pmatrix} a \\ b \end{pmatrix}$ under A.

Exercise 4. Let B be a linear transformation such that whenever $\mathbf{X}$ is a nonzero vector, then $B(\mathbf{X}) \neq \mathbf{0}$. Show that for every straight line L through the origin, the image of L under B is a straight line through the origin.

If $\mathbf{X}$ is a nonzero vector then the set of vectors $\{r\mathbf{X} \mid 0 \leqslant r \leqslant 1\}$ is the segment from $\mathbf{0}$ to the point $\mathbf{X}$. If $\mathbf{X} = \mathbf{0}$, then the collection $\{r\mathbf{X} \mid 0 \leqslant r \leqslant 1\}$ contains only the zero vector, and in this case we say that the segment *degenerates to a point.*

If T is any linear transformation, then $T(r\mathbf{X}) = rT(\mathbf{X})$, so the image of the segment $\{r\mathbf{X} \mid 0 \leqslant r \leqslant 1\}$ is the segment $\{r(T(\mathbf{X})) \mid 0 \leqslant r \leqslant 1\}$, possibly degenerate if $T(\mathbf{X}) = \mathbf{0}$.

The set of points $\{\mathbf{U} + r\mathbf{X} \mid 0 \leqslant r \leqslant 1\}$ is also a segment, from $\mathbf{U}$ to $\mathbf{U} + \mathbf{X}$.

If $\mathbf{X}$ and $\mathbf{U}$ are linearly independent vectors, then the set of vectors $\{r\mathbf{X} + s\mathbf{U} \mid 0 \leqslant r \leqslant 1, 0 \leqslant s \leqslant 1\}$ describes the *parallelogram* determined by $\mathbf{X}$ and $\mathbf{U}$ (see Fig. 2.27). The sets $\{r\mathbf{X} \mid 0 \leqslant r \leqslant 1\}$ and $\{s\mathbf{U} \mid 0 \leqslant s \leqslant 1\}$ form two edges of the parallelogram and the other two edges are $\{r\mathbf{X} + \mathbf{U} \mid 0 \leqslant r \leqslant 1\}$ and $\{\mathbf{X} + s\mathbf{U} \mid 0 \leqslant s \leqslant 1\}$. The four corners of the parallelogram are, in order: $\mathbf{U}, \mathbf{0}, \mathbf{X}, \mathbf{U} + \mathbf{X}$.

If $\mathbf{X}$ and $\mathbf{U}$ are linearly dependent, but not both $\mathbf{0}$, then the four points $\mathbf{U}, \mathbf{0}, \mathbf{X}$, and $\mathbf{U} + \mathbf{X}$ all lie on the same line and the set $\{r\mathbf{X} + s\mathbf{U} \mid 0 \leqslant r \leqslant 1, 0 \leqslant s \leqslant 1\}$ is then a *degenerate* or *collapsed* parallelogram.

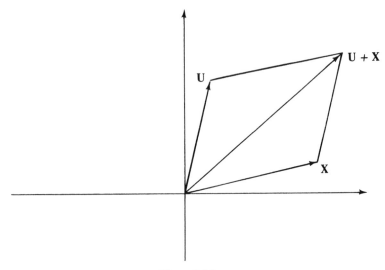

Figure 2.27

If **X** and **U** are both **0**, then $\{r\mathbf{X} + s\mathbf{U} \mid 0 \leqslant r, s \leqslant 1\}$ is also just the point **0**, so the parallelogram degenerates to a single point.

If T is a linear transformation, then $T(r\mathbf{X} + s\mathbf{U}) = rT(\mathbf{X}) + sT(\mathbf{U})$, so the image of the parallelogram $\Pi = \{r\mathbf{X} + s\mathbf{U} \mid 0 \leqslant r, s \leqslant 1\}$ is the parallelogram $T(\Pi) = \{rT(\mathbf{X}) + sT(\mathbf{U}) \mid 0 \leqslant r \leqslant 1, \ 0 \leqslant s \leqslant 1\}$. Even if **X**, **U** is a linearly independent pair, the parallelogram $T(\Pi)$ might be degenerate.

Exercise 5. Describe the parallelograms determined by the following pairs of vectors:

(a) $\begin{pmatrix} 2 \\ 1 \end{pmatrix}, \begin{pmatrix} 1 \\ 1 \end{pmatrix}$,

(b) $\begin{pmatrix} 2 \\ 1 \end{pmatrix}, \begin{pmatrix} 4 \\ 2 \end{pmatrix}$,

(c) $\begin{pmatrix} 2 \\ 1 \end{pmatrix}, \begin{pmatrix} -2 \\ -1 \end{pmatrix}$,

(d) $\begin{pmatrix} 0 \\ 0 \end{pmatrix}, \begin{pmatrix} 2 \\ 1 \end{pmatrix}$,

(e) $\begin{pmatrix} 0 \\ 0 \end{pmatrix}, \begin{pmatrix} 0 \\ 0 \end{pmatrix}$.

Exercise 6. Describe the images of each of the preceding parallelograms under the projection to the first coordinate axis:

$$P\begin{pmatrix} x \\ y \end{pmatrix} = \begin{pmatrix} x \\ 0 \end{pmatrix}.$$

Exercise 7. Do the same for the linear transformation with matrix $\begin{pmatrix} -1 & -2 \\ 2 & 4 \end{pmatrix}$.

Products of Linear Transformations

Let A and B be two linear transformations. We define the transformation C which consists of A followed by B, i.e., if $\mathbf{X}$ is any vector

$$C(\mathbf{X}) = B(A(\mathbf{X})).$$

We write $C = BA$ and we call C *the product B times A*.

Associating to A and B their product BA is in some ways analogous to multiplying two numbers, and we shall pursue this analogy later on.

EXAMPLE 1. B is reflection in the x-axis and A is reflection in the y-axis (see Fig. 2.28). Find BA.

Choose $\mathbf{X} = \begin{pmatrix} x \\ y \end{pmatrix}$. Then

$$A(\mathbf{X}) = \begin{pmatrix} -x \\ y \end{pmatrix} \quad \text{and} \quad B(A(\mathbf{X})) = \begin{pmatrix} -x \\ -y \end{pmatrix}.$$

So

$$(BA)(\mathbf{X}) = \begin{pmatrix} -x \\ -y \end{pmatrix} = -\mathbf{X}.$$

Thus, BA sends each vector into its negative. In other words, $BA = R_\pi$, rotation by π radians.

Exercise 1. Show that if S, T are linear transformations, then ST and TS are linear transformations. (Use (7a) and (7b) or (8) in Chapter 2.2.)

Exercise 2. Let A, B have the same meaning as in Example 1. Show that $AB = R_\pi$.

EXAMPLE 2. Let P be projection on the x-axis and Q projection on the y-axis. Find QP. (See Fig. 2.29.)

If $\mathbf{X} = \begin{pmatrix} x \\ y \end{pmatrix}$, then $P(\mathbf{X}) = \begin{pmatrix} x \\ 0 \end{pmatrix}$ and so $(QP)(\mathbf{X}) = Q(P(\mathbf{X})) = Q\begin{pmatrix} x \\ 0 \end{pmatrix} = \begin{pmatrix} 0 \\ 0 \end{pmatrix}.$

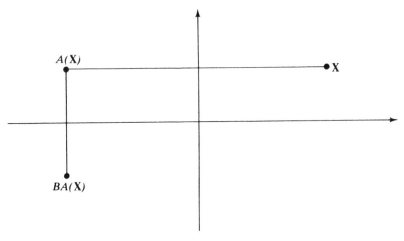

Figure 2.28

Thus QP is the transformation which sends every vector into the origin, i.e., $QP = 0$.

EXAMPLE 3. Let A be a linear transformation and let I denote the identity transformation. Let us find AI and IA.

Fix a vector $\mathbf{X}$

$$(AI)(\mathbf{X}) = A(I(\mathbf{X})) = A(\mathbf{X})$$

and

$$(IA)(\mathbf{X}) = I(A(\mathbf{X})) = A(\mathbf{X}).$$

Hence,

$$AI = A \quad \text{and} \quad IA = A. \tag{1}$$

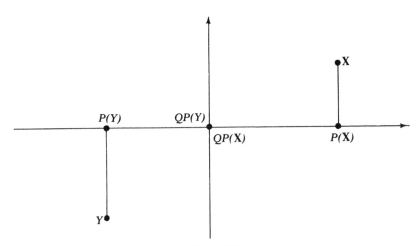

Figure 2.29

Note: The number 1 has the property that

$$a1 = 1a = a$$

for every number a. In view of (1), the identity transformation I plays the same role in multiplying linear transformations as the number 1 does in multiplying numbers.

Exercise 3. Let P be projection on the x-axis and let $R_{\pi/2}$ be rotation by $\pi/2$ radians.

(a) Calculate $PR_{\pi/2}$.
(b) Calculate $(R_{\pi/2})P$.

Observe that your answers for (a) and (b) in Exercise 3 are different. Thus $PR_{\pi/2} \neq R_{\pi/2}P$. The commutative law of multiplication, i.e., the law that $ab = ba$, which is valid for every pair of numbers a, b is false for the product of linear transformations. That is, if A, B are linear transformations, then sometimes $AB = BA$ and sometimes $AB \neq BA$. If $AB = BA$, we say that A and B *commute*. For instance, if A is any linear transformation and I is the identity, then A and I commute.

Now suppose that A and B are two linear transformations having matrices $\begin{pmatrix} a & b \\ c & d \end{pmatrix}$ and $\begin{pmatrix} \bar{a} & \bar{b} \\ \bar{c} & \bar{d} \end{pmatrix}$, respectively. What is the matrix of the transformation AB?

Let $\mathbf{X} = \begin{pmatrix} x \\ y \end{pmatrix}$. Then

$$B(\mathbf{X}) = \begin{pmatrix} \bar{a} & \bar{b} \\ \bar{c} & \bar{d} \end{pmatrix} \begin{pmatrix} x \\ y \end{pmatrix} = \begin{pmatrix} \bar{a}x + \bar{b}y \\ \bar{c}x + \bar{d}y \end{pmatrix}.$$

So

$$AB(\mathbf{X}) = A(B(\mathbf{X})) = \begin{bmatrix} a & b \\ c & d \end{bmatrix} \begin{bmatrix} \bar{a}x + \bar{b}y \\ \bar{c}x + \bar{d}y \end{bmatrix}$$

$$= \begin{bmatrix} a(\bar{a}x + \bar{b}y) + b(\bar{c}x + \bar{d}y) \\ c(\bar{a}x + \bar{b}y) + d(\bar{c}x + \bar{d}y) \end{bmatrix}$$

$$= \begin{bmatrix} (a\bar{a} + b\bar{c})x + (a\bar{b} + b\bar{d})y \\ (c\bar{a} + d\bar{c})x + (c\bar{b} + d\bar{d})y \end{bmatrix}$$

$$= \begin{bmatrix} a\bar{a} + b\bar{c} & a\bar{b} + b\bar{d} \\ c\bar{a} + d\bar{c} & c\bar{b} + d\bar{d} \end{bmatrix} \begin{bmatrix} x \\ y \end{bmatrix}.$$

So the matrix of AB is

$$\begin{pmatrix} a\bar{a} + b\bar{c} & a\bar{b} + b\bar{d} \\ c\bar{a} + d\bar{c} & c\bar{b} + d\bar{d} \end{pmatrix}. \tag{2}$$

We define the *product of matrices* $m(A)$ *and* $m(B)$ to be the matrix $m(AB)$ of AB. Thus

$$m(A)m(B) = m(AB). \tag{3}$$

In other words,

$$\begin{pmatrix} a & b \\ c & d \end{pmatrix}\begin{pmatrix} \bar{a} & \bar{b} \\ \bar{c} & \bar{d} \end{pmatrix} = \begin{pmatrix} a\bar{a} + b\bar{c} & a\bar{b} + b\bar{d} \\ c\bar{a} + d\bar{c} & c\bar{b} + d\bar{d} \end{pmatrix}. \tag{4}$$

Note that on the right-hand side of (4), the upper left-hand entry is the dot product

$$\begin{pmatrix} a \\ b \end{pmatrix} \cdot \begin{pmatrix} \bar{a} \\ \bar{c} \end{pmatrix},$$

the upper right-hand entry is

$$\begin{pmatrix} a \\ b \end{pmatrix} \cdot \begin{pmatrix} \bar{b} \\ \bar{d} \end{pmatrix},$$

the lower left-hand entry is

$$\begin{pmatrix} c \\ d \end{pmatrix} \cdot \begin{pmatrix} \bar{a} \\ \bar{c} \end{pmatrix},$$

and the lower right-hand entry is

$$\begin{pmatrix} c \\ d \end{pmatrix} \cdot \begin{pmatrix} \bar{b} \\ \bar{d} \end{pmatrix}.$$

EXAMPLE 4. Find the product $\begin{pmatrix} 1 & 0 \\ 2 & 3 \end{pmatrix}\begin{pmatrix} 5 & 4 \\ -1 & 7 \end{pmatrix}$. By formula (4),

$$\begin{pmatrix} 1 & 0 \\ 2 & 3 \end{pmatrix}\begin{pmatrix} 5 & 4 \\ -1 & 7 \end{pmatrix} = \begin{pmatrix} 1 \cdot 5 + 0 \cdot -1 & 1 \cdot 4 + 0 \cdot 7 \\ 2 \cdot 5 + 3 \cdot -1 & 2 \cdot 4 + 3 \cdot 7 \end{pmatrix} = \begin{pmatrix} 5 & 4 \\ 7 & 29 \end{pmatrix}.$$

Exercise 4. In each case, calculate the indicated product of two matrices:

(a) $\begin{pmatrix} -1 & 0 \\ 0 & 0 \end{pmatrix}\begin{pmatrix} -1 & 0 \\ 0 & 0 \end{pmatrix}$,

(b) $\begin{pmatrix} 0 & 1 \\ 0 & 0 \end{pmatrix}\begin{pmatrix} 0 & 1 \\ 0 & 0 \end{pmatrix}$,

(c) $\begin{pmatrix} 1 & 2 \\ 3 & 4 \end{pmatrix}\begin{pmatrix} 5 & 6 \\ 7 & 8 \end{pmatrix}$.

Exercise 5. Let U be the linear transformation having matrix $\begin{pmatrix} 0 & 1 \\ 1 & 0 \end{pmatrix}$.

(a) Interpret U geometrically.
(b) Show that $UU = I$.

Exercise 6. Let V be the linear transformation having matrix $\begin{pmatrix} 0 & -1 \\ 1 & 0 \end{pmatrix}$. Show that $VV = R_\pi$, rotation by π radians.

Exercise 7. Let R_θ and R_ϕ be rotation by angles of θ radians and Q radians, respectively. Show that

$$R_\theta R_\phi = R_{\theta + \phi},$$

rotation by $\theta + \phi$ radians.

Exercise 8. Exhibit a linear transformation N such that $N \neq 0$, while $NN = 0$.

Exercise 9. Let A be the linear transformation with matrix $\begin{pmatrix} 1 & -1 \\ 2 & -2 \end{pmatrix}$.

(a) Show that if $\mathbf{X}$ is any vector which lies on the line along $\begin{pmatrix} 1 \\ 1 \end{pmatrix}$, then $A(\mathbf{X}) = 0$.

(b) Show that if $\mathbf{X}$ is any vector, then $A(\mathbf{X})$ lies on the line along $\begin{pmatrix} 1 \\ 2 \end{pmatrix}$.

(c) Find a linear transformation B with $B \neq 0$ such that $BA = 0$.
(d) Find a linear transformation C with $C \neq 0$ such that $AC = 0$.

If a, b, c are three numbers, then the *associative law* holds, i.e., $(ab)c = a(bc)$.

If A, B, C are three linear transformations, then

$$(AB)C = A(BC), \tag{5a}$$

and

$$(m(A)m(B))m(C) = m(A)(m(B)m(C)). \tag{5b}$$

Note: (5a) says that the associative law holds for multiplication of linear transformations, while (5b) says that it holds for multiplication of matrices.

PROOF OF (5a): Let $\mathbf{X}$ be any vector. Then

$$((AB)C)(\mathbf{X}) = AB(C(\mathbf{X})) = A(B(C(\mathbf{X})))$$
$$= A((BC)(\mathbf{X})) = (A(BC))(\mathbf{X}).$$

Hence

$$(AB)C = A(BC).$$

So (5a) holds.

PROOF OF (5b): By definition of multiplication of matrices, if S and T are linear transformation, then $m(S)m(T) = m(ST)$. Hence, using (5a), we get

$$(m(A)m(B))m(C) = m(AB)m(C)$$
$$= m((AB)C) = m(A(BC))$$
$$= m(A)m(BC) = m(A)(m(B)m(C)).$$

Thus (5b) holds.

Note: If $\begin{pmatrix} a & b \\ c & d \end{pmatrix}$, $\begin{pmatrix} x & y \\ u & v \end{pmatrix}$, and $\begin{pmatrix} p & q \\ r & s \end{pmatrix}$ are three matrices, (5b) yields

$$\left(\begin{pmatrix} a & b \\ c & d \end{pmatrix}\begin{pmatrix} x & y \\ u & v \end{pmatrix}\right)\begin{pmatrix} p & q \\ r & s \end{pmatrix} = \begin{pmatrix} a & b \\ c & d \end{pmatrix}\left(\begin{pmatrix} x & y \\ u & v \end{pmatrix}\begin{pmatrix} p & q \\ r & s \end{pmatrix}\right).$$

We could obtain this equation directly, using formula (4), but that would require more effort.

Let A and B be two linear transformations. By the *sum of A and B*, $A + B$, we mean the transformation which assigns to each vector $\mathbf{X}$ the vector $A(\mathbf{X}) + B(\mathbf{X})$ so

$$(A + B)(\mathbf{X}) = A(\mathbf{X}) + B(\mathbf{X}), \qquad \text{for each } \mathbf{X}.$$

If the matrix $m(A) = \begin{pmatrix} a & b \\ c & d \end{pmatrix}$ and the matrix $m(B) = \begin{pmatrix} \bar{a} & \bar{b} \\ \bar{c} & \bar{d} \end{pmatrix}$, then

$$(A + B)\left[\begin{pmatrix} x \\ y \end{pmatrix}\right] = A\begin{pmatrix} x \\ y \end{pmatrix} + B\begin{pmatrix} x \\ y \end{pmatrix} = \begin{pmatrix} a & b \\ c & d \end{pmatrix}\begin{pmatrix} x \\ y \end{pmatrix} + \begin{pmatrix} \bar{a} & \bar{b} \\ \bar{c} & \bar{d} \end{pmatrix}\begin{pmatrix} x \\ y \end{pmatrix}$$

$$= \begin{pmatrix} ax + by \\ cx + dy \end{pmatrix} + \begin{pmatrix} \bar{a}x + \bar{b}y \\ \bar{c}x + \bar{d}y \end{pmatrix} = \begin{bmatrix} (a + \bar{a})x + (b + \bar{b})y \\ (c + \bar{c})x + (d + \bar{d})y \end{bmatrix}$$

$$= \begin{pmatrix} a + \bar{a} & b + \bar{b} \\ c + \bar{c} & d + \bar{d} \end{pmatrix}\begin{pmatrix} x \\ y \end{pmatrix}.$$

Thus, $A + B$ is a linear transformation and its matrix is $\begin{pmatrix} a + \bar{a} & b + \bar{b} \\ c + \bar{d} & d + \bar{d} \end{pmatrix}$.
We define the *sum of the matrices* $m(A)$ and $m(B)$, denoted $m(A) + m(B)$, as $m(A + B)$. Thus

$$\begin{pmatrix} a & b \\ c & d \end{pmatrix} + \begin{pmatrix} \bar{a} & \bar{b} \\ \bar{c} & \bar{d} \end{pmatrix} = \begin{pmatrix} a + \bar{a} & b + \bar{b} \\ c + \bar{c} & d + \bar{d} \end{pmatrix}.$$

Similarly, if A is as above and t is a scalar, we denote by tA the transformation defined by

$$(tA)(\mathbf{X}) = tA(\mathbf{X}), \qquad \text{for every vector } \mathbf{X},$$

and we define

$$tm(A) = m(tA).$$

It follows that

$$t\begin{pmatrix} a & b \\ c & d \end{pmatrix} = \begin{pmatrix} ta & tb \\ tc & td \end{pmatrix}.$$

EXAMPLE 5.

$$\begin{pmatrix} 1 & 2 \\ 3 & 4 \end{pmatrix} + 2\begin{pmatrix} -1 & 0 \\ 0 & 1 \end{pmatrix} = \begin{pmatrix} 1 + 2(-1) & 2 \\ 3 & 4 + 2(1) \end{pmatrix} = \begin{pmatrix} -1 & 2 \\ 3 & 6 \end{pmatrix}.$$

As an application of the notion of the sum of two linear transformations, let us do the following example.

EXAMPLE 6. Let L be a straight line through the origin and denote by S the transformation which reflects each vector in $\mathbf{X}$. Let P denote the transformation of projection to L. If $\mathbf{X}$ is any vector, then $\frac{1}{2}(\mathbf{X} + S(\mathbf{X})) = P(\mathbf{X})$. Hence

$$\mathbf{X} + S(\mathbf{X}) = 2P(\mathbf{X}),$$

and so

$$S(\mathbf{X}) = 2P(\mathbf{X}) - \mathbf{X} = (2P - I)(\mathbf{X}).$$

Since this holds for every vector $\mathbf{X}$, we get

$$S = 2P - I. \tag{6}$$

EXAMPLE 7. Let L be a straight line through the origin and denote by θ the angle from the positive x-axis to L. Find the matrix of the transformation S which reflects each vector in L. (See Fig. 2.30.) The vector $\mathbf{U} = \begin{pmatrix} \cos\theta \\ \sin\theta \end{pmatrix}$ is a unit vector and lies on L. By (10) of Chapter 2.2, if P is the transformation which projects to L, then

$$m(P) = \begin{pmatrix} \cos^2\theta & \sin\theta\cos\theta \\ \cos\theta\sin\theta & \sin^2\theta \end{pmatrix}.$$

By Example 6, $S = 2P - I$, so

$$m(S) = 2m(P) - m(I) = \begin{pmatrix} 2\cos^2\theta & 2\cos\theta\sin\theta \\ 2\cos\theta\sin\theta & 2\sin^2\theta \end{pmatrix} - \begin{pmatrix} 1 & 0 \\ 0 & 1 \end{pmatrix}$$

$$= \begin{pmatrix} 2\cos^2\theta - 1 & 2\cos\theta\sin\theta \\ 2\cos\theta\sin\theta & 2\sin^2\theta - 1 \end{pmatrix}.$$

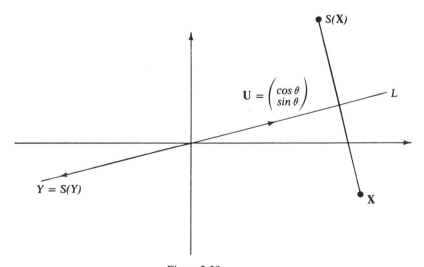

Figure 2.30

By the double-angle formulae from trigonometry, we have

$$\cos 2\theta = 2\cos^2\theta - 1, \qquad \sin 2\theta = 2\sin\theta\cos\theta.$$

So

$$m(S) = \begin{pmatrix} \cos 2\theta & \sin 2\theta \\ \sin 2\theta & -\cos 2\theta \end{pmatrix}. \tag{7}$$

Exercise 10.

(a) Using (7) show that $(m(S))^2 = m(I)$.
(b) Give a geometric explanation of the result of part (a).

Exercise 11. Using formula (7), find the matrix of the transformation which reflects each vector in the line along $\begin{pmatrix} 1 \\ 2 \end{pmatrix}$.

Exercise 12. Let H_k be the transformation with matrix $\begin{pmatrix} 1 & 0 \\ k & 1 \end{pmatrix}$. Show $\begin{pmatrix} 1 & 0 \\ k & 1 \end{pmatrix}\begin{pmatrix} 1 & 0 \\ m & 1 \end{pmatrix} = \begin{pmatrix} 1 & 0 \\ k+m & 1 \end{pmatrix}$, and conclude that $H_k H_m = H_{k+m}$.

Exercise 13. Let J_k be the transformation with matrix $\begin{pmatrix} 1 & k \\ 0 & 1 \end{pmatrix}$. Show that $J_k J_m = J_{k+m}$.

Exercise 14. Conclude $H_k J_m$ and $J_m H_k$ for a given pair of scalars k, m.

Exercise 15. Describe the images of the unit square under the transformations H_1, H_2, H_{-1}.

Exercise 16. Describe the images of the unit square under the transformations J_1, J_2, J_{-1}.

Note: The transformations H_k and J_m are called *shear transformations*. The transformation K with matrix

$$m(K) = \begin{pmatrix} 0 & 1 \\ 1 & 0 \end{pmatrix}$$

is called a *permutation* and its matrix is called a *permutation matrix*.

Exercise 17. Let $\begin{pmatrix} a & b \\ c & d \end{pmatrix}$ be a matrix. Show that

$$\begin{pmatrix} 0 & 1 \\ 1 & 0 \end{pmatrix}\begin{pmatrix} a & b \\ c & d \end{pmatrix} = \begin{pmatrix} c & d \\ a & b \end{pmatrix}$$

and

$$\begin{pmatrix} a & b \\ c & d \end{pmatrix}\begin{pmatrix} 0 & 1 \\ 1 & 0 \end{pmatrix} = \begin{pmatrix} b & a \\ d & c \end{pmatrix}.$$

Note that multiplying a matrix on the left by a permutation matrix *interchanges the rows*, while the corresponding multiplication on the right *interchanges the columns*.

Permutation matrices and shear matrices, as well as the identity matrix, are called *elementary matrices*. A matrix

$$\begin{pmatrix} d_1 & 0 \\ 0 & d_2 \end{pmatrix}$$

with entries 0 except on the diagonal is called a *diagonal matrix*.

Theorem 2.3. *Let* $\begin{pmatrix} a & b \\ c & d \end{pmatrix}$ *be an arbitrary matrix. We can find elementary matrices* e_1, e_2, e_3, e_4 *and a diagonal matrix* $\begin{pmatrix} d_1 & 0 \\ 0 & d_2 \end{pmatrix}$, *such that*

$$e_1 e_2 \begin{pmatrix} a & b \\ c & d \end{pmatrix} e_3 e_4 = \begin{pmatrix} d_1 & 0 \\ 0 & d_2 \end{pmatrix}. \tag{8}$$

PROOF. Suppose $a \neq 0$. We have

$$\begin{pmatrix} 1 & 0 \\ k & 1 \end{pmatrix} \begin{pmatrix} a & b \\ c & d \end{pmatrix} = \begin{pmatrix} a & b \\ ka + c & kb + d \end{pmatrix}.$$

Taking $k = -c/a$, we get

$$\begin{pmatrix} 1 & 0 \\ k & 1 \end{pmatrix} \begin{pmatrix} a & b \\ c & d \end{pmatrix} = \begin{pmatrix} a & b \\ 0 & x \end{pmatrix},$$

where $x = -(c/a)b + d$. Also,

$$\begin{pmatrix} a & b \\ 0 & x \end{pmatrix} \begin{pmatrix} 1 & m \\ 0 & 1 \end{pmatrix} = \begin{pmatrix} a & ma + b \\ 0 & x \end{pmatrix}.$$

Taking $m = -b/a$, we get

$$\begin{pmatrix} a & b \\ 0 & x \end{pmatrix} \begin{pmatrix} 1 & m \\ 0 & 1 \end{pmatrix} = \begin{pmatrix} a & 0 \\ 0 & x \end{pmatrix}.$$

Thus we have

$$\begin{pmatrix} 1 & 0 \\ k & 1 \end{pmatrix} \begin{pmatrix} a & b \\ c & d \end{pmatrix} \begin{pmatrix} 1 & m \\ 0 & 1 \end{pmatrix} = \begin{pmatrix} a & 0 \\ 0 & x \end{pmatrix},$$

and so (8) holds.

What if $a = 0$? Either $\begin{pmatrix} a & b \\ c & d \end{pmatrix}$ is the zero matrix, or some entry is nonzero, say $c \neq 0$. Then

$$\begin{pmatrix} 0 & 1 \\ 1 & 0 \end{pmatrix} \begin{pmatrix} a & b \\ c & d \end{pmatrix} = \begin{pmatrix} c & d \\ a & b \end{pmatrix}.$$

Since $c \neq 0$, the preceding reasoning applies to $\begin{pmatrix} c & d \\ a & b \end{pmatrix}$ and we can choose shear matrices e_1 and e_3 such that

$$e_1 \begin{pmatrix} c & d \\ a & b \end{pmatrix} e_3 = e_1 e_2 \begin{pmatrix} a & b \\ c & d \end{pmatrix} e_3$$

is a diagonal matrix, where $e_2 = \begin{pmatrix} 0 & 1 \\ 1 & 0 \end{pmatrix}$. So again (8) holds. If $b \neq 0$ or $d \neq 0$, we proceed in a similar way to obtain (8).

EXAMPLE 8. Let us find formula (8) for the matrix $\begin{pmatrix} 0 & 5 \\ 2 & 3 \end{pmatrix}$.

$$\begin{pmatrix} 0 & 1 \\ 1 & 0 \end{pmatrix}\begin{pmatrix} 0 & 5 \\ 2 & 3 \end{pmatrix} = \begin{pmatrix} 2 & 3 \\ 0 & 5 \end{pmatrix},$$

$$\begin{pmatrix} 0 & 1 \\ 1 & 0 \end{pmatrix}\begin{pmatrix} 0 & 5 \\ 2 & 3 \end{pmatrix}\begin{pmatrix} 1 & -\frac{3}{2} \\ 0 & 1 \end{pmatrix} = \begin{pmatrix} 2 & 3 \\ 0 & 5 \end{pmatrix}\begin{pmatrix} 1 & -\frac{3}{2} \\ 0 & 1 \end{pmatrix} = \begin{pmatrix} 2 & 0 \\ 0 & 5 \end{pmatrix}.$$

Thus $\begin{pmatrix} 2 & 0 \\ 0 & 5 \end{pmatrix}$ is the diagonal matrix of formula (8) here.

Exercise 18. Find elementary matrices e_1 and e_3 and a diagonal matrix $\begin{pmatrix} d_1 & 0 \\ 0 & d_2 \end{pmatrix}$ such that

$$e_1 \begin{pmatrix} 2 & 2 \\ 3 & 3 \end{pmatrix} e_2 = \begin{pmatrix} d_1 & 0 \\ 0 & d_2 \end{pmatrix}.$$

Figures 2.31a–f indicate the effects of elementary transformations and diagonal matrices. Figure 2.31a shows the identity transformation I. Figure 2.31b shows the matrix with diagonal entries 2 and 1, Figure 2.31c shows the shear transformation H_2, Figure 2.31d shows the permutation matrix K, Figure 2.31e shows the shear J_2, and Figure 2.31f shows the diagonal matrix with entries 1 and 2.

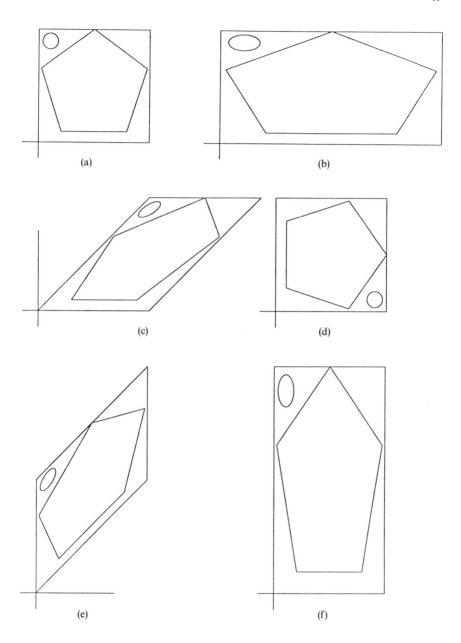

Figure 2.31

Inverses and Systems of Equations

§1 Inverses

If a, x, y are numbers, then

$$a(x + y) = ax + ay.$$

If A is a linear transformation and $\mathbf{X}$ and $\mathbf{Y}$ are vectors, then by Theorem 2.1 of Chapter 2.2,

$$A(\mathbf{X} + \mathbf{Y}) = A(\mathbf{X}) + A(\mathbf{Y}).$$

Thus we see that the operation which takes a number x into the number ax is somehow similar to the operation which takes a vector $\mathbf{X}$ into the vector $A(\mathbf{X})$, where A is a linear transformation.

Next, consider the equation:

$$ax = y \tag{1}$$

where a and y are given numbers, $a \neq 0$, and x is an unknown number. We solve (1) by taking the *reciprocal* $1/a$ of a, and multiplying both sides by it, arriving at

$$\frac{1}{a}(ax) = \frac{1}{a} y, \quad \text{and so} \quad x = \frac{1}{a} y.$$

As an analogue of equation (1) for vectors, we may consider a linear transformation A and a vector $\mathbf{Y}$ and look for a vector $\mathbf{X}$ such that

$$A(\mathbf{X}) = \mathbf{Y}. \tag{2}$$

To solve (2) we should like to have an analogue of the reciprocal for the transformation A. Now the reciprocal $1/a$ satisfies

$$\frac{1}{a} \cdot a = 1 \quad \text{and} \quad a \cdot \frac{1}{a} = 1.$$

A reasonable analogue would be a linear transformation B such that

$$BA = I \quad \text{and} \quad AB = I. \qquad (3)$$

Suppose we have found such a B. Then we can solve Eq. (2) by applying B to both sides. This gives

$$B(A(\mathbf{X})) = B(\mathbf{Y}).$$

But

$$B(A(\mathbf{X})) = (BA)(\mathbf{X}) = I(\mathbf{X}) = \mathbf{X},$$

so we get

$$\mathbf{X} = B(\mathbf{Y}). \qquad (4)$$

We can verify that (4) really gives a solution to (2) by applying A to both sides of (4). This gives

$$A(\mathbf{X}) = A(B(\mathbf{Y})) = (AB)(\mathbf{Y}) = I(\mathbf{Y}) = \mathbf{Y},$$

and so (2) is valid.

The problem of solving Eq. (2) will thus be resolved, provided we can find a linear transformation B satisfying $BA = I$ and $AB = I$. Such a linear transformation B is called an *inverse* of A.

Note that there is exactly one number which fails to have a reciprocal, namely the number 0. It turns out that there are many linear transformations which have no inverse, and later on in this chapter we shall see how we can decide whether or not a given linear transformation has an inverse.

Let A be a linear transformation. If B is an inverse of A, then B *undoes* the effect of A on a vector in the following sense: if A sends the vector $\mathbf{X}$ to the vector $\mathbf{Y}$, then B sends the vector $\mathbf{Y}$ to the vector $\mathbf{X}$.

To see that this is so, consider a vector $\mathbf{X}$. Define $\mathbf{Y} = A(\mathbf{X})$. By (3),

$$(BA)(\mathbf{X}) = I(\mathbf{X}) \quad \text{or} \quad B(A(\mathbf{X})) = \mathbf{X}.$$

So $B(\mathbf{Y}) = \mathbf{X}$, as we have claimed (see Fig. 2.32).

EXAMPLE 1. Fix $r \neq 0$. Find the inverse of D_r, i.e., of stretching by r, where D_r takes the vector $\mathbf{X}$ into the vector $r\mathbf{X}$. To undo this, we must multiply by the scalar $1/r$. Thus, we set $B = D_{1/r}$. Then if $\mathbf{X}$ is any vector,

$$(BD_r)(\mathbf{X}) = B(D_r(\mathbf{X})) = B(r\mathbf{X}) = \frac{1}{r}(r\mathbf{X}) = \mathbf{X}.$$

Hence, $BD_r = I$. Also,

$$(D_r B)(\mathbf{X}) = D_r(B(\mathbf{X})) = D_r\left(\frac{1}{r}\mathbf{X}\right) = r\left(\frac{1}{r}\mathbf{X}\right) = \mathbf{X},$$

Thus, B satisfies (3) and so $D_{1/r} = B$ is an inverse of D.

EXAMPLE 2. R_θ denotes rotation by θ radians. Find the inverse of $R_{\pi/2}$.

Let $\mathbf{X}$ be a vector. $R_{\pi/2}$ rotates $\mathbf{X}$ by $\pi/2$ radians counterclockwise around $\mathbf{0}$. To undo the effect of $R_{\pi/2}$, we can rotate by $-\pi/2$ radians.

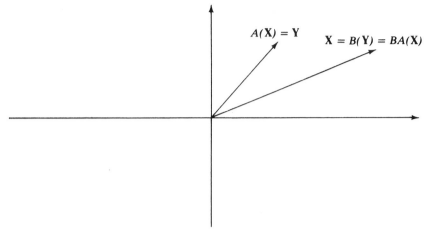

Figure 2.32

Thus, we set $B = R_{-\pi/2}$. If we prefer, we can write $B = R_{3\pi/2}$, since rotation by $-\pi/2$ radians and rotation by $3\pi/2$ radians have the same effect on each vector, and so $R_{-\pi/2} = R_{3\pi/2}$. Then, if $\mathbf{X}$ is a vector,

$$(BR_{\pi/2})(\mathbf{X}) = B(R_{\pi/2}(\mathbf{X})) = R_{-\pi/2}(R_{\pi/2}(\mathbf{X})) = \mathbf{X}.$$

Thus, $BR_{\pi/2} = I$. Also,

$$(R_{\pi/2}B)(\mathbf{X}) = R_{\pi/2}(B(\mathbf{X})) = R_{\pi/2}(R_{-\pi/2}(\mathbf{X})) = \mathbf{X}.$$

Thus, $R_{\pi/2}B = I$. So B satisfies (3), and $R_{-\pi/2} = B$ is an inverse of $R_{\pi/2}$.

Exercise 1. Find an inverse of $R_{3\pi/4}$.

EXAMPLE 3. Let L be a straight line through the origin. Let S be reflection in the line L. Find an inverse to S (see Fig. 2.33).

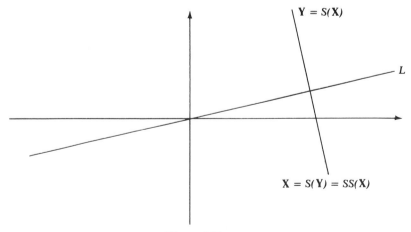

Figure 2.33

If we start with a vector $\mathbf{X}$, then reflect $\mathbf{X}$ in L and then reflect in L again, we return to $\mathbf{X}$. In other words,

$$S(S(\mathbf{X})) = \mathbf{X} \quad \text{or} \quad (SS)(\mathbf{X}) = \mathbf{X}.$$

Thus, $SS = I$. Hence, if we take $A = S$ and $B = S$, then (3) is satisfied. So S is an inverse of itself.

Note: For numbers, the analogous situation occurs when a number a is its own reciprocal, as in case $a = 1$ or $a = -1$.

Now let A be an arbitrary linear transformation. Can A have *more than one* inverse? Assume that B and C are two linear transformations each of which satisfies (3), i.e., assume

$$AB = I \quad \text{and} \quad BA = I \tag{5}$$

and, also,

$$AC = I \quad \text{and} \quad CA = I. \tag{6}$$

By (5), $BA = I$.
Hence, $(BA)C = IC = C$.
By the associative property, $(BA)C = B(AC)$. So

$$B(AC) = C.$$

By (6), $AC = I$, so $B(AC) = BI = B$. Hence,

$$B = C.$$

We have seen, then, that if B and C each is an inverse of A, then $B = C$. In other words, A can have only *one* inverse. Thus we can speak of *the* inverse of A, and we denote this inverse, provided it exists, by A^{-1}. Thus, $A \cdot A^{-1} = I$ and $A^{-1} \cdot A = I$. Examples 1, 2, and 3 can then be expressed as follows:

$$(D_r)^{-1} = D_{1/r},$$
$$(R_{\pi/2})^{-1} = R_{-\pi/2},$$
$$S^{-1} = S.$$

Exercise 2. Let T be the transformation with matrix $\begin{pmatrix} 2 & 0 \\ 0 & 5 \end{pmatrix}$, so that $T\begin{pmatrix} x \\ y \end{pmatrix} = \begin{pmatrix} 2x \\ 5y \end{pmatrix}$ for every vector $\begin{pmatrix} x \\ y \end{pmatrix}$. Find the matrix of T^{-1}.

Exercise 3. Let T be the transformation with matrix $\begin{pmatrix} 2 & 1 \\ 0 & 5 \end{pmatrix}$. Find the matrix of T^{-1}.

EXAMPLE 4. Let P be projection on the x-axis. Suppose B is an inverse of P. Then, $BP = I$, so $(BP)(\mathbf{X}) = \mathbf{X}$ for every vector $\mathbf{X}$.

Choose a vector $\mathbf{X} = \begin{pmatrix} 0 \\ y \end{pmatrix}$ with $y \neq 0$. Then $P(\mathbf{X}) = \begin{pmatrix} 0 \\ 0 \end{pmatrix}$, and so

$$(BP)(\mathbf{X}) = B(P(\mathbf{X})) = B\begin{pmatrix} 0 \\ 0 \end{pmatrix} = \mathbf{0}.$$

Hence

$$\mathbf{X} = (BP)(\mathbf{X}) = \mathbf{0}.$$

But $\mathbf{X}$ was not the zero vector, and so we have reached a contradiction. From this we are forced to conclude that there is no linear transformation B satisfying $BP = I$. So P has no inverse.

Next we observe that if A is a linear transformation which has an inverse A^{-1}, then A satisfies the following condition:

The only vector $\mathbf{X}$ with $A(\mathbf{X}) = \mathbf{0}$ is the vector $\mathbf{X} = \mathbf{0}$.   (7)

To see this, choose $\mathbf{X}$ with $A(\mathbf{X}) = \mathbf{0}$. Then $A^{-1}(A(\mathbf{X})) = A^{-1}(\mathbf{0}) = \mathbf{0}$, and also $A^{-1}(A(\mathbf{X})) = (A^{-1}A)(\mathbf{X}) = I(\mathbf{X}) = \mathbf{X}$. So $\mathbf{X} = \mathbf{0}$, and so (7) is true.

Now let A be a linear transformation with matrix $\begin{pmatrix} a & b \\ c & d \end{pmatrix}$. We shall prove:

Proposition 1. *Condition* (7) *holds if and only if* $ad - bc \neq 0$.

PROOF. Suppose $ad - bc = 0$. Then we have

$$A\begin{pmatrix} -b \\ a \end{pmatrix} = \begin{pmatrix} a & b \\ c & d \end{pmatrix}\begin{pmatrix} -b \\ a \end{pmatrix} = \begin{pmatrix} 0 \\ -cb + ad \end{pmatrix} = \begin{pmatrix} 0 \\ 0 \end{pmatrix},$$

$$A\begin{pmatrix} d \\ -c \end{pmatrix} = \begin{pmatrix} a & b \\ c & d \end{pmatrix}\begin{pmatrix} d \\ -c \end{pmatrix} = \begin{pmatrix} ad - bc \\ 0 \end{pmatrix} = \begin{pmatrix} 0 \\ 0 \end{pmatrix}.$$

If (7) holds, it follows that $\begin{pmatrix} -b \\ a \end{pmatrix} = \begin{pmatrix} 0 \\ 0 \end{pmatrix}$ and $\begin{pmatrix} d \\ -c \end{pmatrix} = \begin{pmatrix} 0 \\ 0 \end{pmatrix}$. Hence a, b, c, d are all zero, so $A(\mathbf{X}) = \mathbf{0}$ for every $\mathbf{X}$. But then (7) is false, so we have a contradiction. Hence if $ad - bc = 0$, then (7) does not hold.

Conversely, suppose $ad - bc \neq 0$. Let $\mathbf{X} = \begin{pmatrix} x \\ y \end{pmatrix}$ be a vector with $A(\mathbf{X})$ $= \mathbf{0}$. Then $\begin{pmatrix} 0 \\ 0 \end{pmatrix} = A(\mathbf{X}) = \begin{pmatrix} a & b \\ c & d \end{pmatrix}\begin{pmatrix} x \\ y \end{pmatrix} = \begin{pmatrix} ax + by \\ cx + dy \end{pmatrix}$. So

$$ax + by = 0,$$
$$cx + dy - 0.$$

Multiplying the first equation by d and the second by b and subtracting, we get

$$(ad - bc)x = 0,$$

and hence $x = 0$. Similarly, we get $y = 0$. Hence $\mathbf{X} = \begin{pmatrix} x \\ y \end{pmatrix} = \begin{pmatrix} 0 \\ 0 \end{pmatrix}$. Thus

$X = 0$ is the only vector with $A(X) = 0$, so (7) holds. The proposition is proved.

We saw earlier that if A has an inverse, then (7) holds and so $ad - bc \neq 0$. Let us now proceed in the converse direction.

Consider a linear transformation A with matrix $\begin{pmatrix} a & b \\ c & d \end{pmatrix}$. Assume

$$ad - bc \neq 0. \tag{8}$$

We seek an inverse B for A. Set $m(B) = \begin{pmatrix} p & q \\ r & s \end{pmatrix}$, where p, q, r, s are unknown numbers. We must have

$$\begin{pmatrix} a & b \\ c & d \end{pmatrix}\begin{pmatrix} p & q \\ r & s \end{pmatrix} = \begin{pmatrix} 1 & 0 \\ 0 & 1 \end{pmatrix},$$

so

$$\begin{aligned} ap + br &= 1, \\ cp + dr &= 0, \end{aligned} \tag{9}$$

and

$$\begin{aligned} aq + bs &= 0, \\ cq + ds &= 1. \end{aligned}$$

Hence

$$\begin{aligned} dap + dbr &= d, \\ bcp + bdr &= 0, \end{aligned}$$

and so

$$(ad - bc)p = d,$$

and, since (8) holds, we get

$$p = \frac{d}{ad - bc}.$$

To simplify the notation, we set $\Delta = ad - bc$.

Exercise 4. Using the system (9), show that

$$r = \frac{-c}{\Delta}. \tag{10}$$

Exercise 5. Using the relations

$$\begin{aligned} aq + bs &= 0, \\ cq + ds &= 1, \end{aligned}$$

show that

$$q = \frac{-b}{\Delta} \quad \text{and} \quad s = \frac{a}{\Delta}. \tag{11}$$

We have obtained

$$
\begin{bmatrix} p & q \\ r & s \end{bmatrix} = \begin{bmatrix} \dfrac{d}{\Delta} & \dfrac{-b}{\Delta} \\ \dfrac{-c}{\Delta} & \dfrac{a}{\Delta} \end{bmatrix}.
$$

Exercise 6. Calculate

$$
\begin{bmatrix} \dfrac{d}{\Delta} & \dfrac{-b}{\Delta} \\ \dfrac{-c}{\Delta} & \dfrac{a}{\Delta} \end{bmatrix} \begin{bmatrix} a & b \\ c & d \end{bmatrix} \quad \text{and} \quad \begin{bmatrix} a & b \\ c & d \end{bmatrix} \begin{bmatrix} \dfrac{d}{\Delta} & \dfrac{-b}{\Delta} \\ \dfrac{-c}{\Delta} & \dfrac{a}{\Delta} \end{bmatrix}.
$$

Show that both products equal $\begin{pmatrix} 1 & 0 \\ 0 & 1 \end{pmatrix}$.

Earlier we saw that if A has an inverse, then $ad - bc \neq 0$. Combining this with Exercise 6, we have

Theorem 2.4. *Let A be a linear transformation with matrix* $\begin{pmatrix} a & b \\ c & d \end{pmatrix}$.

  (i) *If A has an inverse, then $ad - bc \neq 0$;*
 (ii) *If $ad - bc \neq 0$, then A has an inverse B and*

$$
m(B) = \begin{bmatrix} \dfrac{d}{\Delta} & \dfrac{-b}{\Delta} \\ \dfrac{-c}{\Delta} & \dfrac{a}{\Delta} \end{bmatrix}, \tag{12}
$$

where Δ denotes $ad - bc$.

EXAMPLE 5. Let A have the matrix $\begin{pmatrix} 1 & 2 \\ 3 & 4 \end{pmatrix}$. Since $1 \cdot 4 - 2 \cdot 3 = -2 \neq 0$, A has an inverse A^{-1}. The matrix of A^{-1} is

$$
\begin{bmatrix} \dfrac{4}{-2} & \dfrac{-2}{-2} \\ \dfrac{-3}{-2} & \dfrac{1}{-2} \end{bmatrix} = \begin{bmatrix} -2 & 1 \\ \dfrac{3}{2} & -\dfrac{1}{2} \end{bmatrix}.
$$

EXAMPLE 6. Solve the system

$$
\begin{aligned} x + 2y &= 4, \\ 3x + 4y &= 0 \end{aligned} \tag{13}
$$

for x and y.
   We write the system in the form

$$
A\begin{pmatrix} x \\ y \end{pmatrix} = \begin{pmatrix} 1 & 2 \\ 3 & 4 \end{pmatrix}\begin{pmatrix} x \\ y \end{pmatrix} = \begin{pmatrix} 4 \\ 0 \end{pmatrix}.
$$

By the preceding example, A^{-1} has the matrix $\begin{pmatrix} -2 & 1 \\ 3/2 & -1/2 \end{pmatrix}$. Hence,

$$\begin{pmatrix} x \\ y \end{pmatrix} = A^{-1}\begin{pmatrix} 4 \\ 0 \end{pmatrix} = \begin{pmatrix} -2 & 1 \\ 3/2 & -1/2 \end{pmatrix}\begin{pmatrix} 4 \\ 0 \end{pmatrix} = \begin{pmatrix} -8 \\ 6 \end{pmatrix}.$$

Hence, the solution is

$$x = -8, \qquad y = 6.$$

§2. Systems of Linear Equations

We consider the following system of two equations in two unknowns:

$$ax + by = u, \tag{14}$$

$$cx + dy = v.$$

For each choice of numbers u, v, we may ask: Does the system (14) have a solution x, y? And if (14) has a solution, is this solution unique?

We may write the above system in matrix form by introducing the linear transformation A with matrix $\begin{pmatrix} a & b \\ c & d \end{pmatrix}$. Then the system (14) may be written

$$A(\mathbf{X}) = \mathbf{U}, \tag{15}$$

where $\mathbf{X}$ is the vector $\begin{pmatrix} x \\ y \end{pmatrix}$ and $\mathbf{U} = \begin{pmatrix} u \\ v \end{pmatrix}$.

Suppose that the transformation A has an inverse A^{-1}. For given vector $\mathbf{U}$,

$$A(A^{-1}(\mathbf{U})) = (AA^{-1})(\mathbf{U}) = \mathbf{U},$$

so $\mathbf{X} = A^{-1}(\mathbf{U})$ is a solution of (15). Conversely, if $\mathbf{X}$ is a solution of (15), then

$$\mathbf{X} = A^{-1}(A(\mathbf{X})) = A^{-1}(\mathbf{U}).$$

So (15) has the unique solution $\mathbf{X} = A^{-1}(\mathbf{U})$.

In particular, if $\mathbf{U} = 0$, we find that $\mathbf{X} = \begin{pmatrix} 0 \\ 0 \end{pmatrix} = A^{-1}\begin{pmatrix} 0 \\ 0 \end{pmatrix}$ is the unique solution of the system

$$ax + by = 0,$$
$$cx + dy = 0. \tag{16}$$

This system, with zero on the right-hand side, is called the *homogeneous system* associated with the system (14).

No matter what the matrix $\begin{pmatrix} a & b \\ c & d \end{pmatrix}$ is, the homogeneous system has at least one solution, the solution $\mathbf{X} = \begin{pmatrix} 0 \\ 0 \end{pmatrix}$. This is called the *trivial solution* of the homogeneous system, and we have seen above that if A has an inverse, then the trivial solution is the only solution of the homogeneous system. If A does not have an inverse, then by Theorem 2.4, $ad - bc = 0$, and so by Proposition 1 in §1 of this chapter, there is a nonzero vector $\mathbf{X} = \begin{pmatrix} x \\ y \end{pmatrix}$ with $A(\mathbf{X}) = \mathbf{0}$. Then x, y is a nontrivial solution of the homogeneous system (16).

What is the totality of solutions of (16)? If a, b, c, d are all 0, then every vector $\begin{pmatrix} x \\ y \end{pmatrix}$ in the plane is a solution. If a and b are both 0, but c and d are not both 0, then the solutions are all $\begin{pmatrix} x \\ y \end{pmatrix}$ with $cx + dy = 0$, and so the totality of solutions is the line $cx + dy = 0$. A similar statement holds if c and d are both 0, but a and b are not both 0.

Finally, if A does not have an inverse and either $a \neq 0$ or $b \neq 0$ and, also, either $c \neq 0$ or $d \neq 0$, we may conclude that the totality of solutions of (16) is the line through the origin orthogonal to $\begin{pmatrix} a \\ b \end{pmatrix}$.

We can summarize what we have found so far in the following two propositions.

Proposition 2. *The system* (14) *has a unique solution for every vector* $\begin{pmatrix} u \\ v \end{pmatrix}$ *if and only if the transformation A has an inverse.*

Proposition 3. *The homogeneous system* (16) *has a nontrivial solution if and only if A fails to have an inverse. In this case the totality of solutions of* (16) *is either the whole plane or a line through the origin.*

Now suppose that A fails to have an inverse, and $A \neq 0$. Then the solutions of (16) form a line through the origin, or, in other words, if we fix one nonzero solution $\mathbf{X}^h$ of (16), then every solution of (16) equals $t\mathbf{X}^h$ for some scalar t. If $\mathbf{X}$ and $\overline{\mathbf{X}}$ are two solutions of the system $A(\mathbf{X}) = \mathbf{U}$, then

$$A(\mathbf{X} - \overline{\mathbf{X}}) = A(\mathbf{X}) - A(\overline{\mathbf{X}}) = \mathbf{U} - \mathbf{U} = \mathbf{0},$$

so $\mathbf{X} - \overline{\mathbf{X}}$ is a solution of (3). Hence $\mathbf{X} - \overline{\mathbf{X}} = t\mathbf{X}^h$ and so for some t,

$$\mathbf{X} = \overline{\mathbf{X}} + t\mathbf{X}^h.$$

We can therefore describe all solutions of the nonhomogeneous system (15) in the following way:

Proposition 4. *Assume A is not the zero transformation. If A does not have an inverse, then if $\overline{\mathbf{X}}$ is a particular solution of* (15), *so that $A(\overline{\mathbf{X}}) = \mathbf{U}$, we may express every solution of* (15) *in the form $\overline{\mathbf{X}} + t\mathbf{X}^h$, where $\mathbf{X}^h$ is a non-trivial solution of the homogeneous system* (16).

EXAMPLE 7. Find all solutions of the system

$$x + 2y = 3,$$
$$-2x - 4y = -6. \tag{17}$$

The corresponding homogeneous system is

$$x + 2y = 0,$$
$$-2x - 4y = 0.$$

and the solutions to this system are the multiples $t\begin{pmatrix} -2 \\ 1 \end{pmatrix}$ of a vector $\mathbf{X}^h$ perpendicular to $\begin{pmatrix} 1 \\ 2 \end{pmatrix}$ and $\begin{pmatrix} -2 \\ -4 \end{pmatrix}$. We observe that $\begin{pmatrix} x \\ y \end{pmatrix} = \begin{pmatrix} 1 \\ 1 \end{pmatrix}$ is a particular solution of the system (17), so it follows, by the above proposition, that the set of all solutions is given by

$$\mathbf{X} = \begin{pmatrix} 1 \\ 1 \end{pmatrix} + t\begin{pmatrix} -2 \\ 1 \end{pmatrix} = \begin{pmatrix} 1 - 2t \\ 1 + t \end{pmatrix}.$$

EXAMPLE 8. Find all solutions of the system

$$x + 2y = 3,$$
$$-2x + 4y = -6. \tag{18}$$

In this case the matrix $\begin{pmatrix} 1 & 2 \\ -2 & 4 \end{pmatrix}$ has an inverse, $\frac{1}{8}\begin{pmatrix} 4 & -2 \\ 2 & 1 \end{pmatrix}$, so the unique solution to the system (18) is given by

$$\frac{1}{8}\begin{pmatrix} 4 & -2 \\ 2 & 1 \end{pmatrix}\begin{pmatrix} 3 \\ -6 \end{pmatrix} = \begin{pmatrix} 3 \\ 0 \end{pmatrix}.$$

EXAMPLE 9. Find all solutions of the system

$$x + 2y = 3,$$
$$-2x - 4y = 5. \tag{19}$$

In this case the system (19) has no solution. If we had a solution to the first equation, we could multiply both sides of the equation by -2, to get

$$-2x - 4y = -6,$$

and this is inconsistent with the second equation,

$$-2x - 4y = 5.$$

More generally, we can get a solution of the system

$$x + 2y = u,$$
$$-2x - 4y = v$$

if and only if

$$-2x - 4y = -2u$$

and

$$-2x - 4y = v$$

are consistent, i.e., if $-2u = v$. For example, in the system (17), we have $u = 3, v = -6$.

Exercise 7. Find all solutions of the following systems.

(a) $2x + y = 0$,
    $3x - y = 0$.
(b) $2x + y = 0$,
    $-4x - 2y = 0$.

Exercise 8. Find all solutions of the following systems.

(a) $2x + y = 1$,
    $3x - y = 1$.
(b) $2x + y = 1$,
    $-4x - 2y = 1$.

Exercise 9. Find all solutions of the system

$$2x + y = 1,$$
$$-4x - 2y = -2.$$

Exercise 10. Find all solutions of the system

$$x + y = 10,$$
$$5x + 5y = 50.$$

Exercise 11. For what choices of the numbers u, v does the system

$$x + y = u,$$
$$5x + 5y = v$$

have a solution?

§3. Inverses of Shears and Permutations

Recall the elementary matrices, H_k, J_k, and K which we discussed at the end of Chapter 2.3. We had

$$m(H_k) = \begin{pmatrix} 1 & 0 \\ k & 1 \end{pmatrix}, \quad m(J_k) = \begin{pmatrix} 1 & k \\ 0 & 1 \end{pmatrix}, \quad m(K) = \begin{pmatrix} 0 & 1 \\ 1 & 0 \end{pmatrix}.$$

Exercise 12. Show that

$$H_k^{-1} = H_{-k}, \qquad J_k^{-1} = J_{-k}, \qquad K^{-1} = K.$$

Determinants

Let A be a linear transformation with matrix $\begin{pmatrix} a & b \\ c & d \end{pmatrix}$. The quantity

$$ad - bc$$

is called the *determinant* of the matrix $\begin{pmatrix} a & b \\ c & d \end{pmatrix}$ and is denoted

$$\begin{vmatrix} a & b \\ c & d \end{vmatrix}. \tag{1}$$

Expressed in these terms, Theorem 2.4 states that A has an inverse if and only if $\begin{vmatrix} a & b \\ c & d \end{vmatrix} \neq 0$. We shall see that the determinant gives us further information about the behavior of A.

Consider a pair of vectors $\mathbf{X}_1, \mathbf{X}_2$ regarded as an *ordered* pair with $\mathbf{X}_1$ first and $\mathbf{X}_2$ second. Denote by α the angle from $\mathbf{X}_1$ to $\mathbf{X}_2$, measured counterclockwise, and assume that $\alpha \neq 0$ and $\alpha \neq \pi$.

If $\sin \alpha > 0$, we say that the pair $\mathbf{X}_1, \mathbf{X}_2$ is *positively oriented*. This holds exactly when α lies between 0 and π (see Fig. 2.34). If $\sin \alpha < 0$, we say the pair $\mathbf{X}_1, \mathbf{X}_2$ is *negatively oriented*. This holds if α is between π and 2π (see Fig. 2.35).

EXAMPLE 1. The pair $\mathbf{E}_1, \mathbf{E}_2$ is positively oriented (see Fig. 2.36). The pair $\mathbf{E}_1, -\mathbf{E}_2$ is negatively oriented (see Fig. 2.37). The pair $\mathbf{E}_2, \mathbf{E}_1$ is negatively oriented (see Fig. 2.38). The pair $\begin{pmatrix} 1 \\ 2 \end{pmatrix}, \begin{pmatrix} -1 \\ 2 \end{pmatrix}$ is positively oriented (see Fig. 2.39).

We saw in (20), Chapter 2.0, that if $\begin{pmatrix} x \\ y \end{pmatrix}$ and $\begin{pmatrix} u \\ v \end{pmatrix}$ are two vectors and if α

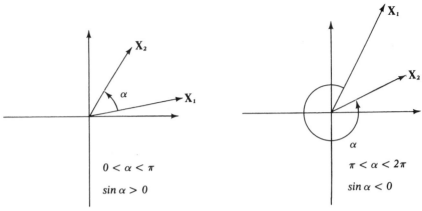

Figure 2.34   Figure 2.35

is the angle from $\begin{pmatrix} x \\ y \end{pmatrix}$ to $\begin{pmatrix} u \\ v \end{pmatrix}$, then

$$\sin \alpha = \frac{xv - yu}{\sqrt{x^2 + y^2} \sqrt{u^2 + v^2}} .$$

Now let $\mathbf{X_1} = \begin{pmatrix} x_1 \\ y_1 \end{pmatrix}$, $\mathbf{X_2} = \begin{pmatrix} x_2 \\ y_2 \end{pmatrix}$ be a given pair of vectors. How can we tell from the numbers x_1, y_1, x_2, y_2 whether or not the pair $\mathbf{X_1}, \mathbf{X_2}$ is positively oriented? Let α denote the angle from $\mathbf{X_1}$ to $\mathbf{X_2}$, measured counterclockwise. By the preceding,

$$\sin \alpha = \frac{x_1 y_2 - y_1 x_2}{\sqrt{x_1^2 + y_1^2} \sqrt{x_2^2 + y_2^2}} . \tag{2}$$

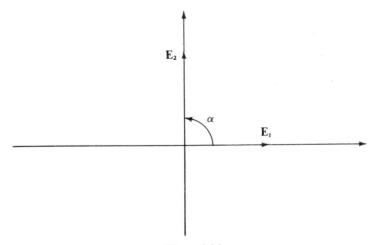

Figure 2.36

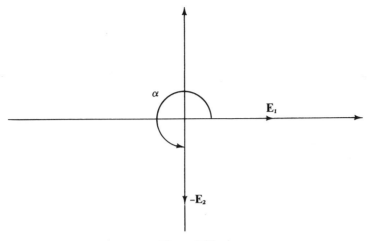

Figure 2.37

Hence, $\sin \alpha > 0$ if and only if $x_1 y_2 - y_1 x_2 > 0$. But $x_1 y_2 - y_1 x_2 = \begin{vmatrix} x_1 & x_2 \\ y_1 & y_2 \end{vmatrix}$. So, we conclude:

The pair X_1, X_2 is positively oriented if

and only if the determinant $\begin{vmatrix} x_1 & x_2 \\ y_1 & y_2 \end{vmatrix} > 0.$ (3)

Next let A be a linear transformation which has an inverse. We say that A *preserves orientation* if whenever X_1, X_2 is a positively oriented pair of

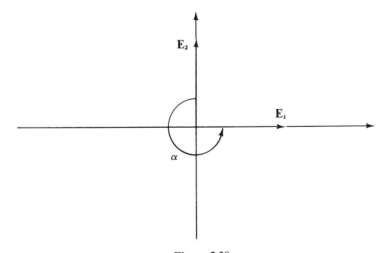

Figure 2.38

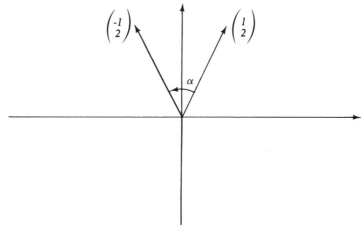

Figure 2.39

vectors, then the pair $A(\mathbf{X}_1), A(\mathbf{X}_2)$ of image-vectors is again positively oriented.

EXAMPLE 2.

(a) Rotation by $R_{\pi/2}$ preserves orientation;
(b) D_3, stretching by 3, preserves orientation;
(c) reflection in the x-axis does not preserve orientation.

Let A be a linear transformation which preserves orientation and let $\begin{pmatrix} a & b \\ c & d \end{pmatrix}$ be its matrix. Set $\mathbf{E}_1 = \begin{pmatrix} 1 \\ 0 \end{pmatrix}$, $\mathbf{E}_2 = \begin{pmatrix} 0 \\ 1 \end{pmatrix}$. The pair $\mathbf{E}_1, \mathbf{E}_2$ is positively oriented. Hence, the pair $A(\mathbf{E}_1), A(\mathbf{E}_2)$ is positively oriented. $A(\mathbf{E}_1) = \begin{pmatrix} a \\ c \end{pmatrix}$, $A(\mathbf{E}_2) = \begin{pmatrix} b \\ d \end{pmatrix}$. So by (3), we have

$$\begin{vmatrix} a & b \\ c & d \end{vmatrix} > 0.$$

Thus, if A preserves orientation, then the determinant is positive. Conversely, suppose $\begin{pmatrix} a & b \\ c & d \end{pmatrix} > 0$ and let us see whether it follows that A preserves orientation. Let $\mathbf{X}_1 = \begin{pmatrix} x_1 \\ y_1 \end{pmatrix}$, $\mathbf{X}_2 = \begin{pmatrix} x_2 \\ y_2 \end{pmatrix}$ be a positively oriented pair of vectors. Then

$$A(\mathbf{X}_1) = \begin{pmatrix} a & b \\ c & d \end{pmatrix} \begin{pmatrix} x_1 \\ y_1 \end{pmatrix} = \begin{pmatrix} ax_1 + by_1 \\ cx_1 + dy_1 \end{pmatrix}$$

and

$$A(\mathbf{X_2}) = \begin{pmatrix} a & b \\ c & d \end{pmatrix} \begin{pmatrix} x_2 \\ y_2 \end{pmatrix} = \begin{pmatrix} ax_2 + by_2 \\ cx_2 + dy_2 \end{pmatrix}.$$

The pair $A(\mathbf{X_1}), A(\mathbf{X_2})$ is positively oriented, by (3), if and only if the determinant

$$\begin{vmatrix} ax_1 + by_1 & ax_2 + by_2 \\ cx_1 + dy_1 & cx_2 + dy_2 \end{vmatrix} > 0.$$

This determinant equals

$$(ax_1 + by_1)(cx_2 + dy_2) - (ax_2 + by_2)(cx_1 + dy_1)$$
$$= acx_1x_2 + bdy_1y_2 + adx_1y_2 + bcy_1x_2$$
$$\quad - acx_2x_1 - bdy_2y_1 - adx_2y_1 - bcy_2x_1$$
$$= ad(x_1y_2 - x_2y_1) - bc(x_1y_2 - x_2y_1)$$
$$= (ad - bc)(x_1y_2 - x_2y_1).$$

So we have found

$$\begin{vmatrix} ax_1 + by_1 & ax_2 + by_2 \\ cx_1 + dy_1 & cx_2 + dy_2 \end{vmatrix} = \begin{vmatrix} a & b \\ c & d \end{vmatrix} \cdot \begin{vmatrix} x_1 & x_2 \\ y_1 & y_2 \end{vmatrix}. \tag{4}$$

Since $\mathbf{X_1}, \mathbf{X_2}$ is positively oriented, the determinant $\begin{vmatrix} x_1 & x_2 \\ y_1 & y_2 \end{vmatrix} > 0$. By hypothesis, $\begin{vmatrix} a & b \\ c & d \end{vmatrix} > 0$. So

$$\begin{vmatrix} ax_1 + by_1 & ax_2 + by_2 \\ cx_1 + dy_1 & cx_2 + dy_2 \end{vmatrix} > 0,$$

and so the pair $A(\mathbf{X_1}), A(\mathbf{X_2})$ is positively oriented. If $\begin{vmatrix} a & b \\ c & d \end{vmatrix} < 0$, the same calculation shows that $A(\mathbf{X_1}), A(\mathbf{X_2})$ is negatively oriented. In the preceding, the pair $\mathbf{X_1}, \mathbf{X_2}$ could be any given positively oriented pair of vectors. So we have proved the following:

Theorem 2.5. *Let A be a linear transformation with matrix $\begin{pmatrix} a & b \\ c & d \end{pmatrix}$. If $\begin{vmatrix} a & b \\ c & d \end{vmatrix} > 0$, then A preserves orientation. If $\begin{vmatrix} a & b \\ c & d \end{vmatrix} < 0$, then A does not preserve orientation.*

Let us say that A *reverses orientation* if whenever $\mathbf{X_1}, \mathbf{X_2}$ is a positively oriented pair, then $A(\mathbf{X_1}), A(\mathbf{X_2})$ is a negatively oriented pair. If we look back over our preceding argument, we see that, in fact, we have shown: if $\begin{vmatrix} a & b \\ c & d \end{vmatrix} < 0$, then A reverses orientation.

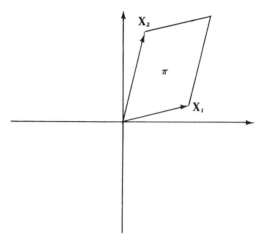

Figure 2.40

Next, we shall calculate the effect of a transformation A on *area*. Let A have the matrix $\begin{pmatrix} a & b \\ c & d \end{pmatrix}$ and assume $\begin{vmatrix} a & b \\ c & d \end{vmatrix} > 0$. Let π be a parallelogram, two of whose sides are the vectors $\mathbf{X}_1 = \begin{pmatrix} x_1 \\ y_1 \end{pmatrix}$ and $\mathbf{X}_2 = \begin{pmatrix} x_2 \\ y_2 \end{pmatrix}$, such that the pair $\mathbf{X}_1, \mathbf{X}_2$ is positively oriented.

Let $A(\pi)$ be the image of π under A, i.e., $A(\pi) = \{A(\mathbf{X}) \mid \mathbf{X} \text{ is a vector in }$ $\pi\}$ (see Figs. 2.40 and 2.41). By (3), $\begin{vmatrix} x_1 & x_2 \\ y_1 & y_2 \end{vmatrix} > 0$. By (25), Chapter 2.0,

$$\text{area}(\pi) = \begin{vmatrix} x_1 & x_2 \\ y_1 & y_2 \end{vmatrix}. \tag{5}$$

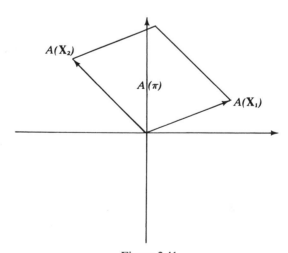

Figure 2.41

By Theorem 2.5 $A(\mathbf{X}_1), A(\mathbf{X}_2)$ is a positively oriented pair. By (5), with $A(\pi)$ replacing π,

$$\text{area}(A(\pi)) = \begin{vmatrix} ax_1 + by_1 & ax_2 + by_2 \\ cx_1 + dy_1 & cx_2 + dy_2 \end{vmatrix}.$$

By calculation (4), this determinant equals $\begin{vmatrix} a & b \\ c & d \end{vmatrix} \cdot \begin{vmatrix} x_1 & x_2 \\ y_1 & y_2 \end{vmatrix}$. Thus, we have

$$\text{area}(A(\pi)) = \begin{vmatrix} a & b \\ c & d \end{vmatrix} \text{area}(\pi). \tag{6}$$

If we instead assume that $\begin{vmatrix} a & b \\ c & d \end{vmatrix} < 0$ and perform the corresponding calculation, we get

$$\text{area}(A(\pi)) = -\begin{vmatrix} a & b \\ c & d \end{vmatrix} \text{area}(\pi). \tag{7}$$

We thus have:

Theorem 2.6. *Let A be a linear transformation with matrix $\begin{pmatrix} a & b \\ c & d \end{pmatrix}$ such that $\begin{vmatrix} a & b \\ c & d \end{vmatrix} \neq 0$. If π is any parallelogram with one vertex at 0, then*

$$\text{area}(A(\pi)) = \left(\text{absolute value of} \begin{vmatrix} a & b \\ c & d \end{vmatrix}\right) \text{area}(\pi).$$

We can derive an interesting consequence from Theorem 2.6. If C is a linear transformation, we write $\det C$ for the determinant of the matrix of C. Now let A, B be two linear transformations. Assume $\det A > 0$, $\det B > 0$. Then A preserves orientation and B preserves orientation. It follows that BA preserves orientation. Let Q be the unit square $Q = \{(x, y) \mid 0 \leqslant x \leqslant 1, 0 \leqslant y \leqslant 1\}$.

$$(BA)(Q) = B(A(Q)).$$

so

$$\text{area}((BA)((Q))) = \text{area}(B(A(Q))) = (\det B)\text{area}(A(Q)),$$

by Theorem 2.6. Hence,

$$\det(BA)\,\text{area}(Q) = (\det B)(\det A)\,\text{area}(Q).$$

It follows that

$$\det(BA) = (\det B)(\det A). \tag{8}$$

We have obtained this under the assumption $\det A > 0$ and $\det B > 0$. Recall the earlier result:

$$\begin{vmatrix} ax_1 + by_1 & ax_2 + by_2 \\ cx_1 + dy_1 & cx_2 + dy_2 \end{vmatrix} = \begin{vmatrix} a & b \\ c & d \end{vmatrix} \cdot \begin{vmatrix} x_1 & x_2 \\ y_1 & y_2 \end{vmatrix}.$$

For this formula, $\begin{pmatrix} a & b \\ c & d \end{pmatrix}$ and $\begin{pmatrix} x_1 & x_2 \\ y_1 & y_2 \end{pmatrix}$ are any two matrices, and on the left-hand side we have the matrix of $\begin{pmatrix} a & b \\ c & d \end{pmatrix}\begin{pmatrix} x_1 & x_2 \\ y_1 & y_2 \end{pmatrix}$. So (8) is true without restriction. We have:

Theorem 2.7. *If A, B are two linear transformations, then*

$$\det(BA) = (\det B)(\det A).$$

Exercise 1. Calculate the product of the matrices $\begin{pmatrix} 1 & 2 \\ 3 & 4 \end{pmatrix}$ and $\begin{pmatrix} -1 & 1 \\ 0 & 5 \end{pmatrix}$ and verify Theorem 2.7 when $A = \begin{pmatrix} 1 & 2 \\ 3 & 4 \end{pmatrix}$ and $B = \begin{pmatrix} -1 & 1 \\ 0 & 5 \end{pmatrix}$.

Exercise 2. Let Q be the square of side 1 whose edges are parallel to the coordinate axes and whose lower left-hand corner is at $\begin{pmatrix} 2 \\ 5 \end{pmatrix}$. If A is a linear transformation, define

$$A(Q) = \{ A(\mathbf{X}) \mid \text{the vector } \mathbf{X} \text{ is in } Q \}.$$

In each of the following cases, sketch $A(Q)$ and find area($A(Q)$).

(i) Matrix of A is $\begin{pmatrix} 1 & 0 \\ 1 & 1 \end{pmatrix}$.

(ii) Matrix of A is $\begin{pmatrix} 2 & 0 \\ 1 & 3 \end{pmatrix}$.

(iii) Matrix of A is $\begin{pmatrix} a & b \\ c & d \end{pmatrix}$.

Exercise 3. Show that the conclusion of Theorem 2.6 remains valid when π is any parallelogram, not necessarily with one vertex at 0.

Exercise 4. In this exercise Q_1, Q_2, etc., are rectangles with sides parallel to the axes. Q_1 is the square of side 10 with lower left-hand corner at $\begin{pmatrix} 0 \\ 0 \end{pmatrix}$. Q_2 and Q_3 are squares of side 2 with lower left-hand corners at $\begin{pmatrix} 2 \\ 6 \end{pmatrix}$ and $\begin{pmatrix} 6 \\ 6 \end{pmatrix}$, respectively. Q_4 is a square of side 1 with lower left-hand corner at $\begin{pmatrix} 4.5 \\ 4 \end{pmatrix}$. Q_5 is the rectangle of height 1, base 4, with lower left-hand corner at $\begin{pmatrix} 3 \\ 1 \end{pmatrix}$. We denote by W the region obtained by removing from Q_1 the figures Q_2, Q_3, Q_4, Q_5.

(a) Draw W on graph paper.

(b) Let A be the linear transformation having matrix $\begin{pmatrix} 1 & 0 \\ 1 & 1 \end{pmatrix}$.

(c) Draw the image $A(W)$ on graph paper.

(d) What is the area of $A(W)$?

§1. Isometries of the Plane

Let us find all linear transformation T which *preserve length*, i.e., such that for every segment, the length of the image of the segment under T equals the length of the segment, or, in other words, whenever $\mathbf{X}_1$ and $\mathbf{X}_2$ are two vectors, then

$$|T(\mathbf{X}_1) - T(\mathbf{X}_2)| = |\mathbf{X}_1 - \mathbf{X}_2|. \tag{9}$$

Such a transformation is called an *isometry*.

We know that $T(\mathbf{X}_1) - T(\mathbf{X}_2) = T(\mathbf{X}_1 - \mathbf{X}_2)$. So (9) says that

$$|T(\mathbf{X}_1 - \mathbf{X}_2)| = |\mathbf{X}_1 - \mathbf{X}_2|.$$

Hence (9) holds, provided we have

$$|T(\mathbf{X})| = |\mathbf{X}| \qquad \text{for every vector } \mathbf{X}. \tag{10}$$

Conversely, if (9) holds, we get (10) by setting $\mathbf{X}_1 = \mathbf{X}$, $\mathbf{X}_2 = \mathbf{0}$. So (9) and (10) are equivalent conditions.

Let T be a linear transformation satisfying (10) and denote by $\begin{pmatrix} a & b \\ c & d \end{pmatrix}$ the matrix of T. What consequences follow for the entries a, b, c, d from the fact that T preserves length, i.e., that (10) is true?

Set $\mathbf{X} = \begin{pmatrix} x \\ y \end{pmatrix}$. Then

$$T(\mathbf{X}) = \begin{pmatrix} a & b \\ c & d \end{pmatrix} \begin{pmatrix} x \\ y \end{pmatrix} = \begin{pmatrix} ax + by \\ cx + dy \end{pmatrix},$$

and so

$$|T(\mathbf{X})| = \sqrt{(ax + by)^2 + (cx + dy)^2}\;.$$

Since $|T(\mathbf{X})| = |\mathbf{X}| = \sqrt{x^2 + y^2}$, we have

$$\sqrt{(ax + by)^2 + (cx + dy)^2} = \sqrt{x^2 + y^2}\;,$$

and so, simplifying, we get

$$a^2x^2 + 2abxy + b^2y^2 + c^2x^2 + 2cdxy + d^2y^2 = x^2 + y^2,$$

i.e.,

$$(a^2 + c^2)x^2 + (b^2 + d^2)y^2 + (2ab + 2cd)xy = x^2 + y^2. \tag{11}$$

This holds for every vector $\mathbf{X} = \begin{pmatrix} x \\ y \end{pmatrix}$. Setting $\mathbf{X} = \begin{pmatrix} 1 \\ 0 \end{pmatrix}$, we get

(i) $a^2 + c^2 = 1$

and setting $\mathbf{X} = \begin{pmatrix} 0 \\ 1 \end{pmatrix}$, we get

(ii) $b^2 + d^2 = 1.$

Inserting this information in (11) and simplifying, we get

$$(2ab + 2cd)xy = 0.$$

Setting $x = 1$, $y = 1$ and dividing by 2, we get

(iii) $ab + cd = 0.$

Thus relations (i), (ii), and (iii) are consequences of (10). We can interpret these relations geometrically. Set

$$\mathbf{U} = \begin{pmatrix} a \\ c \end{pmatrix}, \qquad \mathbf{V} = \begin{pmatrix} b \\ d \end{pmatrix}.$$

Then (i), (ii), and (iii) say that:

$$|\mathbf{U}| = 1, \quad |\mathbf{V}| = 1, \quad \text{and} \quad \mathbf{U} \cdot \mathbf{V} = 0.$$

Since $|\mathbf{U}| = 1$, we can write $\begin{pmatrix} a \\ c \end{pmatrix} = \mathbf{U} = \begin{pmatrix} \cos\theta \\ \sin\theta \end{pmatrix}$, where θ is the polar angle of $\mathbf{U}$, so $a = \cos\theta$, $c = \sin\theta$.

Since $\mathbf{U} \cdot \mathbf{V} = 0$ and $|\mathbf{V}| = 1$, $\mathbf{V}$ is obtained from $\mathbf{U}$ either by a positive or a negative rotation by $\pi/2$. In the first case, $\begin{pmatrix} b \\ d \end{pmatrix} = \mathbf{V} = R_{\pi/2}\begin{pmatrix} \cos\theta \\ \sin\theta \end{pmatrix}$
$= \begin{pmatrix} -\sin\theta \\ \cos\theta \end{pmatrix}$, so $b = -\sin\theta$, $d = \cos\theta$. Hence,

$$\begin{pmatrix} a & b \\ c & d \end{pmatrix} = \begin{pmatrix} \cos\theta & -\sin\theta \\ \sin\theta & \cos\theta \end{pmatrix}. \tag{12}$$

In the second case, $\begin{pmatrix} b \\ d \end{pmatrix} = \mathbf{V} = R_{-\pi/2}\begin{pmatrix} \cos\theta \\ \sin\theta \end{pmatrix} = \begin{pmatrix} \sin\theta \\ -\cos\theta \end{pmatrix}$, so $b = \sin\theta$, $d = -\cos\theta$. Hence,

$$\begin{pmatrix} a & b \\ c & d \end{pmatrix} = \begin{pmatrix} \cos\theta & \sin\theta \\ \sin\theta & -\cos\theta \end{pmatrix}. \tag{13}$$

We recognize the matrix (12) as the matrix of the rotation R_θ (see Fig. 2.42). Also, we recall that in Chapter 2.3, we saw that $\begin{pmatrix} \cos 2\theta & \sin 2\theta \\ \sin 2\theta & -\cos 2\theta \end{pmatrix}$ is the matrix of the reflection in the line through the origin in which forms an angle θ with the positive x-axis. It follows that the matrix $\begin{pmatrix} \cos\theta & \sin\theta \\ \sin\theta & -\cos\theta \end{pmatrix}$ in (13) is the matrix of reflection through the line forming an angle $\frac{1}{2}\theta$ with the positive x-axis. So we have:

Theorem 2.8. *Let T be a length-preserving linear transformation, i.e., assume that T satisfies (10). Then either the matrix of T is $\begin{pmatrix} \cos\theta & -\sin\theta \\ \sin\theta & \cos\theta \end{pmatrix}$ for some number θ and then T is rotation R_θ, or else the matrix of T is $\begin{pmatrix} \cos\theta & \sin\theta \\ \sin\theta & -\cos\theta \end{pmatrix}$, and then T is reflection through the line through the origin which forms an angle of $\theta/2$ with the positive x-axis.*

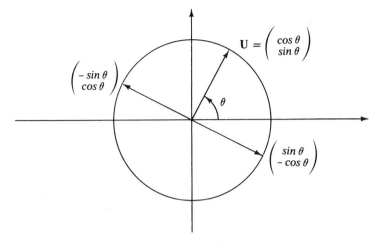

Figure 2.42

Exercise 5. For each of the following matrices, the corresponding transformation is either a rotation or a reflection. Decide which case occurs for each matrix. When it is a rotation, find the angle of rotation, and when it is a reflection, find the line in which it reflects.

(a) $\begin{pmatrix} -1/\sqrt{2} & 1/\sqrt{2} \\ 1/\sqrt{2} & 1/\sqrt{2} \end{pmatrix}$,

(b) $\begin{pmatrix} 3/5 & 4/5 \\ -4/5 & 3/5 \end{pmatrix}$,

(c) $\begin{pmatrix} 0 & 1 \\ 1 & 0 \end{pmatrix}$.

Exercise 6. A transformation T is length preserving and the matrix of T is $\begin{pmatrix} a & b \\ c & d \end{pmatrix}$.

(i) Show that the determinant $\begin{vmatrix} a & b \\ c & d \end{vmatrix}$ is either 1 or -1.

(ii) Show that T is a rotation when the determinant is 1 and a reflection when the determinant is -1.

Exercise 7. Let T_1, T_2 be two length-preserving transformations.

(a) Show that $T_1 T_2$ is length preserving.
(b) Show that if T_1 and T_2 are both reflections, then $T_1 T_2$ is a rotation.
(c) Show that if T_1 is a rotation and T_2 is a reflection, then $T_1 T_2$ is a reflection.

Exercise 8. Let T_1 be reflection through the line along $\begin{pmatrix} 1 \\ 1 \end{pmatrix}$ and let T_2 be reflection through the line along $\begin{pmatrix} 1 \\ 2 \end{pmatrix}$. Write $T_1 T_2$ in the form $T_1 T_2 = R_\theta$ and find the number θ.

Having studied the effect of a linear transformation on area and length, we can ask what happens to *angles*. Let L_1, L_2 be two rays beginning at the origin and let θ be the angle from L_1 to L_2, measured counterclockwise. Let

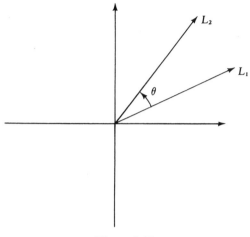

Figure 2.43

A be a linear transformation having an inverse. The images $A(L_1)$ and $A(L_2)$ are two new rays beginning at $\mathbf{0}$. Denote by θ' the angle from $A(L_1)$ to $A(L_2)$ (see Figs. 2.43 and 2.44). If $\theta' = \theta$ for each pair of rays L_1, L_2, then we say that A *preserves angle*.

We devote the next set of exercises to studying those linear transformations which preserve angles.

Exercise 9. Let A and B be two linear transformations which preserve angles. Show that the transformation AB and BA preserve angles.

Exercise 10.

(a) Show that each rotation R_θ preserves angles.
(b) Show that each stretching D_r preserves angles.

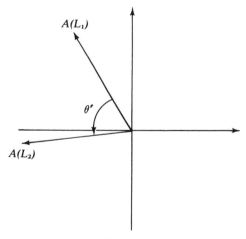

Figure 2.44

Exercise 11. Let R_θ and D_r be rotation by θ and stretching by t, respectively. Then

$$m(R_\theta) = \begin{pmatrix} \cos\theta & -\sin\theta \\ \sin\theta & \cos\theta \end{pmatrix}$$

and

$$m(D_r) = \begin{pmatrix} r & 0 \\ 0 & r \end{pmatrix}.$$

(a) Show that $D_r R_\theta$ preserves angles.
(b) Find the matrix $m(D_r R_\theta)$.
(c) Show there exist numbers a, b, not both 0, such that $m(D_r R_\theta) = \begin{pmatrix} a & b \\ -b & a \end{pmatrix}$.

Exercise 12. Show that if a, b are any two numbers not both 0, then the transformation whose matrix is $\begin{pmatrix} a & b \\ -b & a \end{pmatrix}$ preserves angles.

Exercise 13. Let A have the matrix $\begin{pmatrix} t_1 & 0 \\ 0 & t_2 \end{pmatrix}$. Show that A preserves angles if and only if $t_1 = t_2$.

In the following exercises, A denotes a linear transformation which preserves angles and $\begin{pmatrix} a & b \\ c & d \end{pmatrix} = m(A)$.

Exercise 14. Set $E_1 = \begin{pmatrix} 1 \\ 0 \end{pmatrix}$, $E_2 = \begin{pmatrix} 0 \\ 1 \end{pmatrix}$. Then $A(E_1) = \begin{pmatrix} a \\ c \end{pmatrix}$, $A(E_2) = \begin{pmatrix} b \\ d \end{pmatrix}$. Show that the angle from $\begin{pmatrix} a \\ c \end{pmatrix}$ to $\begin{pmatrix} b \\ d \end{pmatrix}$ is $\pi/2$.

Exercise 15. Write $\begin{pmatrix} a \\ c \end{pmatrix}$ in the form $\begin{pmatrix} a \\ c \end{pmatrix} = \begin{pmatrix} t_1\cos\theta \\ t_1\sin\theta \end{pmatrix}$ where $t_1 > 0$. Show that $\begin{pmatrix} b \\ d \end{pmatrix}$ can be expressed in the form $\begin{pmatrix} b \\ d \end{pmatrix} = \begin{pmatrix} -t_2\sin\theta \\ t_2\cos\theta \end{pmatrix}$, where $t_2 > 0$.

Exercise 16. By the preceding,

$$\begin{pmatrix} a & b \\ c & d \end{pmatrix} = \begin{pmatrix} t_1\cos\theta & -t_2\sin\theta \\ t_1\sin\theta & t_2\cos\theta \end{pmatrix}.$$

Show that

$$\begin{pmatrix} a & b \\ c & d \end{pmatrix} = \begin{pmatrix} \cos\theta & -\sin\theta \\ \sin\theta & \cos\theta \end{pmatrix}\begin{pmatrix} t_1 & 0 \\ 0 & t_2 \end{pmatrix}.$$

Exercise 17. Denoting by B the transformation with matrix $\begin{pmatrix} t_1 & 0 \\ 0 & t_2 \end{pmatrix}$, by the preceding exercise, we get $A = R_\theta B$.

(a) Show that B preserves angles.
(b) Using Exercise 13, show that $t_1 = t_2$ and deduce that B equals the stretching D_r, where $r = t_1$.

Exercise 18. Use the preceding exercise to prove the following result: if A is a linear transformation which preserves angles, then A is the product of a stretching and a rotation.

Exercise 19. Show that if A is a linear transformation which preserves angles, then the matrix of A has the form $\begin{pmatrix} a & b \\ -b & a \end{pmatrix}$.

§2. Determinants of Shears and Permutations

Recall the elementary matrices

$$m(H_k) = \begin{pmatrix} 1 & 0 \\ k & 1 \end{pmatrix}, \quad m(J_k) = \begin{pmatrix} 1 & k \\ 0 & 1 \end{pmatrix}, \quad m(K) = \begin{pmatrix} 0 & 1 \\ 1 & 0 \end{pmatrix}.$$

Exercise 20. Find the determinants for $m(H_k)$, $m(J_k)$, and $m(K)$.

Eigenvalues

EXAMPLE 1. Let L be a line through the origin and let S be the transformation which reflects each vector in L. If X is on the line L, then $S(X) = X$. If X is on the line L' which goes through the origin and is perpendicular to L, then $S(X) = -X$.

Let T be a linear transformation. Fix a scalar t. If there is a vector $X \neq 0$ such that $T(X) = tX$, then we say that t is an *eigenvalue* of T. If t is an eigenvalue of T, then each vector Y such that $T(Y) = tY$ is called an *eigenvector* corresponding to t.

In the preceding example, $t = 1$ and $t = -1$ are eigenvalues of the reflection S. Every vector Y on L is an eigenvector of S corresponding to $t = 1$, since $S(Y) = Y = 1 \cdot Y$. Every vector Y on L' is an eigenvector of S corresponding to $t = -1$, since $S(Y) = -Y = (-1) \cdot Y$ (see Fig. 2.45).

Let T be a linear transformation. A vector $X \neq 0$ is an eigenvector of T, corresponding to some eigenvalue, if and only if T takes X into a scalar multiple of itself. In other words, X is an eigenvector of T if and only if X and $T(X)$ lie on the same straight line through the origin (see Fig. 2.46).

EXAMPLE 2. Let D_r be stretching by r. Then for every vector X, $D_r(X) = rX$. Hence, r is an eigenvalue of D_r. Every vector X is an eigenvector of D corresponding to the eigenvalue r.

EXAMPLE 3. Let $R_{\pi/2}$ be rotation by $\pi/2$ radians. If X is any vector $\neq 0$, it is clear that X and $R_{\pi/2}(X)$ do not lie on the same straight line through the origin. It follows that $R_{\pi/2}$ has *no* eigenvalue (see Fig. 2.47).

Exercise 1. Let L be a straight line through the origin and let P be the transformation which projects each vector X to L (see Fig. 2.48).

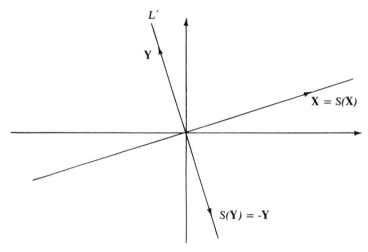

Figure 2.45

(a) Show that 0 and 1 are eigenvalues of P.
(b) Find all the eigenvectors which correspond to each of these eigenvalues.
(c) Show that P has no eigenvalues except for 0 and 1.

Exercise 2. Find all eigenvalues and corresponding eigenvectors for each of the following transformations:

(a) Rotation by π radians,
(b) I,
(c) 0.

Let A be a linear transformation and $\begin{pmatrix} a & b \\ c & d \end{pmatrix}$ its matrix. Assume t is an eigenvalue of A and $\begin{pmatrix} x \\ y \end{pmatrix}$ is a corresponding eigenvector with $\begin{pmatrix} x \\ y \end{pmatrix} \neq \begin{pmatrix} 0 \\ 0 \end{pmatrix}$.

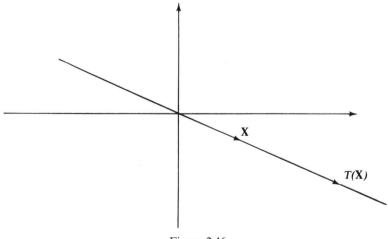

Figure 2.46

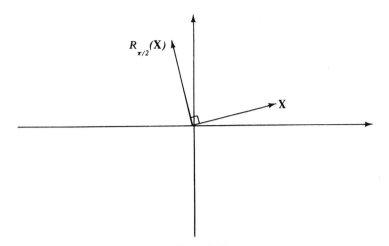

Figure 2.47

Then $A\begin{pmatrix} x \\ y \end{pmatrix} = t\begin{pmatrix} x \\ y \end{pmatrix}$, so

$$\begin{pmatrix} a & b \\ c & d \end{pmatrix}\begin{pmatrix} x \\ y \end{pmatrix} = t\begin{pmatrix} x \\ y \end{pmatrix},$$

or

$$ax + by = tx,$$
$$cx + dy = ty,$$

and so

$$(a - t)x + by = 0,$$
$$cx + (d - t)y = 0.$$

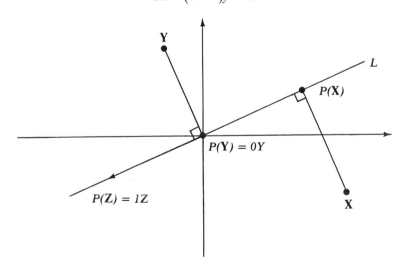

Figure 2.48

But x, y are not both 0. So we can apply Proposition 1, Chapter 2.4, and conclude that

$$(a - t)(d - t) - bc = 0. \tag{1}$$

Equation (1) can be expressed in the equivalent forms:

$$\begin{vmatrix} a - t & b \\ c & d - t \end{vmatrix} = 0, \tag{2}$$

and

$$t^2 - (a + d)t + (ad - bc) = 0. \tag{3}$$

We call Eq. (3) the *characteristic equation* of A. We have seen that if t is an eigenvalue of A, then t is a root of Eq. (3). Furthermore, since t is a real number by definition, t is a real root.

Conversely, suppose that t is a real root of (3). Let us show that t is then an eigenvalue of A. By assumption,

$$\begin{vmatrix} a - t & b \\ c & d - t \end{vmatrix} = 0.$$

By Proposition 1, Chapter 2.4, there exists a pair of scalars x, y with $\begin{pmatrix} x \\ y \end{pmatrix} \neq \begin{pmatrix} 0 \\ 0 \end{pmatrix}$ such that $\begin{pmatrix} a - t & b \\ c & d - t \end{pmatrix}\begin{pmatrix} x \\ y \end{pmatrix} = \begin{pmatrix} 0 \\ 0 \end{pmatrix}$, and so

$$(a - t)x + by = 0$$

and

$$cx + (d - t)y = 0.$$

This implies that

$$ax + by = tx,$$
$$cx + dy = ty,$$

so

$$\begin{pmatrix} a & b \\ c & d \end{pmatrix}\begin{pmatrix} x \\ y \end{pmatrix} = t\begin{pmatrix} x \\ y \end{pmatrix}.$$

Thus, t is an eigenvalue of A, and $\begin{pmatrix} x \\ y \end{pmatrix}$ is an eigenvector corresponding to t. In summary, we now know:

Theorem 2.9. *A real number t is an eigenvalue of the linear transformation A if and only if t is a root of the characteristic equation of A.*

Exercise 3. Find the characteristic equation and calculate its roots for each of the following transformations:

(a) stretching D_r;
(b) rotation R_θ;
(c) reflection in the y-axis;
(d) reflection in the line along $\begin{pmatrix} 1 \\ 1 \end{pmatrix}$.

Note. By an eigenvalue (or eigenvector) *of a matrix* we shall mean an eigenvalue (or eigenvector) of the corresponding linear transformation.

EXAMPLE 4. Find all eigenvalues and eigenvectors of the matrix $\begin{pmatrix} 3 & 4 \\ 4 & -3 \end{pmatrix}$.

SOLUTION. The characteristic equation here is

$$\begin{vmatrix} 3-t & 4 \\ 4 & -3-t \end{vmatrix} = 0,$$

or

$$t^2 - 25 = 0.$$

Its roots are $t = 5$ and $t = -5$. So, the values 5 and -5 are the eigenvalues. Let us find the eigenvectors which correspond to $t = 5$. We seek $\begin{pmatrix} x \\ y \end{pmatrix}$ with

$$\begin{pmatrix} 3 & 4 \\ 4 & -3 \end{pmatrix}\begin{pmatrix} x \\ y \end{pmatrix} = 5\begin{pmatrix} x \\ y \end{pmatrix}.$$

Thus,

$$3x + 4y = 5x,$$
$$4x - 3y = 5y,$$

so

$$-2x + 4y = 0,$$
$$4x - 8y = 0.$$

It follows that $x = 2y$. Thus, an eigenvector with eigenvalue 5 has the form

$$\begin{pmatrix} 2y \\ y \end{pmatrix}.$$

Conversely, *every* vector of this form is an eigenvector for

$$\begin{pmatrix} 3 & 4 \\ 4 & -3 \end{pmatrix}\begin{pmatrix} 2y \\ y \end{pmatrix} = \begin{pmatrix} 10y \\ 5y \end{pmatrix} = 5\begin{pmatrix} 2y \\ y \end{pmatrix}.$$

Notice that the eigenvectors we have found fill up a straight line through the origin. Try to find the eigenvectors of $\begin{pmatrix} 3 & 4 \\ 4 & -3 \end{pmatrix}$ which have -5 as their eigenvalue.

We now turn our attention to a class of matrices which occur in many applications of linear algebra, the *symmetric matrices*. A matrix $\begin{pmatrix} a & b \\ c & d \end{pmatrix}$ is called *symmetric* if $b = c$, i.e., if the matrix has the form $\begin{pmatrix} s & t \\ t & u \end{pmatrix}$. The matrix $\begin{pmatrix} 3 & 4 \\ 4 & -3 \end{pmatrix}$, which we studied in Example 4, is symmetric.

Let A be a linear transformation and assume $m(A) = \begin{pmatrix} a & b \\ b & c \end{pmatrix}$, so that $m(A)$ is symmetric. The characteristic equation of A is

$$\begin{vmatrix} a - t & b \\ b & c - t \end{vmatrix} = 0,$$

or

$$(a - t)(c - t) - b^2 = t^2 - (a + c)t + (ac - b^2) = 0. \tag{4}$$

The roots of (4) are

$$t = \frac{a + c \pm \sqrt{(a + c)^2 - 4(ac - b^2)}}{2}.$$

Simplifying, we obtain

$$(a + c)^2 - 4(ac - b^2) = a^2 + 2ac + c^2 - 4ac + 4b^2$$
$$= a^2 - 2ac + c^2 + 4b^2 = (a - c)^2 + 4b^2.$$

Since $(a - c)^2 + 4b^2 \geqslant 0$, its square root is a real number. We consider two possible cases:

(i) $(a - c)^2 + 4b^2 = 0$;

then $a = c$ and $b = 0$, so $m(A) = \begin{pmatrix} a & b \\ b & c \end{pmatrix} = \begin{pmatrix} a & 0 \\ 0 & a \end{pmatrix}$, and so A is stretching, $A = aI$.

(ii) $(a - c)^2 + 4b^2 > 0$;
then (4) has the two distinct real roots

$$t_1 = \frac{(a + c) + \sqrt{(a - c)^2 + 4b^2}}{2}$$

and

$$t_2 = \frac{(a + c) - \sqrt{(a - c)^2 + 4b^2}}{2}.$$

By Theorem 2.9, t_1 and t_2 are eigenvalues of A. We have proved:

Proposition 1. *Let A be a linear transformation with symmetric matrix* $\begin{pmatrix} a & b \\ b & c \end{pmatrix}$. *Then either $A = aI$ or A has two distinct eigenvalues t_1, t_2 where*

$$t_1 = \tfrac{1}{2}\left((a + c) + \sqrt{(a - c)^2 + 4b^2}\right),$$

$$t_2 = \tfrac{1}{2}\left((a + c) - \sqrt{(a - c)^2 + 4b^2}\right).$$

Exercise 4. For each of the following matrices, find two eigenvalues of the corresponding linear transformation:

(i) $\begin{pmatrix} 3 & 0 \\ 0 & -5 \end{pmatrix}$,

(ii) $\begin{pmatrix} 1 & 1 \\ 1 & 1 \end{pmatrix}$,

(iii) $\begin{pmatrix} 0 & \sqrt{2} \\ \sqrt{2} & \pi \end{pmatrix}$.

Exercise 5. For each of the linear transformations in the preceding exercise, find one nonzero eigenvector corresponding to each eigenvalue. Show that in each case, if X_1, X_2 are eigenvectors corresponding to distinct eigenvalues, then X_1 and X_2 are orthogonal.

Exercise 5 suggests that the following theorem may be true.

Theorem 2.10. *Let A be a linear transformation with symmetric matrix* $\begin{pmatrix} a & b \\ b & c \end{pmatrix}$ *and let t_1, t_2 be distinct eigenvalues of A. Choose nonzero eigenvectors* X_1, X_2 *corresponding to t_1 and t_2, respectively. Then X_1 and X_2 are orthogonal.*

PROOF. First assume that $b \neq 0$. Set $X_1 = \begin{pmatrix} x_1 \\ y_1 \end{pmatrix}$, $X_2 = \begin{pmatrix} x_2 \\ y_2 \end{pmatrix}$. Then $\begin{pmatrix} a & b \\ b & c \end{pmatrix} \begin{pmatrix} x_1 \\ y_1 \end{pmatrix} = t_1 \begin{pmatrix} x_1 \\ y_1 \end{pmatrix}$. So

$$ax_1 + by_1 = t_1 x_1$$

or

$$by_1 = (t_1 - a)x_1 .$$

If $x_1 = 0$, then $by_1 = 0$, and since $b \neq 0$ by assumption, then $y_1 = 0$, so $X_1 = \begin{pmatrix} 0 \\ 0 \end{pmatrix}$, contrary to hypothesis. So, $x_1 \neq 0$ and

$$\frac{y_1}{x_1} = \frac{t_1 - a}{b} .$$

Similarly, $x_2 \neq 0$ and

$$\frac{y_2}{x_2} = \frac{t_2 - a}{b} .$$

Since y_1/x_1 and y_2/x_2 are the slopes of X_1 and X_2, to prove that X_1 and X_2 are orthogonal amounts to showing that

$$\left(\frac{y_1}{x_1} \right) \left(\frac{y_2}{x_2} \right) = -1, \quad \text{i.e.,} \quad \left(\frac{t_1 - a}{b} \right) \left(\frac{t_2 - a}{b} \right) = -1. \quad (5)$$

Try showing that (5) is true, using the values of t_1, t_2 obtained in the last theorem, before reading the rest of the proof.

By Proposition 1, in this chapter,

$$t_1 = \tfrac{1}{2}a + \tfrac{1}{2}c + \tfrac{1}{2}\sqrt{(a-c)^2 + 4b^2}$$

and

$$t_2 = \tfrac{1}{2}a + \tfrac{1}{2}c - \tfrac{1}{2}\sqrt{(a-c)^2 + 4b^2} \ .$$

So,

$$t_2 - a = \tfrac{1}{2}(c-a) - \tfrac{1}{2}\sqrt{(a-c)^2 + 4b^2}$$

and

$$t_1 - a = \tfrac{1}{2}(c-a) + \tfrac{1}{2}\sqrt{(a-c)^2 + 4b^2} \ .$$

Then

$$(t_1 - a)(t_2 - a) = \tfrac{1}{4}(c-a)^2 - \tfrac{1}{4}\left[(a-c)^2 + 4b^2\right] = -b^2.$$

Hence,

$$\left(\frac{t_1 - a}{b}\right)\left(\frac{t_2 - a}{b}\right) = \frac{-b^2}{b^2} = -1,$$

as desired. Thus, $\mathbf{X}_1$ and $\mathbf{X}_2$ are orthogonal.

Now if $b = 0$, then $\begin{pmatrix} a & b \\ b & c \end{pmatrix} = \begin{pmatrix} a & 0 \\ 0 & c \end{pmatrix}$. So $t_1 = a$, $\mathbf{X}_1 = \begin{pmatrix} x_1 \\ 0 \end{pmatrix}$ and $t_2 = c$, $\mathbf{X}_2 = \begin{pmatrix} 0 \\ y_2 \end{pmatrix}$. Since $\begin{pmatrix} x_1 \\ 0 \end{pmatrix}$ and $\begin{pmatrix} 0 \\ y_2 \end{pmatrix}$ are orthogonal, the desired conclusion holds here as well.

An alternative proof of Theorem 2.10 can be obtained from the following exercises.

Exercise 6. Let A be the linear transformation which occurs in Theorem 2.10. Let $\mathbf{X}, \mathbf{Y}$ be any two vectors. Show that

$$A(\mathbf{X}) \cdot \mathbf{Y} = \mathbf{X} \cdot A(\mathbf{Y}).$$

Exercise 7. Let A be as in the preceding exercise and let t_1, t_2 be distinct eigenvalues of A, and $\mathbf{X}_1, \mathbf{X}_2$ the corresponding eigenvectors.

(a) Show $A(\mathbf{X}_1) \cdot \mathbf{X}_2 = t_1(\mathbf{X}_1 \cdot \mathbf{X}_2)$ and $A(\mathbf{X}_2) \cdot \mathbf{X}_1 = t_2(\mathbf{X}_1 \cdot \mathbf{X}_2)$.
(b) Using Exercise 6, deduce from (a) that $t_1(\mathbf{X}_1 \cdot \mathbf{X}_2) = t_2(\mathbf{X}_1 \cdot \mathbf{X}_2)$.
(c) Using the fact that $t_1 \neq t_2$, conclude that $\mathbf{X}_1 \cdot \mathbf{X}_2 = 0$. Thus, Theorem 2.10 holds.

Exercise 8. For each of the following matrices, find all eigenvalues and eigenvectors:

(a) $\begin{pmatrix} 1 & 2 \\ 3 & 4 \end{pmatrix}$,

(b) $\begin{pmatrix} 4 & 3 \\ 2 & 1 \end{pmatrix}$,

(c) $\begin{pmatrix} p & 0 \\ 0 & q \end{pmatrix}$,

(d) $\begin{pmatrix} 0 & 7 \\ 0 & 0 \end{pmatrix}$,

(e) $\begin{pmatrix} 0 & 5 \\ -1 & 1 \end{pmatrix}$,

(f) $\begin{pmatrix} \cos\theta & -\sin\theta \\ \sin\theta & \cos\theta \end{pmatrix}$,

(g) $\begin{pmatrix} \cos\theta & \sin\theta \\ \sin\theta & -\cos\theta \end{pmatrix}$.

Exercise 9. Given numbers a, b, c, d, show that the matrices $\begin{pmatrix} a & b \\ c & d \end{pmatrix}$ and $\begin{pmatrix} d & c \\ b & a \end{pmatrix}$

have the same eigenvalues.

Exercise 10. Denote by N a linear transformation such that $N^2 = 0$. Show that 0 is the only eigenvalue of N.

Exercise 11. Let B be a linear transformation such that B has the eigenvalue 0 and no other eigenvalue. Show that $B^2 = 0$.

Exercise 12. Let E be a linear transformation such that $E^2 = E$. What are the eigenvalues of E?

Exercise 13. Let C be a linear transformation such that C has eigenvalues 0 and 1. Show that $C^2 = C$.

Exercise 14. Let T be a linear transformation with nonzero eigenvectors X_1, X_2 and corresponding eigenvalues t_1, t_2, where $t_1 \neq t_2$.
   Set $S = (T - t_1 I)(T - t_2 I)$.

(a) Show that $S(X_2) = 0$.
(b) Show that $S = (T - t_2 I)(T - t_1 I)$.
(c) Show that $S(X_1) = 0$.
(d) Show that $S(c_1 X_1 + c_2 X_2) = 0$, where c_1, c_2 are given constants.
(e) Show that $S = 0$, i.e.,

$$(T - t_1 I)(T - t_2 I) = 0.$$

Exercise 15. Let T be a linear transformation and let $\begin{pmatrix} a & b \\ c & d \end{pmatrix}$ be its matrix. Assume T has eigenvalues t_1, t_2 with $t_1 \neq t_2$.
   Using part (e) of Exercise 14, show that

$$T^2 - (a + d)T + (ad - bc)I = 0. \tag{6}$$

Exercise 16. Let T be a linear transformation and let $\begin{pmatrix} a & b \\ c & d \end{pmatrix}$ be its matrix. Do not assume that T has any eigenvalues. Show by direct calculation that (6) is still true.

Exercise 17. Verify formula (6) when T has matrix

(a) $\begin{pmatrix} 1 & 1 \\ 1 & 1 \end{pmatrix}$,

(b) $\begin{pmatrix} 2 & 0 \\ 0 & 3 \end{pmatrix}$,

(c) $\begin{pmatrix} 0 & 5 \\ 0 & 0 \end{pmatrix}$.

Classification of Conic Sections

We can use matrix multiplication to keep track of the action of a transformation A on a pair of vectors $\begin{pmatrix} x_1 \\ y_1 \end{pmatrix}$ and $\begin{pmatrix} x_2 \\ y_2 \end{pmatrix}$. Let $m(A)$ be the matrix of A, and consider the matrix $\begin{pmatrix} x_1 & x_2 \\ y_1 & y_2 \end{pmatrix}$ whose columns are $\begin{pmatrix} x_1 \\ y_1 \end{pmatrix}$ and $\begin{pmatrix} x_2 \\ y_2 \end{pmatrix}$. If we set $\begin{pmatrix} x_1' \\ y_1' \end{pmatrix} = A\begin{pmatrix} x_1 \\ y_1 \end{pmatrix}$ and $\begin{pmatrix} x_2' \\ y_2' \end{pmatrix} = A\begin{pmatrix} x_2 \\ y_2 \end{pmatrix}$, then, as we shall prove,

$$m(A)\begin{pmatrix} x_1 & x_2 \\ y_1 & x_1 \end{pmatrix} = \begin{pmatrix} x_1' & x_2' \\ y_1' & y_2' \end{pmatrix}. \tag{1}$$

EXAMPLE 1. $m(A) = \begin{pmatrix} 1 & 2 \\ 3 & 4 \end{pmatrix}$, $\begin{pmatrix} x_1 & x_2 \\ y_1 & y_2 \end{pmatrix} = \begin{pmatrix} 0 & -1 \\ 1 & 0 \end{pmatrix}$. Then $A\begin{pmatrix} 0 \\ 1 \end{pmatrix} = \begin{pmatrix} 2 \\ 4 \end{pmatrix}$, $A\begin{pmatrix} -1 \\ 0 \end{pmatrix} = \begin{pmatrix} -1 \\ -3 \end{pmatrix}$. By (1),

$$\begin{pmatrix} 1 & 2 \\ 3 & 4 \end{pmatrix}\begin{pmatrix} 0 & -1 \\ 1 & 0 \end{pmatrix} = \begin{pmatrix} 2 & -1 \\ 4 & -3 \end{pmatrix}.$$

Direct computation verifies this equation.

To prove (1) in general, we write

$$m(A) = \begin{pmatrix} a & b \\ c & d \end{pmatrix}.$$

Then

$$A\begin{pmatrix} x_1 \\ y_1 \end{pmatrix} = \begin{pmatrix} ax_1 + by_1 \\ cx_1 + dy_1 \end{pmatrix}, \qquad A\begin{pmatrix} x_2 \\ y_2 \end{pmatrix} = \begin{pmatrix} ax_2 + by_2 \\ cx_2 + dy_2 \end{pmatrix}.$$

Formula (1) states that

$$m(A)\begin{pmatrix} x_1 & x_2 \\ y_1 & y_2 \end{pmatrix} = \begin{pmatrix} ax_1 + by_1 & ax_2 + by_2 \\ cx_1 + dy_1 & cx_2 + dy_2 \end{pmatrix},$$

which is true because of the way we have defined multiplication of matrices. So formula (1) holds in general.

Now suppose that

$$A\begin{pmatrix} x_1 \\ y_1 \end{pmatrix} = t_1\begin{pmatrix} x_1 \\ y_1 \end{pmatrix}, \qquad A\begin{pmatrix} x_2 \\ y_2 \end{pmatrix} = t_2\begin{pmatrix} x_2 \\ y_2 \end{pmatrix},$$

or, in other words, that $\begin{pmatrix} x_1 \\ y_1 \end{pmatrix}$ and $\begin{pmatrix} x_2 \\ y_2 \end{pmatrix}$ are eigenvectors of A. Then (1) gives

$$m(A)\begin{pmatrix} x_1 & x_2 \\ y_1 & y_2 \end{pmatrix} = \begin{pmatrix} t_1 x_1 & t_2 x_2 \\ t_1 y_1 & t_2 y_2 \end{pmatrix}.$$

The matrix on the right-hand side equals

$$\begin{pmatrix} x_1 & x_2 \\ y_1 & y_2 \end{pmatrix}\begin{pmatrix} t_1 & 0 \\ 0 & t_2 \end{pmatrix},$$

so we have found the following result.

Let the linear transformation A have eigenvectors $\begin{pmatrix} x_1 \\ y_1 \end{pmatrix}$ and $\begin{pmatrix} x_2 \\ y_2 \end{pmatrix}$

corresponding to eigenvalues t_1, t_2. Then

$$m(A)\begin{pmatrix} x_1 & x_2 \\ y_1 & y_2 \end{pmatrix} = \begin{pmatrix} x_1 & x_2 \\ y_1 & y_2 \end{pmatrix}\begin{pmatrix} t_1 & 0 \\ 0 & t_2 \end{pmatrix}. \tag{2}$$

In addition, now suppose that the eigenvectors $\begin{pmatrix} x_1 \\ y_1 \end{pmatrix}$ and $\begin{pmatrix} x_2 \\ y_2 \end{pmatrix}$ are not *linearly dependent*. It follows by (25), Chapter 2.0, that the determinant $\begin{vmatrix} x_1 & x_2 \\ y_1 & y_2 \end{vmatrix}$ is different from 0, and so, by Theorem 2.4, the matrix $\begin{pmatrix} x_1 & x_2 \\ y_1 & y_2 \end{pmatrix}$ possesses an inverse $\begin{pmatrix} x_1 & x_2 \\ y_1 & y_2 \end{pmatrix}^{-1}$. We now multiply both sides of Eq. (2) on the right-hand side by $\begin{pmatrix} x_1 & x_2 \\ y_1 & y_2 \end{pmatrix}^{-1}$. This yields

$$m(A) = m(A)\begin{pmatrix} x_1 & x_2 \\ y_1 & y_2 \end{pmatrix}\begin{pmatrix} x_1 & x_2 \\ y_1 & y_2 \end{pmatrix}^{-1} = \begin{pmatrix} x_1 & x_2 \\ y_1 & y_2 \end{pmatrix}\begin{pmatrix} t_1 & 0 \\ 0 & t_2 \end{pmatrix}\begin{pmatrix} x_1 & x_2 \\ y_1 & y_2 \end{pmatrix}^{-1}.$$

We introduce the linear transformations P and D with $m(P) = \begin{pmatrix} x_1 & x_2 \\ y_1 & y_2 \end{pmatrix}$, $m(D) = \begin{pmatrix} t_1 & 0 \\ 0 & t_2 \end{pmatrix}$. The last equation can now be written

$$m(A) = m(P)m(D)m(P)^{-1}.$$

It follows that $A = PDP^{-1}$. We have proved:

Theorem 2.11. *Let A be a linear transformation with linearly independent eigenvectors* $\begin{pmatrix} x_1 \\ y_1 \end{pmatrix}$ *and* $\begin{pmatrix} x_2 \\ y_2 \end{pmatrix}$, *corresponding to the eigenvalues* t_1 *and* t_2. *Then*

$$A = PDP^{-1} \tag{3}$$

where $m(P) = \begin{pmatrix} x_1 & x_2 \\ y_1 & y_2 \end{pmatrix}$ *and* $m(D) = \begin{pmatrix} t_1 & 0 \\ 0 & t_2 \end{pmatrix}$.

Note: Assume t_1 and t_2 are eigenvalues of A and $\mathbf{X}_1, \mathbf{X}_2$ are corresponding nonzero eigenvectors. If $t_1 \neq t_2$, then it follows that $\mathbf{X}_1$ and $\mathbf{X}_2$ are linearly independent. To see this, suppose the contrary, i.e., suppose $\mathbf{X}_2 = s\mathbf{X}_1$ for some scalar s with $s \neq 0$. Then

$$A(\mathbf{X}_2) = A(s\mathbf{X}_1) = sA(\mathbf{X}_1) = st_1\mathbf{X}_1 .$$

and so $st_1\mathbf{X}_1 = A(\mathbf{X}_2) = t_2\mathbf{X}_2 = t_2s\mathbf{X}_1$. It follows that $st_1 = t_2s$, and so $t_1 = t_2$, which contradicts our assumption. So $\mathbf{X}_1$ and $\mathbf{X}_2$ are linearly independent, as claimed, as long as t_1 and t_2 are distinct.

Whenever A is a linear transformation whose characteristic polynomial has distinct real roots, then formula (3) is valid.

Note: Recall that a matrix whose entries are 0 except for those on the diagonal, i.e., whose form is $\begin{pmatrix} s & 0 \\ 0 & t \end{pmatrix}$, is called a *diagonal* matrix. The matrix of the transformation D above is a diagonal matrix.

It is easy to compute the powers of a diagonal matrix.

$$\begin{pmatrix} s & 0 \\ 0 & t \end{pmatrix}^2 = \begin{pmatrix} s & 0 \\ 0 & t \end{pmatrix}\begin{pmatrix} s & 0 \\ 0 & t \end{pmatrix} = \begin{pmatrix} s^2 & 0 \\ 0 & t^2 \end{pmatrix},$$

$$\begin{pmatrix} s & 0 \\ 0 & t \end{pmatrix}^3 = \begin{pmatrix} s & 0 \\ 0 & t \end{pmatrix}^2\begin{pmatrix} s & 0 \\ 0 & t \end{pmatrix} = \begin{pmatrix} s^2 & 0 \\ 0 & t^2 \end{pmatrix}\begin{pmatrix} s & 0 \\ 0 & t \end{pmatrix} = \begin{pmatrix} s^3 & 0 \\ 0 & t^3 \end{pmatrix}.$$

Continuing in this way, we see that the nth power of the diagonal matrix $\begin{pmatrix} s & 0 \\ 0 & t \end{pmatrix}$ is the diagonal matrix $\begin{pmatrix} s^n & 0 \\ 0 & t^n \end{pmatrix}$ whose entries on the diagonal are the nth powers of the original entries.

Exercise 1. For each of the following matrices, find the nth power of the matrix when $n = 2, 3, 7, 100$.

(i) $\begin{pmatrix} -1 & 0 \\ 0 & 1 \end{pmatrix}$,

(ii) $\begin{pmatrix} 2 & 0 \\ 0 & 10 \end{pmatrix}$,

(iii) $\begin{pmatrix} 0 & 0 \\ 0 & 4 \end{pmatrix}$.

Now let A and D be the linear transformations which occur in Theorem 2.11. By (3), $A = PDP^{-1}$. Hence,

$$A^2 = (PDP^{-1})(PDP^{-1}) = PD(P^{-1}P)DP^{-1} = PDIDP^{-1}$$

$$= PDDP^{-1} = PD^2P^{-1},$$

$$A^3 = AA^2 = (PDP^{-1})(PD^2P^{-1}) = PD(P^{-1}P)D^2P^{-1}$$

$$= PDID^2P^{-1} = PD^3P^{-1}.$$

Continuing in this way, we find that

$$A^4 = PD^4P^{-1}, \qquad A^5 = PD^5P^{-1},$$

and in general,

$$A^n = PD^nP^{-1}, \tag{4}$$

where n is a positive integer. It follows that

$$m(A^n) = m(PD^nP^{-1}) = m(P)m(D^n)m(P^{-1}) = m(P)(m(D))^n m(P^{-1}).$$

Since $m(D) = \begin{pmatrix} t_1 & 0 \\ 0 & t_2 \end{pmatrix}$, we know that $(m(D))^n = \begin{pmatrix} t_1^n & 0 \\ 0 & t_2^n \end{pmatrix}$. So we have

$$(m(A))^n = m(A^n) = m(P)\begin{pmatrix} t_1^n & 0 \\ 0 & t_2^n \end{pmatrix} m(P^{-1}). \tag{5}$$

Note: Formula (5) allows us to calculate the nth power of $m(A)$ in a practical way, as the following example and exercises illustrate.

EXAMPLE 2. Let us find the nth power of the matrix $\begin{pmatrix} 3 & 4 \\ 4 & -3 \end{pmatrix}$ for $n = 1, 2, 3 \ldots$.

In Example 4, Chapter 2.6, we found that the eigenvalues are $t_1 = 5$, $t_2 = -5$. As corresponding eigenvectors, we can take

$$\mathbf{X}_1 = \begin{pmatrix} 2 \\ 1 \end{pmatrix} \quad \text{and} \quad \mathbf{X}_2 = \begin{pmatrix} -1 \\ 2 \end{pmatrix}.$$

The transformation P of Theorem 2.11 then has matrix $m(P) = \begin{pmatrix} 2 & -1 \\ 1 & 2 \end{pmatrix}$.

Then

$$m(P^{-1}) = (m(P))^{-1} = \begin{pmatrix} 2/5 & 1/5 \\ -1/5 & 2/5 \end{pmatrix}.$$

By (5),

$$\begin{pmatrix} 3 & 4 \\ 4 & -3 \end{pmatrix}^n = \begin{pmatrix} 2 & -1 \\ 1 & 2 \end{pmatrix}\begin{pmatrix} 5^n & 0 \\ 0 & (-5)^n \end{pmatrix}\begin{pmatrix} 2/5 & 1/5 \\ -1/5 & 2/5 \end{pmatrix}.$$

For instance, taking $n = 3$, we get

$$\begin{pmatrix} 3 & 4 \\ 4 & -3 \end{pmatrix}^3 = \begin{pmatrix} 2 & -1 \\ 1 & 2 \end{pmatrix} \begin{pmatrix} 125 & 0 \\ 0 & -125 \end{pmatrix} \begin{pmatrix} 2/5 & 1/5 \\ -1/5 & 2/5 \end{pmatrix}$$

$$= \begin{pmatrix} 2 & -1 \\ 1 & 2 \end{pmatrix} \begin{pmatrix} 50 & 25 \\ 25 & -50 \end{pmatrix} = \begin{pmatrix} 75 & 100 \\ 100 & -75 \end{pmatrix}.$$

Exercise 2. Let A be the linear transformation whose matrix is $\begin{pmatrix} 1 & 3 \\ 3 & -1 \end{pmatrix}$. Find a linear transformation D with diagonal matrix and find a linear transformation P, using Theorem 2.11 such that $A = PDP^{-1}$.

Exercise 3. Let A be as in the preceding exercise. Calculate the matrix $m(A^{10})$ $= \begin{pmatrix} 1 & 3 \\ 3 & -1 \end{pmatrix}^{10}$.

Exercise 4. Using Theorem 2.11, calculate $\begin{pmatrix} 1 & 1 \\ 1 & 1 \end{pmatrix}^8$.

Exercise 5. Using Theorem 2.11, calculate $\begin{pmatrix} 3 & 0 \\ 4 & 2 \end{pmatrix}^6$.

Exercise 6. Fix scalars a, b. Show that if n is an even integer, then $\begin{pmatrix} a & b \\ b & -a \end{pmatrix}^n$ is a diagonal matrix.

The second application of eigenvalues which we shall discuss in this chapter concerns quadratic forms. Let a, b, c be given scalars. For every pair of numbers x, y, we define

$$H(x, y) = ax^2 + 2bxy + cy^2.$$

H is called a *quadratic form*, i.e., a polynomial in x and y, each of whose terms is of the second degree.

Associated with H, we consider the curve whose equation is

$$ax^2 + 2bxy + cy^2 = 1. \tag{6}$$

We denote this curve by C_H.

EXAMPLE 3.

(i) $a = 1$, $b = 0$, $c = 1$. Then C_H is the circle: $x^2 + y^2 = 1$.
(ii) $a = 1$, $b = 0$, $c = -1$. Then C_H is the hyperbola: $x^2 - y^2 = 1$.

Question. Given numbers a, b, c, how can we decide what kind of curve C_H is?

Let us introduce new coordinate axes, to be called the u-axis and the v-axis, by rotating the x- and y-axes about the origin (see Fig. 2.49).

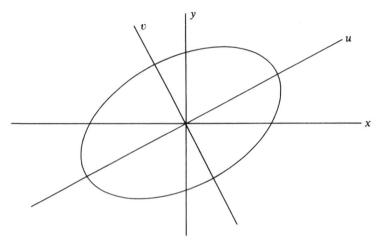

Figure 2.49

Expressed in terms of the new u and v coordinates, the equation of C_H may look more familiar. This will happen if C_H has an axis of symmetry and we manage to choose the u-axis along that axis of symmetry.

EXAMPLE 4. $a = 0$, $2b = 3$, $c = 0$. C_H has the equation: $3xy = 1$ or $xy = \frac{1}{3}$. Evidently the line $x = y$ is an axis of symmetry of C_H. Let us choose the u-axis along this line. Then the v-axis falls on the line $x = -y$.

Suppose a point $\mathbf{X}$ has old coordinates $\begin{pmatrix} x \\ y \end{pmatrix}$ and new coordinates $\begin{pmatrix} u \\ v \end{pmatrix}$. Denote by α the polar angle of $\mathbf{X}$ in the (u, v)-system. Then the polar angle of $\mathbf{X}$ in the (x, y)-system is $\alpha + \pi/4$ (see Fig. 2.50). Hence,

$$\begin{bmatrix} x \\ y \end{bmatrix} = |\mathbf{X}| \begin{bmatrix} \cos(\alpha + \pi/4) \\ \sin(\alpha + \pi/4) \end{bmatrix} = |\mathbf{X}| \begin{bmatrix} (\cos \alpha)\sqrt{2}/2 - (\sin \alpha)\sqrt{2}/2 \\ (\sin \alpha)\sqrt{2}/2 + (\cos \alpha)\sqrt{2}/2 \end{bmatrix}$$

$$= \frac{\sqrt{2}}{2} \begin{bmatrix} |\mathbf{X}|\cos \alpha - |\mathbf{X}|\sin \alpha \\ |\mathbf{X}|\sin \alpha + |\mathbf{X}|\cos \alpha \end{bmatrix}.$$

Also,

$$\begin{pmatrix} u \\ v \end{pmatrix} = |\mathbf{X}| \begin{pmatrix} \cos \alpha \\ \sin \alpha \end{pmatrix},$$

so

$$u = |\mathbf{X}|\cos \alpha,$$
$$v = |\mathbf{X}|\sin \alpha.$$

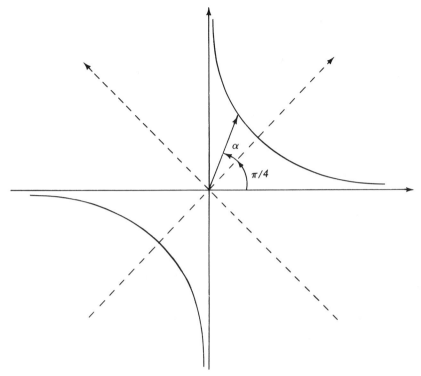

Figure 2.50

Thus,

$$\begin{pmatrix} x \\ y \end{pmatrix} = \frac{\sqrt{2}}{2} \begin{pmatrix} u - v \\ u + v \end{pmatrix},$$

$$\begin{cases} x = \dfrac{\sqrt{2}}{2}\, u - \dfrac{\sqrt{2}}{2}\, v, \\[2mm] y = \dfrac{\sqrt{2}}{2}\, u + \dfrac{\sqrt{2}}{2}\, v. \end{cases} \tag{7}$$

Now suppose $\mathbf{X}$ is a point on C_H. Then $3xy = 1$. Hence,

$$3\left(\frac{\sqrt{2}}{2}\, u - \frac{\sqrt{2}}{2}\, v \right)\left(\frac{\sqrt{2}}{2}\, u + \frac{\sqrt{2}}{2}\, v \right) = 1,$$

or

$$3\left(\frac{\sqrt{2}}{2} \right)^2 (u - v)(u + v) = 1,$$

or

$$3 \cdot \tfrac{1}{2}(u^2 - v^2) = 1.$$

We have found: if $\mathbf{X}$ is a point on C_H with new coordinates $\begin{pmatrix} u \\ v \end{pmatrix}$, then

$$\tfrac{3}{2}u^2 - \tfrac{3}{2}v^2 = 1. \tag{8}$$

Equation (8) is an equation for C_H in the (u,v)-system. We recognize (8) as describing a hyperbola with one axis along the u-axis.

Now let a, b, c be given numbers with $b \neq 0$ and let $H(x, y) = ax^2 + 2bxy + cy^2$. We can express $H(x, y)$ as the dot product

$$\begin{pmatrix} ax + by \\ bx + cy \end{pmatrix} \cdot \begin{pmatrix} x \\ y \end{pmatrix} = (ax + by)x + (bx + cy)y = H(x, y).$$

Also,

$$\begin{pmatrix} ax + by \\ bx + cy \end{pmatrix} = \begin{pmatrix} a & b \\ b & c \end{pmatrix} \begin{pmatrix} x \\ y \end{pmatrix}.$$

Thus,

$$H(x, y) = \left(\begin{pmatrix} a & b \\ b & c \end{pmatrix} \begin{pmatrix} x \\ y \end{pmatrix} \right) \cdot \begin{pmatrix} x \\ y \end{pmatrix}. \tag{9}$$

Now, $\begin{pmatrix} a & b \\ b & c \end{pmatrix}$ is a symmetric matrix, so Proposition 1, Chapter 2.6, tells us that $\begin{pmatrix} a & b \\ b & c \end{pmatrix}$ has eigenvalues t_1, t_2. Since $b \neq 0$, $t_1 \neq t_2$. Let $\mathbf{X}_1$ be an eigenvector corresponding to t_1 such that $|\mathbf{X}_1| = 1$.

We choose new coordinate axes as follows: The u-axis passes through $\mathbf{X}_1$, directed so that $\mathbf{X}_1$ points in the positive direction. The v-axis is chosen orthogonal to the u-axis and oriented so that the positive u-direction goes over into the positive v-direction by a counterclockwise rotation of $\pi/2$ radians.

Set $\mathbf{X}_1 = \begin{pmatrix} x_1 \\ y_1 \end{pmatrix}$ and set $\mathbf{X}_2 = \begin{pmatrix} -y_1 \\ x_1 \end{pmatrix}$. Then $\mathbf{X}_2$ lies on the v-axis. We know that each eigenvector corresponding to t_2 is orthogonal to $\mathbf{X}_1$ and, hence, lies on the v-axis. Hence, every vector lying on the v-axis is an eigenvector corresponding to t_2. In particular, $\mathbf{X}_2$ is such an eigenvector (see Fig. 2.51).

Let X be any point, with $\begin{pmatrix} x \\ y \end{pmatrix}$ its old coordinates and $\begin{pmatrix} u \\ v \end{pmatrix}$ its new coordinates. Then

$$\mathbf{X} = u\mathbf{X}_1 + v\mathbf{X}_2.$$

Then

$$\begin{cases} u = \mathbf{X} \cdot \mathbf{X}_1 = \begin{pmatrix} x \\ y \end{pmatrix} \cdot \begin{pmatrix} x_1 \\ y_1 \end{pmatrix} = xx_1 + yy_1, \\[2mm] v = \mathbf{X} \cdot \mathbf{X}_2 = \begin{pmatrix} x \\ y \end{pmatrix} \cdot \begin{pmatrix} -y_1 \\ x_1 \end{pmatrix} = -xy_1 + yx_1. \end{cases} \tag{10}$$

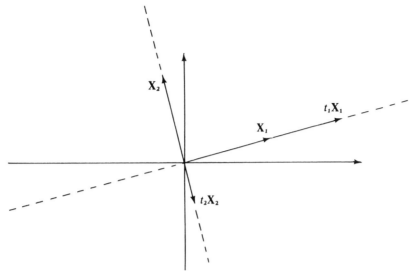

Figure 2.51

Using (9), we have

$$H(x, y) = \begin{pmatrix} a & b \\ b & c \end{pmatrix}(\mathbf{X}) \cdot \mathbf{X} = \left[\begin{pmatrix} a & b \\ b & c \end{pmatrix}(u\mathbf{X}_1 + v\mathbf{X}_2) \right] \cdot [\, u\mathbf{X}_1 + v\mathbf{X}_2]$$

$$= (ut_1\mathbf{X}_1 + vt_2\mathbf{X}_2) \cdot (u\mathbf{X}_1 + v\mathbf{X}_2)$$

$$= u^2 t_1 + v^2 t_2 .$$

Thus, we have found:

Theorem 2.12. *Let* t_1, t_2 *be the eigenvalues of the matrix* $\begin{pmatrix} a & b \\ b & c \end{pmatrix}$, *where* $b \neq 0$. *Let* $\mathbf{X}$ *be any point, and let* $\begin{pmatrix} x \\ y \end{pmatrix}$ *be its old coordinates and* $\begin{pmatrix} u \\ v \end{pmatrix}$ *be its new coordinates. Then*

$$ax^2 + 2bxy + cy^2 = t_1 u^2 + t_2 v^2. \tag{11}$$

Now let $\mathbf{X}$ be a point on the curve C_H whose equation is $ax^2 + 2bxy + cy^2 = 1$. Let $\begin{pmatrix} x \\ y \end{pmatrix}$ and $\begin{pmatrix} u \\ v \end{pmatrix}$ be, respectively, the old coordinates and the new coordinates of $\mathbf{X}$. Then using (11), we get

$$t_1 u^2 + t_2 v^2 = ax^2 + 2bxy + cy^2 = 1,$$

since $\begin{pmatrix} x \\ y \end{pmatrix}$ lies on C_H. So

$$t_1 u^2 + t_2 v^2 = 1. \tag{12}$$

Equation (12) is valid for every point on C_H and only for such points. This means that (12) is *an equation for C_H in the (u, v)-system.*

EXAMPLE 5. Describe and sketch the curve C_H: $3x^2 + 8xy - 3y^2 = 1$. Here $a = 3$, $b = 4$, $c = -3$. The matrix $\begin{pmatrix} a & b \\ b & c \end{pmatrix} = \begin{pmatrix} 3 & 4 \\ 4 & -3 \end{pmatrix}$. The eigenvalues are $t_1 = 5$ and $t_2 = -5$.

$$\begin{pmatrix} 3 & 4 \\ 4 & -3 \end{pmatrix} \begin{pmatrix} 2 \\ 1 \end{pmatrix} = 5 \begin{pmatrix} 2 \\ 1 \end{pmatrix}.$$

Since we need an eigenvector $\mathbf{X}_1 = \begin{pmatrix} x_1 \\ y_1 \end{pmatrix}$ of length 1, we set $\mathbf{X}_1 = (1/\sqrt{5}) \begin{pmatrix} 2 \\ 1 \end{pmatrix} = \begin{bmatrix} 2/\sqrt{5} \\ 1/\sqrt{5} \end{bmatrix}$. So

$$x_1 = \frac{2}{\sqrt{5}}, \qquad y_1 = \frac{1}{\sqrt{5}}.$$

We choose the u-axis to pass through $\mathbf{X}_1$, so it is the line $y = 2x$, and the v-axis is the line $y = -\frac{1}{2}x$. In the new system, the equation of C_H is

$$5u^2 - 5v^2 = 1, \tag{13}$$

where we have used (12) with $t_1 = 5$, $t_2 = -5$.

We can check Eq. (13) by using the relations between u, v and x, y. By (10) we know

$$u = x_1 x + y_1 y = \frac{2}{\sqrt{5}} x + \frac{1}{\sqrt{5}} y,$$

$$v = -y_1 x + x_1 y = -\frac{1}{\sqrt{5}} x + \frac{2}{\sqrt{5}} y.$$

Hence,

$$5u^2 - 5v^2 = 5 \left(\frac{2}{\sqrt{5}} x + \frac{1}{\sqrt{5}} y \right)^2 - 5 \left(-\frac{1}{\sqrt{5}} x + \frac{2}{\sqrt{5}} y \right)^2$$

$$= 5 \cdot \tfrac{1}{5} (2x + y)^2 - 5 \cdot \tfrac{1}{5} (-x + 2y)^2$$

$$= 4x^2 + 4xy + y^2 - x^2 + 4xy - 4y^2$$

$$= 3x^2 + 8xy - 3y^2.$$

Thus, for every point $\mathbf{X}$ in the plane, if $\begin{pmatrix} x \\ y \end{pmatrix}$ are the old and $\begin{pmatrix} u \\ v \end{pmatrix}$ are the new coordinates of $\mathbf{X}$, then we have

$$5u^2 - 5v^2 = 3x^2 + 8xy - 3y^2. \tag{14}$$

By definition of C_H, $\mathbf{X}$ lies on C_H if and only if $3x^2 + 8xy - 3y^2 = 1$ and so if and only if $5u^2 - 5v^2 = 1$. So (13) is verified.

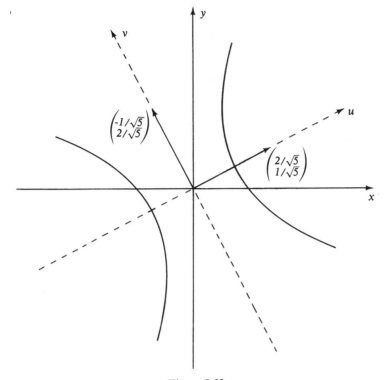

Figure 2.52

From Eq. (13), we find that C_H is a hyperbola with its axes of symmetry along the u- and v-axes (see Fig. 2.52).

Now let $H(x, y) = ax^2 + 2bxy + cy^2$ be a given quadratic form. Assume $b \neq 0$. Let t_1, t_2 denote the eigenvalues of the matrix $\begin{pmatrix} a & b \\ b & c \end{pmatrix}$. We have:

Theorem 2.13. *Let C_H denote the curve $ax^2 + 2bxy + cy^2 = 1$.*

(i) *If t_1, t_2 are both > 0, then C_H is an ellipse.*
(ii) *If $t_1, t_2 < 0$, then C_H is empty.*
(iii) *If t_1, t_2 have opposite signs, then C_H is a hyperbola.*

Exercise 7. Using the fact that in the (u, v)-system, C_H has equation

$$t_1 u^2 + t_2 v^2 = 1,$$

prove Theorem 2.13.

EXAMPLE 6. Describe the curve C_H,

$$x^2 + 2xy + y^2 = 1. \tag{15}$$

Here $\begin{pmatrix} a & b \\ b & c \end{pmatrix} = \begin{pmatrix} 1 & 1 \\ 1 & 1 \end{pmatrix}$ and t_1, t_2 are the roots of the polynomial

$$\begin{vmatrix} 1-t & 1 \\ 1 & 1-t \end{vmatrix} = t^2 - 2t.$$

So $t_1 = 0$, $t_2 = 2$. In the (u, v)-system, C_H has the equation

$$0u^2 + 2v^2 = 1 \quad \text{or} \quad v^2 = \tfrac{1}{2}.$$

This equation describes a locus consisting of two parallel straight lines $v = 1/\sqrt{2}$ and $v = 1/\sqrt{2}$. Thus, here C_H consists of two straight lines. We can also see this directly by writing (15) in the form

$$(x + y)^2 = 1 \quad \text{or} \quad (x + y)^2 - 1 = 0$$

or, equivalently, $((x + y) + 1)((x + y) - 1) = 0$. So C_H consists of the two lines $x + y + 1 = 0$ and $x + y - 1 = 0$.

Thus, in addition to the possibilities of an ellipse, hyperbola, and empty locus, noted in Theorem 2.13, C_H may consist of two lines.

Exercise 8. Classify and sketch the curve $2xy - y^2 = 1$.

Exercise 9. Classify and sketch the curve $4x^2 + 2\sqrt{2}\, xy + 3y^2 = 1$.

Exercise 10. Classify and sketch the curve $x^2 - 2xy + y^2 = 1$.

The quadratic form $H(x, y) = ax^2 + 2bxy + cy^2$ is called *positive definite* if $H(x, y) > 0$ whenever $(x, y) \neq (0, 0)$. For instance, $H(x, y) = 2x^2 + 3y^2$ is positive definite and $H(x, y) = x^2 - y^2$ is not positive definite.

Exercise 11. Give conditions on the coefficients a, b, c in order that $H(x, y)$ is positive definite. *Hint*: Make use of formula (11).

Next, instead of the curve C_H with the equation,

$$H(x, y) = 1,$$

we consider the locus defined by the equation,

$$H(x, y) = 0 \quad \text{or} \quad ax^2 + 2bxy + cy^2 = 0. \tag{16}$$

This locus is not always a curve in the ordinary sense of the word; for instance, with $a = c = 1$, $b = 0$, we get the equation $x^2 + y^2 = 0$, which defines a single point, the origin. If $a = b = c = 0$, then the locus is the entire plane.

If $a = 0$, equation (16) becomes

$$2bxy + cy^2 = 0 \quad \text{or} \quad y(2bx + c) = 0. \tag{17}$$

Exercise 12. Describe the locus defined by equation (17).

If $a \neq 0$, the only point (x, y) that satisfies equation (16), with $y = 0$, is the origin. Let us consider a point (x, y) on the locus defined by (16) with

$y \neq 0$. Then,

$$H(x, y) = y^2 \left[a \left(\frac{x^2}{y^2} \right) + 2b \left(\frac{x}{y} \right) + c \right].$$

We may write $H(x, y) = y^2 P \left(\frac{x}{y} \right)$, where $P(t) = at^2 + 2bt + c$. If $b^2 -$

$ac < 0$, then P has no real roots, so $P \left(\frac{x}{y} \right) \neq 0$ and $H(x, y) \neq 0$. Hence, in

this case, the locus consists of the origin.

If $b^2 - ac \geqslant c$, the P has real roots t_1, t_2, and we can write for all t
$P(t) = a(t - t_1)(t - t_2)$. Therefore,

$$H(x, y) = y^2 P \left(\frac{x}{y} \right) = y^2 a \left[\frac{x}{y} - t_1 \right] \left[\frac{x}{y} - t_2 \right] \quad \text{or}$$

$$H(x, y) = a(x - t_1 y)(x - t_2 y).$$

So (x, y) belongs to the locus if and only if $x - t_1 y = 0$ or $x - t_2 y = 0$. Thus, the locus consists of the two lines with equations: $x - t_1 y = 0$ and $x - t_2 y = 0$. If $t_1 = t_2$, the two lines coincide. To sum up, we have *the locus defined by the equation $ax^2 + 2bxy + cy^2 = 0$ is either a single point, (the origin), a pair of straight lines meeting at the origin, a single line passing through the origin, or the entire plane.*

Exercise 13. Classify and sketch the following loci:

(a) $x^2 - 3y^2 = 0$,
(b) $xy + y^2 = 0$,
(c) $x^2 + 2xy + y^2 = 0$,
(d) $x^2 + xy + y^2 = 0$.

Note: A student who has arrived at this point in the book and who is familiar with elementary calculus is now ready to learn one of the applications of linear algebra given in Chapter 8, namely, the study of systems of differential equations in two dimensions.

CHAPTER 3.0

Vector Geometry in 3-Space

Just as in the plane, we may use vectors to express the analytic geometry of 3-dimensional space.

We define a *vector* in 3-space as a triplet of numbers $\begin{bmatrix} x_1 \\ x_2 \\ x_3 \end{bmatrix}$ written in column form, with x_1, x_2, and x_3 as the *first*, *second*, and *third coordinates*. We designate this vector by a single capital letter $\mathbf{X}$, i.e., we write $\mathbf{X} = \begin{bmatrix} x_1 \\ x_2 \\ x_3 \end{bmatrix}$. We can picture the vector $\mathbf{X}$ as an arrow or directed segment, starting at the origin and ending at the point $\begin{bmatrix} x_1 \\ x_2 \\ x_3 \end{bmatrix}$. We denote by $\mathbb{R}^3$ the set of all vectors in 3-space, and we denote by $\mathbb{R}^2$ the set of all vectors in the plane.

We *add* two vectors by adding their components, so if $\mathbf{X} = \begin{bmatrix} x_1 \\ x_2 \\ x_3 \end{bmatrix}$ and $\mathbf{U} = \begin{bmatrix} u_1 \\ u_2 \\ u_3 \end{bmatrix}$, then

$$\mathbf{X} + \mathbf{U} = \begin{bmatrix} x_1 + u_1 \\ x_2 + u_2 \\ x_3 + u_3 \end{bmatrix}.$$

We *multiply* a vector by a scalar c by multiplying each of the coordinates by c, so

$$c\mathbf{X} = c\begin{bmatrix} x_1 \\ x_2 \\ x_3 \end{bmatrix} = \begin{bmatrix} cx_1 \\ cx_2 \\ cx_3 \end{bmatrix}.$$

We set $\mathbf{E}_1 = \begin{bmatrix} 1 \\ 0 \\ 0 \end{bmatrix}$, $\mathbf{E}_2 = \begin{bmatrix} 0 \\ 1 \\ 0 \end{bmatrix}$, and $\mathbf{E}_3 = \begin{bmatrix} 0 \\ 0 \\ 1 \end{bmatrix}$, and we call these the *basis vectors*
of 3-space. The first *coordinate axis* is then obtained by taking all multiples
$x_1\mathbf{E}_1 = x_1 \begin{bmatrix} 1 \\ 0 \\ 0 \end{bmatrix} = \begin{bmatrix} x_1 \\ 0 \\ 0 \end{bmatrix}$, of $\mathbf{E}_1$, and the second and third coordinate axes are
defined similarly. Any vector $\mathbf{X}$ may be expressed uniquely as a sum of
vectors on the three coordinate axes:

$$\mathbf{X} = \begin{bmatrix} x_1 \\ x_2 \\ x_3 \end{bmatrix} = \begin{bmatrix} x_1 \\ 0 \\ 0 \end{bmatrix} + \begin{bmatrix} 0 \\ x_2 \\ 0 \end{bmatrix} + \begin{bmatrix} 0 \\ 0 \\ x_3 \end{bmatrix} = x_1\mathbf{E}_1 + x_2\mathbf{E}_2 + x_3\mathbf{E}_3 .$$

Geometrically, we may think of $\mathbf{X}$ as a diagonal segment in a rectangular
prism with edges parallel to the coordinate axes (see Fig. 3.1).

Let $\mathbf{X} = \begin{bmatrix} x_1 \\ x_2 \\ x_3 \end{bmatrix}$ and $\mathbf{U} = \begin{bmatrix} u_1 \\ u_2 \\ u_3 \end{bmatrix}$ be two vectors. What is the geometric
description of the vector $\mathbf{X} + \mathbf{U}$? (see Fig. 3.2). By analogy with the
situation in Section 2.0, we expect to obtain $\mathbf{X} + \mathbf{U}$ by moving the segment
$\mathbf{U}$ parallel to itself so that its starting point lands at $\mathbf{X}$, and then taking its
endpoint. To see that this expectation is correct, we can reason as follows:
If we move the segment $\mathbf{U}$ first in the x-direction by x_1 units, then in the

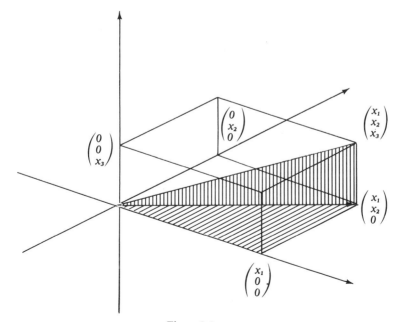

Figure 3.1

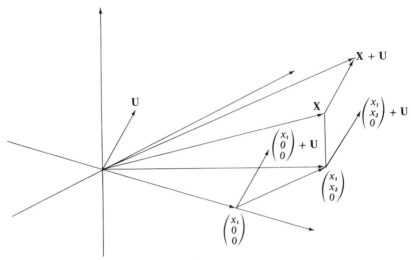

Figure 3.2

y-direction by x_2 units, and finally in the z-direction by x_3 units, it will at all times remain parallel to its original position and in its final position it will start at $\mathbf{X}$. If we move the segment $\mathbf{U}$ by x_1 units in the x-direction, its

new endpoint is $\begin{bmatrix} u_1 + x_1 \\ u_2 \\ u_3 \end{bmatrix}$. Taking the corresponding steps in the y- and

z-directions, we get $\begin{bmatrix} u_1 + x_1 \\ u_2 + x_2 \\ u_3 + x_3 \end{bmatrix} = \mathbf{U} + \mathbf{X}$ for the final position of the end-

point.

By the *difference* of two vectors $\mathbf{X}$ and $\mathbf{U}$, we mean the vector $\mathbf{X} + (-\mathbf{U})$, which we add to $\mathbf{U}$ to get $\mathbf{X}$ (see Fig. 3.3). We may think of $\mathbf{X} + (-\mathbf{U})$ as the vector from the origin which is parallel to the segment from the endpoint of $\mathbf{U}$ to the endpoint of $\mathbf{X}$ and has the same length and direction. We often write $\mathbf{X} - \mathbf{U}$ for the sum $\mathbf{X} + (-\mathbf{U})$.

As in the case of the plane, we can establish the following properties of vector addition and scalar multiplication in 3-space: For all vectors $\mathbf{X}$, $\mathbf{U}$, $\mathbf{A}$, and all scalars r, s, we have

(i) $\mathbf{X} + \mathbf{U} = \mathbf{U} + \mathbf{X}$;
(ii) $(\mathbf{X} + \mathbf{U}) + \mathbf{A} = \mathbf{X} + (\mathbf{U} + \mathbf{A})$;
(iii) there is a vector $\mathbf{0}$ such that $\mathbf{X} + \mathbf{0} = \mathbf{X} = \mathbf{0} + \mathbf{X}$ for all $\mathbf{X}$;
(iv) For any $\mathbf{X}$ there is a vector $-\mathbf{X}$ such that $\mathbf{X} + (-\mathbf{X}) = \mathbf{0}$;
(v) $r(\mathbf{X} + \mathbf{U}) = r\mathbf{X} + r\mathbf{U}$;
(vi) $(r + s)(\mathbf{X}) = r\mathbf{X} + s\mathbf{X}$;
(vii) $r(s\mathbf{X}) = (rs)\mathbf{X}$;
(viii) $1 \cdot \mathbf{X} = \mathbf{X}$ for each $\mathbf{X}$.

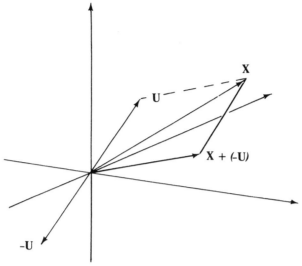

Figure 3.3

Each of these properties can be established by referring to the co-ordinatewise definitions of addition and scalar multiplication.

If **A** and **B** are vectors in $\mathbb{R}^3$ and s and t are scalars, the expression

$$sA + tB$$

is called a *linear combination* of **A** and **B**.

As in two dimensions, the vectors **A** and **B** in $\mathbb{R}^3$ are said to be *linearly dependent* if one is a scalar multiple of the other. A collection of three vectors, **A**, **B**, **C** is said to be *linearly dependent* if one of the vectors is a linear combination of the other two vectors.

EXAMPLE 1. If $A = \begin{bmatrix} 2 \\ 1 \\ 3 \end{bmatrix}$, $B = \begin{bmatrix} -1 \\ 0 \\ -2 \end{bmatrix}$, and $C = \begin{bmatrix} 5 \\ 2 \\ 8 \end{bmatrix}$, then the triplet **A**, **B**, **C** is linearly dependent because

$$C = 2A + (-1)B.$$

If a collection of vectors is not linearly dependent, it is said to be *linearly independent*.

EXAMPLE 2. The vectors $\begin{bmatrix} 1 \\ 0 \\ 0 \end{bmatrix}$, $\begin{bmatrix} 0 \\ 1 \\ 0 \end{bmatrix}$, and $\begin{bmatrix} 0 \\ 0 \\ 1 \end{bmatrix}$ are linearly independent since it is impossible to write $\begin{bmatrix} 0 \\ 0 \\ 1 \end{bmatrix}$ as $s\begin{bmatrix} 1 \\ 0 \\ 0 \end{bmatrix} + t\begin{bmatrix} 0 \\ 1 \\ 0 \end{bmatrix}$, and, similarly, for $\begin{bmatrix} 0 \\ 1 \\ 0 \end{bmatrix}$ and $\begin{bmatrix} 0 \\ 0 \\ 1 \end{bmatrix}$.

Exercise 1. Prove that the set of three vectors **A**, **B**, **C** is linearly dependent if and only if it is possible to find scalars r, s, t *not all zero*, such that

$$r\mathbf{A} + s\mathbf{B} + t\mathbf{C} = \mathbf{0}.$$

Show that under these conditions we can write one of the vectors as a linear combination of the other two and show the converse.

An extremely useful notion which helps to express many of the ideas of the geometry of 3-space is the *dot product*. We define

$$\mathbf{X} \cdot \mathbf{U} = \begin{pmatrix} x_1 \\ x_2 \\ x_3 \end{pmatrix} \cdot \begin{pmatrix} u_1 \\ u_2 \\ u_3 \end{pmatrix} = x_1 u_1 + x_2 u_2 + x_3 u_3 .$$

The dot product of two vectors is a *scalar*, e.g.,

$$\begin{pmatrix} 2 \\ 1 \\ 3 \end{pmatrix} \cdot \begin{pmatrix} 2 \\ 4 \\ -6 \end{pmatrix} = 2 \cdot 2 + 1 \cdot 4 + 3 \cdot (-6) = -10.$$

We have $\mathbf{E}_1 \cdot \mathbf{E}_2 = \begin{pmatrix} 1 \\ 0 \\ 0 \end{pmatrix} \cdot \begin{pmatrix} 0 \\ 1 \\ 0 \end{pmatrix} = 0$ and $\mathbf{E}_1 \cdot \mathbf{E}_1 = \begin{pmatrix} 1 \\ 0 \\ 0 \end{pmatrix} \cdot \begin{pmatrix} 1 \\ 0 \\ 0 \end{pmatrix} = 1.$ Similarly, $\mathbf{E}_2 \cdot \mathbf{E}_3 = 0 = \mathbf{E}_3 \cdot \mathbf{E}_1$, while $\mathbf{E}_2 \cdot \mathbf{E}_2 = 1 = \mathbf{E}_3 \cdot \mathbf{E}_3$.

As in the plane, the dot product behaves somewhat like the ordinary product of numbers. We have the distributive property $(\mathbf{X} + \mathbf{Y}) \cdot \mathbf{U} = \mathbf{X} \cdot \mathbf{U} + \mathbf{Y} \cdot \mathbf{U}$ and the commutative property $\mathbf{X} \cdot \mathbf{U} = \mathbf{U} \cdot \mathbf{X}$. Moreover, for scalar multiplication, we have $(t\mathbf{X}) \cdot \mathbf{U} = t(\mathbf{X} \cdot \mathbf{U})$. To prove this last statement, note that

$$(t\mathbf{X}) \cdot \mathbf{U} = \begin{pmatrix} tx_1 \\ tx_2 \\ tx_3 \end{pmatrix} \cdot \begin{pmatrix} u_1 \\ u_2 \\ u_3 \end{pmatrix} = (tx_1)u_1 + (tx_2)u_2 + (tx_3)u_3$$

$$= t(x_1 u_1 + x_2 u_2 + x_3 u_3) = t(\mathbf{X} \cdot \mathbf{U}).$$

The other properties also have straightforward proofs in terms of coordinates.

By the Pythagorean Theorem in 3-space, the distance from a point $\mathbf{X} = \begin{pmatrix} x_1 \\ x_2 \\ x_3 \end{pmatrix}$ to the origin is $\sqrt{x_1^2 + x_2^2 + x_3^2}$, and we define this number to be the *length* of the vector **X**, written $|\mathbf{X}|$. For example, $\left\| \begin{pmatrix} 3 \\ 4 \\ 12 \end{pmatrix} \right\| = \sqrt{169} = 13,$

while $|\mathbf{E}_i| = 1$ for each i and $|\mathbf{0}| = 0$.

In general, $\mathbf{X} \cdot \mathbf{X} = \begin{pmatrix} x_1 \\ x_2 \\ x_3 \end{pmatrix} \cdot \begin{pmatrix} x_1 \\ x_2 \\ x_3 \end{pmatrix} = x_1^2 + x_2^2 + x_3^2,$ and so this is the square of the length $|\mathbf{X}|$ of the vector **X**. Note that for any scalar c, we have

$|c\mathbf{X}| = \sqrt{c\mathbf{X} \cdot c\mathbf{X}} = \sqrt{c^2 \mathbf{X} \cdot \mathbf{X}} = |c|\sqrt{\mathbf{X} \cdot \mathbf{X}} = |c|\,|\mathbf{X}|$, where $|c| = \sqrt{c^2}$ is the absolute value of the scalar c. We have $|\mathbf{X}| \geqslant 0$ with $|\mathbf{X}| = 0$ if and only if $\mathbf{X} = \mathbf{0}$.

In terms of the notion of dot product, we shall now treat seven basic geometric problems:

   (i) To decide when two given vectors $\mathbf{X}$ and $\mathbf{U}$ are perpendicular.
   (ii) To calculate the angle between two vectors.
   (iii) To find the projection of a given vector on a given line through the origin.
   (iv) To find the projection of a given vector on a given plane through the origin.
   (v) To compute the distance from a given point to a given plane through the origin.
   (vi) To compute the distance from a given point to a given line through the origin.
   (vii) To compute the area of the parallelogram formed by two vectors in 3-space.

(i) As in the case of the plane, we may use the law of cosines to give an interpretation of the dot product in terms of the lengths $|\mathbf{X}|$ and $|\mathbf{U}|$ of the vectors $\mathbf{X}$ and $\mathbf{U}$ and the angle θ between them. The law of cosines states

$$|\mathbf{X} - \mathbf{U}|^2 = |\mathbf{X}|^2 + |\mathbf{U}|^2 - 2|\mathbf{X}|\,|\mathbf{U}| \cos \theta.$$

But

$$|\mathbf{X} - \mathbf{U}|^2 = (\mathbf{X} - \mathbf{U}) \cdot (\mathbf{X} - \mathbf{U}) = \mathbf{X} \cdot \mathbf{X} - 2\mathbf{X} \cdot \mathbf{U} + \mathbf{U} \cdot \mathbf{U}$$

$$= |\mathbf{X}|^2 + |\mathbf{U}|^2 - 2\mathbf{X} \cdot \mathbf{U}.$$

Thus

$$\mathbf{X} \cdot \mathbf{U} = |\mathbf{X}|\,|\mathbf{U}| \cos \theta.$$

The vectors $\mathbf{X}$ and $\mathbf{U}$ will be perpendicular if and only if $\cos \theta = 0$, so if and only if $\mathbf{X} \cdot \mathbf{U} = 0$.

(ii) If $\mathbf{X}$ and $\mathbf{U}$ are not perpendicular, we may use the dot product to compute the cosine of the angle between $\mathbf{X}$ and $\mathbf{U}$. For example, if $\mathbf{X} = \begin{pmatrix} 1 \\ 2 \\ 1 \end{pmatrix}$

and $\mathbf{U} = \begin{pmatrix} -1 \\ 1 \\ 3 \end{pmatrix}$, then $\mathbf{X} \cdot \mathbf{U} = (1)(-1) + (2)(1) + (1)(3) = 4$, $|\mathbf{X}| = \sqrt{6}$,

$|\mathbf{U}| = \sqrt{11}$, so $\cos \theta = 4/\sqrt{6}\,\sqrt{11}$.

(iii) The projection $P(\mathbf{X})$ of a given vector $\mathbf{X}$ to the line of multiples of a given vector $\mathbf{U}$ is the vector $t\mathbf{U}$ such that $\mathbf{X} - t\mathbf{U}$ is perpendicular to $\mathbf{U}$ (see Fig. 3.4). This condition enables us to compute t since $0 = (\mathbf{X} - t\mathbf{U}) \cdot \mathbf{U}$ $= \mathbf{X} \cdot \mathbf{U} - (t\mathbf{U}) \cdot \mathbf{U} = \mathbf{X} \cdot \mathbf{U} - t(\mathbf{U} \cdot \mathbf{U})$, so $t = \mathbf{X} \cdot \mathbf{U}/\mathbf{U} \cdot \mathbf{U}$ and we have a

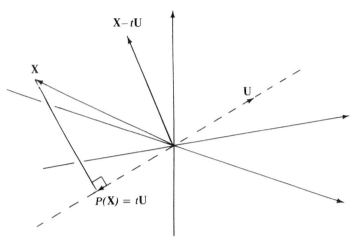

Figure 3.4

formula for the projection:

$$P(\mathbf{X}) = \left(\frac{\mathbf{X} \cdot \mathbf{U}}{\mathbf{U} \cdot \mathbf{U}} \right) \mathbf{U}. \tag{1}$$

Note that this is the same as the formula we obtained in the 2-dimensional case.

Note also that

$$P(\mathbf{X}) \cdot \mathbf{U} = \left(\left(\frac{\mathbf{X} \cdot \mathbf{U}}{\mathbf{U} \cdot \mathbf{U}} \right) \mathbf{U} \right) \cdot \mathbf{U} = \left(\frac{\mathbf{X} \cdot \mathbf{U}}{\mathbf{U} \cdot \mathbf{U}} \right) (\mathbf{U} \cdot \mathbf{U}) = \mathbf{X} \cdot \mathbf{U}. \tag{2}$$

(iv) Fix a nonzero vector $\mathbf{U} = \begin{bmatrix} u_1 \\ u_2 \\ u_3 \end{bmatrix}$ and denote by Π the plane through

the origin which is orthogonal to $\mathbf{U}$. Since $\mathbf{Y} = \begin{bmatrix} y_1 \\ y_2 \\ y_3 \end{bmatrix}$ lies in Π if and only if

$\mathbf{Y} \cdot \mathbf{U} = 0$, an equation of Π is

$$u_1 y_1 + u_2 y_2 + u_3 y_3 = 0. \tag{3}$$

Let $\mathbf{X}$ be a vector. We denote by $Q(\mathbf{X})$ the *projection of* $\mathbf{X}$ *on* Π, i.e., the foot of the perpendicular dropped from $\mathbf{X}$ to Π. Then the segment joining $\mathbf{X}$ and $Q(\mathbf{X})$ is parallel to $\mathbf{U}$, so for some scalar t,

$$\mathbf{X} - Q(\mathbf{X}) = t\mathbf{U}.$$

Then $\mathbf{X} - t\mathbf{U} = Q(\mathbf{X})$. Since $Q(\mathbf{X})$ is in Π, $\mathbf{X} - t\mathbf{U}$ is perpendicular to $\mathbf{U}$. Then, by the discussion of (iii), $\mathbf{X} - Q(\mathbf{X}) = P(\mathbf{X})$ or

$$Q(\mathbf{X}) = \mathbf{X} - P(\mathbf{X}).$$

(v) The *distance* from the point $\mathbf{X}$ to the plane through the origin perpendicular to $\mathbf{U}$ is precisely the length of the projection vector $P(\mathbf{X})$, i.e.,

$$|P(\mathbf{X})| = \left| \frac{\mathbf{X} \cdot \mathbf{U}}{\mathbf{U} \cdot \mathbf{U}} \right| |\mathbf{U}| = \frac{|\mathbf{X} \cdot \mathbf{U}|}{|\mathbf{U}|} .$$

It follows that the distance d from the point $\mathbf{X}$ to the plane through the origin perpendicular to $\mathbf{U}$ with the equation $x_1 u_1 + x_2 u_2 + x_3 u_3 = 0$ is given by

$$d = \frac{|x_1 u_1 + x_2 u_2 + x_3 u_3|}{\sqrt{u_1^2 + u_2^2 + u_3^2}} . \tag{4}$$

(vi) *The distance from the point $\mathbf{X}$ to the line along $\mathbf{U}$ with $\mathbf{U} \neq \mathbf{0}$ is the length of the difference vector $|\mathbf{X} - P(\mathbf{X})|$.* Since $\mathbf{X} - P(\mathbf{X})$ is perpendicular to $P(\mathbf{X})$, we get $|\mathbf{X}|^2 = |P(\mathbf{X})|^2 + |\mathbf{X} - P(\mathbf{X})|^2$, so

$$|\mathbf{X} - P(\mathbf{X})|^2 = |\mathbf{X}|^2 - |P(\mathbf{X})|^2$$

and hence

$$|\mathbf{X} - P(\mathbf{X})|^2 = \mathbf{X} \cdot \mathbf{X} - \frac{|\mathbf{X} \cdot \mathbf{U}|^2}{|\mathbf{U}|^2} ,$$

(see Fig. 3.5) and so the distance is given by

$$\sqrt{\mathbf{X} \cdot \mathbf{X} - \frac{(\mathbf{X} \cdot \mathbf{U})^2}{(\mathbf{U} \cdot \mathbf{U})}} = \frac{\sqrt{(\mathbf{X} \cdot \mathbf{X})(\mathbf{U} \cdot \mathbf{U}) - (\mathbf{X} \cdot \mathbf{U})^2}}{|\mathbf{U}|} . \tag{5}$$

(vii) Let $\mathbf{A}, \mathbf{B}$ be two vectors and let Π denote the parallelogram with two sides along $\mathbf{A}$ and $\mathbf{B}$ (see Fig. 3.6). The area of Π is the product of the base

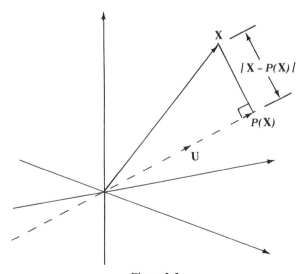

Figure 3.5

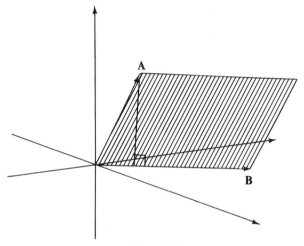

Figure 3.6

$|\mathbf{B}|$, and the altitude on that base which is the distance from $\mathbf{A}$ to $\mathbf{B}$. By (5), this distance $= (1/|\mathbf{B}|)\sqrt{(\mathbf{A} \cdot \mathbf{A})(\mathbf{B} \cdot \mathbf{B}) - (\mathbf{A} \cdot \mathbf{B})^2}$, so

$$\text{area } \Pi = \sqrt{(\mathbf{A} \cdot \mathbf{A})(\mathbf{B} \cdot \mathbf{B}) - (\mathbf{A} \cdot \mathbf{B})^2} . \tag{6}$$

§1. The Cross Product and Systems of Equations

Consider a system of two equations in three unknowns:

$$\begin{aligned} a_1x_1 + a_2x_2 + a_3x_3 &= 0, \\ b_1x_1 + b_2x_2 + b_3x_3 &= 0. \end{aligned} \tag{7}$$

We set $\mathbf{A} = \begin{bmatrix} a_1 \\ a_2 \\ a_3 \end{bmatrix}$ and $\mathbf{B} = \begin{bmatrix} b_1 \\ b_2 \\ b_3 \end{bmatrix}$. A solution vector $\mathbf{X} = \begin{bmatrix} x_1 \\ x_2 \\ x_3 \end{bmatrix}$ for (7) satisfies

$$\mathbf{A} \cdot \mathbf{X} = 0, \qquad \mathbf{B} \cdot \mathbf{X} = 0.$$

We may find such an $\mathbf{X}$ by multiplying the first equation by b_1 and the second by a_1 and subtracting

$$\begin{aligned} a_1b_1x_1 + a_2b_1x_2 + a_3b_1x_3 &= 0, \\ a_1b_1x_1 + a_1b_2x_2 + a_1b_3x_3 &= 0, \\ (a_2b_1 - a_1b_2)x_2 + (a_3b_1 - a_1b_3)x_3 &= 0. \end{aligned} \tag{8a}$$

Similarly, we may multiply the first equation by b_2 and the second by a_2 and subtract to get

$$(a_1b_2 - a_2b_1)x_1 + (a_3b_2 - a_2b_3)x_3 = 0. \tag{8b}$$

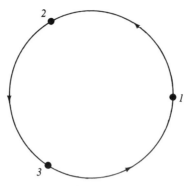

Figure 3.7

We can obtain a solution to the system (8a), (8b) by choosing

$$x_1 = (a_2b_3 - a_3b_2), \qquad x_2 = (a_3b_1 - a_1b_3), \qquad x_3 = (a_1b_2 - a_2b_1). \quad (9)$$

Note that if we think of the subscripts 1, 2, 3 on a wheel, with 1 followed by 2, followed by 3, and three followed by 1 (see Fig. 3.7), then in (9), x_2 is obtained from x_1, and x_3 is obtained from x_2 by following this succession of subscripts of a_i, b_i. We define the *cross product* $\mathbf{A} \times \mathbf{B}$ of $\mathbf{A}$ and $\mathbf{B}$ to be the vector

$$\mathbf{A} \times \mathbf{B} = \begin{bmatrix} a_2b_3 - a_3b_2 \\ a_3b_1 - a_1b_3 \\ a_1b_2 - a_2b_1 \end{bmatrix}. \quad (10)$$

The vector $\mathbf{X} = \mathbf{A} \times \mathbf{B}$ indeed satisfies the conditions

$$\mathbf{A} \cdot \mathbf{X} = 0, \qquad \mathbf{B} \cdot \mathbf{X} = 0,$$

which we set out to satisfy, and this is so since in the expression

$$\mathbf{A} \cdot \mathbf{X} = a_1(a_2b_3 - a_3b_2) + a_2(a_3b_1 - a_1b_3) + a_3(a_1b_2 - a_2b_1)$$

all terms cancel, leaving 0. The same happens for $\mathbf{B} \cdot \mathbf{X}$.

We shall see that the cross product is very useful in solving geometric problems in 3 dimensions.

We may easily verify that the cross product has the following properties:

(i) $\mathbf{A} \times \mathbf{A} = \mathbf{0}$, for every vector $\mathbf{A}$.
(ii) $\mathbf{B} \times \mathbf{A} = -\mathbf{A} \times \mathbf{B}$, for all $\mathbf{A}$, $\mathbf{B}$.
(iii) $\mathbf{A} \times (\mathbf{B} + \mathbf{C}) = \mathbf{A} \times \mathbf{B} + \mathbf{A} \times \mathbf{C}$, for all $\mathbf{A}$, $\mathbf{B}$, $\mathbf{C}$.
(iv) $(t\mathbf{A}) \times \mathbf{B} = t(\mathbf{A} \times \mathbf{B})$ if t is a scalar.

Exercise 2. Show that $\mathbf{E}_1 \times \mathbf{E}_2 = \mathbf{E}_3$, $\mathbf{E}_2 \times \mathbf{E}_3 = \mathbf{E}_1$, and $\mathbf{E}_1 \times \mathbf{E}_3 = -\mathbf{E}_2$.

Exercise 3. Show that $\mathbf{A} \times \mathbf{B} = -\mathbf{B} \times \mathbf{A}$.

Exercise 4. Show that $\mathbf{A} \times (\mathbf{B} + \mathbf{C}) = (\mathbf{A} \times \mathbf{B}) + (\mathbf{A} \times \mathbf{C})$.

Exercise 5. Show that $t\mathbf{A} \times \mathbf{B} = t(\mathbf{A} \times \mathbf{B})$.

Exercise 6. Show that $\mathbf{A} \cdot (\mathbf{B} \times \mathbf{C}) = \mathbf{B} \cdot (\mathbf{C} \times \mathbf{A}) = \mathbf{C} \cdot (\mathbf{A} \times \mathbf{B})$.

We now prove some propositions about the cross product.

(v). *If **A** and **B** are linearly dependent, then* **A** × **B** = **0**.

PROOF. If **B** = **0**, then **A** × **B** = **0** automatically. If **B** ≠ **0** and **A** = *t***B**, then **A** × **B** = (*t***B**) × **B** = *t*(**B** × **B**) = **0**.

(vi). *Conversely, if* **A** × **B** = **0**, *then* **A** *and* **B** *are linearly dependent.*

PROOF. If **A** × **B** = **0**, then either **B** = **0** or at least one of the components of **B** is nonzero. Assume $b_3 \neq 0$. Then $a_3b_2 - a_2b_3 = 0$, so $a_2 = (a_3/b_3)b_2$ and $-a_3b_1 + a_1b_3 = 0$, so $a_1 = (a_3/b_3)b_1$. It follows that

$$\mathbf{A} = \begin{bmatrix} a_1 \\ a_2 \\ a_3 \end{bmatrix} = \begin{bmatrix} (a_3/b_3)b_1 \\ (a_3/b_3)b_2 \\ (a_3/b_3)b_3 \end{bmatrix} = (a_3/b_3)\begin{bmatrix} b_1 \\ b_2 \\ b_3 \end{bmatrix} = (a_3/b_3)\mathbf{B}.$$

Therefore, **A** is a scalar multiple of **B**, so **A** and **B** are linearly dependent. We reason similarly if $b_2 \neq 0$ or $b_1 \neq 0$.

Exercise 7. If **A**, **B**, **C** are linearly dependent, show that **A** · (**B** × **C**) = **B** · (**C** × **A**) = **C** · (**A** × **B**) = **0**.

(vii). *Let* Π *denote the parallelogram in* $\mathbb{R}^3$ *with sides along* **A** *and* **B**, *where* **B** ≠ **0** (*see Fig. 3.8*). *Then*

$$\text{area } \Pi = |\mathbf{A} \times \mathbf{B}|. \tag{11}$$

PROOF. We have already found formula (6) for the area Π:

$$\text{area } \Pi = \sqrt{(\mathbf{A} \cdot \mathbf{A})(\mathbf{B} \cdot \mathbf{B}) - (\mathbf{A} \cdot \mathbf{B})^2} .$$

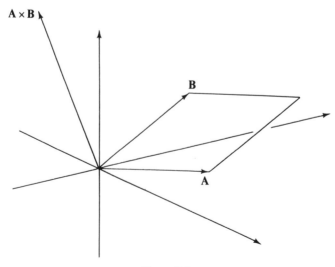

Figure 3.8

In coordinates, this gives

$$(\text{area } \Pi)^2 = (a_1^2 + a_2^2 + a_3^2)(b_1^2 + b_2^2 + b_3^2) - (a_1b_1 + a_2b_2 + a_3b_3)^2$$

$$= a_1^2b_1^2 + a_1^2b_2^2 + a_1^2b_3^2 - (a_1b_1)^2 - 2a_1b_1a_2b_2$$

$$+ a_2^2b_1^2 + a_2^2b_2^2 + a_2^2b_3^2 - (a_2b_2)^2 - 2a_1b_1a_3b_3$$

$$+ a_3^2b_1^2 + a_3^2b_2^2 + a_3^2b_3^2 - (a_3b_3)^2 - 2a_2b_2a_3b_3$$

$$= (a_1b_2 - a_2b_1)^2 + (a_3b_1 - a_1b_3)^2 + (a_2b_3 - a_3b_2)^2$$

$$= |\mathbf{A} \times \mathbf{B}|^2.$$

This establishes formula (11).

Observe that if **A** and **B** are linearly dependent, then the parallelogram Π will be contained in a line so its area will be 0, agreeing with the fact that $\mathbf{A} \times \mathbf{B} = \mathbf{0}$ in this case.

Let us now summarize what we have found.

If **A** and **B** are not linearly dependent, then we may describe the vector $\mathbf{A} \times \mathbf{B}$ by saying that it is perpendicular to **A** and **B** and it has length equal to the area of the parallelogram determined by **A** and **B**. Note that this description applies equally well to *two* vectors, $\mathbf{A} \times \mathbf{B}$ and $-(\mathbf{A} \times \mathbf{B})$ lying on opposite sides of the plane containing **A** and **B**.

In Chapter 3.5 we will go more deeply into the significance of the sign of $\mathbf{A} \times \mathbf{B}$.

Next, we shall give a generalization of the Pythagorean Theorem. If we project **A** and **B** into the x_1x_2 plane, we get the vectors $\begin{bmatrix} a_1 \\ a_2 \\ 0 \end{bmatrix}$ and $\begin{bmatrix} b_1 \\ b_2 \\ 0 \end{bmatrix}$. Note that the area of the parallelogram Π_{12} determined by these two vectors is given by

$$\left| \begin{bmatrix} a_1 \\ a_2 \\ 0 \end{bmatrix} \times \begin{bmatrix} b_1 \\ b_2 \\ 0 \end{bmatrix} \right| = \left| \begin{bmatrix} 0 \\ 0 \\ a_1b_2 - a_2b_1 \end{bmatrix} \right| = |a_1b_2 - a_2b_1|.$$

Similarly, the area of the parallelogram Π_{23} determined by the projections $\begin{bmatrix} 0 \\ a_2 \\ a_3 \end{bmatrix}$ and $\begin{bmatrix} 0 \\ b_2 \\ b_3 \end{bmatrix}$ of **A** and **B** to the x_2x_3 plane is given by area $(\Pi_{23}) = |a_3b_2 - a_2b_3|$, and finally area $\Pi_{13} = |a_3b_1 - a_1b_3|$ (see Fig. 3.9).

Formula (vii) thus yields the following striking result which is a generalization of the Pythagorean Theorem:

(viii) $(\text{area } \Pi)^2 = (\text{area } \Pi_{12})^2 + (\text{area } \Pi_{23})^2 + (\text{area } \Pi_{13})^2.$

This is the analogue of the theorem that the square of the length of a vector is the sum of the squares of its projections to the three coordinate

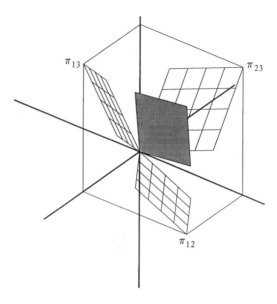

Figure 3.9

axes (see Fig. 3.10):

(ix) Given three vectors **A**, **B**, **C** in 3-space, we shall next obtain a formula for the volume of the parallelepiped determined by these three vectors.

If **A** and **B** are linearly independent, then the distance from the vector **C** to the plane determined by **A** and **B** equals the length of the projection of **C** to the line along **A** × **B**, since the vector **A** × **B** is orthogonal to that plane.

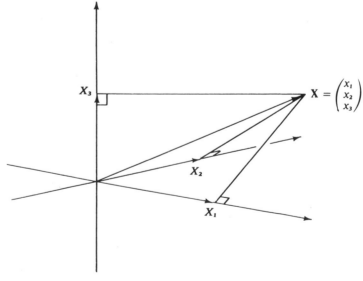

Figure 3.10

So the distance is given by

$$\frac{|\mathbf{C} \cdot (\mathbf{A} \times \mathbf{B})|}{|\mathbf{A} \times \mathbf{B}|}. \tag{12}$$

(x) It follows that the volume of the parallelepiped Π with sides along $\mathbf{A}$, $\mathbf{B}$, and $\mathbf{C}$ is given by the area of the base $|\mathbf{A} \times \mathbf{B}|$ multiplied by the height $|\mathbf{C} \cdot (\mathbf{A} \times \mathbf{B})|/|\mathbf{A} \times \mathbf{B}|$, i.e.,

$$\text{volume } \Pi = |\mathbf{C} \cdot (\mathbf{A} \times \mathbf{B})|. \tag{13}$$

If $\mathbf{A}$, $\mathbf{B}$, and $\mathbf{C}$ are linearly dependent, then the parallelepiped is contained in a plane so its volume will be zero, which agrees with the fact that $\mathbf{C} \cdot (\mathbf{A} \times \mathbf{B}) = 0$ in such a case, as we saw in Exercise 7.

We shall next see how the cross product helps us to study systems of linear equations.

We consider a system of three equations in the three unknowns x_1, x_2, x_3:

$$a_1 x_1 + a_2 x_2 + a_3 x_3 = 0,$$
$$b_1 x_1 + b_2 x_2 + b_3 x_3 = 0, \tag{14}$$
$$c_1 x_1 + c_2 x_2 + c_3 x_3 = 0.$$

We set $\mathbf{A} = \begin{bmatrix} a_1 \\ a_2 \\ a_3 \end{bmatrix}$, $\mathbf{B} = \begin{bmatrix} b_1 \\ b_2 \\ b_3 \end{bmatrix}$, $\mathbf{C} = \begin{bmatrix} c_1 \\ c_2 \\ c_3 \end{bmatrix}$. The solutions $\mathbf{X} = \begin{bmatrix} x_1 \\ x_2 \\ x_3 \end{bmatrix}$ of the system

(14) are the vectors $\mathbf{X}$ such that $\mathbf{X}$ is perpendicular to the three vectors $\mathbf{A}$, $\mathbf{B}$, $\mathbf{C}$. We shall show the following:

(xi). *There exists a nonzero solution vector $\mathbf{X}$ of* (14) *if and only if* $\mathbf{C} \cdot (\mathbf{A} \times \mathbf{B}) = 0$.

PROOF. Assume that $\mathbf{C} \cdot (\mathbf{A} \times \mathbf{B}) = 0$. If $\mathbf{A}$ and $\mathbf{B}$ are linearly dependent, there is some plane Π through the origin which contains $\mathbf{A}$, $\mathbf{B}$, and $\mathbf{C}$. We choose a nonzero vector $\mathbf{X}$ perpendicular to Π. Then $\mathbf{X} \cdot \mathbf{A} = 0$, $\mathbf{X} \cdot \mathbf{B} = 0$, $\mathbf{X} \cdot \mathbf{C} = 0$, so $\mathbf{X}$ is a solution of (14).

If $\mathbf{A}$ and $\mathbf{B}$ are linearly independent, then $\mathbf{A} \times \mathbf{B} \neq 0$. By assumption, $(\mathbf{A} \times \mathbf{B}) \cdot \mathbf{C} = 0$. Also $(\mathbf{A} \times \mathbf{B}) \cdot \mathbf{A} = 0$ and $(\mathbf{A} \times \mathbf{B}) \cdot \mathbf{B} = 0$. So $\mathbf{A} \times \mathbf{B}$ is a nonzero solution of (14).

Conversely, assume that $\mathbf{C} \cdot (\mathbf{A} \times \mathbf{B}) \neq 0$. Then $\mathbf{A}$ and $\mathbf{B}$ are linearly independent, since otherwise $\mathbf{A} \times \mathbf{B} = 0$ and so $\mathbf{C} \cdot (\mathbf{A} \times \mathbf{B}) = 0$, contrary to assumption. Let Π be the plane through the origin containing $\mathbf{A}$ and $\mathbf{B}$. Then $\mathbf{A} \times \mathbf{B}$ is perpendicular to Π, and every vector perpendicular to Π is a scalar multiple of $\mathbf{A} \times \mathbf{B}$. If $\mathbf{X}$ is a solution of (14), it follows that $\mathbf{X} = t(\mathbf{A} \times \mathbf{B})$ for some scalar t. Since $\mathbf{X} \cdot \mathbf{C} = 0$, $(t(\mathbf{A} \times \mathbf{B})) \cdot \mathbf{C} = t((\mathbf{A} \times \mathbf{B}) \cdot \mathbf{C}) = 0$, and so $t = 0$. Hence, $\mathbf{X} = 0$ as claimed.

The following is a fundamental property of the geometry of $\mathbb{R}^3$.

Proposition 1. *If* **A**, **B**, **C** *are three linearly independent vectors in* $\mathbb{R}^3$, *then every vector* **Y** *in* $\mathbb{R}^3$ *can be uniquely expressed as a linear combination of* **A**, **B**, *and* **C**.

PROOF. Since **C** is not a linear combination of **A** and **B**, the line $\{\mathbf{Y} - t\mathbf{C} \mid t$ real$\}$ which goes through **Y** and is parallel to **C** will not be parallel to the plane determined by **A** and **B**. Thus, for some t, $\mathbf{Y} - t\mathbf{C}$ lies in that plane and so

$$\mathbf{Y} - t\mathbf{C} = r\mathbf{A} + s\mathbf{B}$$

for suitable scalars r, s. Therefore,

$$\mathbf{Y} = r\mathbf{A} + s\mathbf{B} + t\mathbf{C}. \tag{15}$$

Expression (15) is unique, for if

$$\mathbf{Y} = r'\mathbf{A} + s'\mathbf{B} + t'\mathbf{C},$$

then $r\mathbf{A} + s\mathbf{B} + t\mathbf{C} = r'\mathbf{A} + s'\mathbf{B} + t'\mathbf{C}$, so

$$(r - r')\mathbf{A} + (s - s')\mathbf{B} + (t - t')\mathbf{C} = \mathbf{0}.$$

Since **A**, **B**, **C** are linearly independent, it follows that $r - r' = 0$, $s - s' = 0$, $t - t' = 0$. So the expression (15) is unique, as claimed.

The vectors of length 1 are called *unit vectors*, and their endpoints form the *unit sphere* in 3-space.

Exercise 8. Show that for any choice of angles θ, ϕ, the vector $\begin{bmatrix} \cos\phi\cos\theta \\ \cos\phi\sin\theta \\ \sin\phi \end{bmatrix}$ is a unit vector. Conversely, show that if $1 = u_1^2 + u_2^2 + u_3^2$ and if $u_3 \neq 0$, then it is possible to find an angle ϕ between $-\pi/2$ and $\pi/2$ so that $u_3 = \sin\phi$. The vector $\begin{pmatrix} u_1 \\ u_2 \\ 0 \end{pmatrix}$ then

has length $\sqrt{u_1^2 + u_2^2} = \sqrt{1 - u_3^2} = \sqrt{1 - \sin^2\phi} = \cos\phi$, so we may write

$$u_1 = \cos\phi\cos\theta, \qquad u_2 = \cos\phi\sin\theta$$

for some θ between 0 and 2π. Hence $\begin{pmatrix} u_1 \\ u_2 \\ u_3 \end{pmatrix} = \begin{bmatrix} \cos\phi\cos\theta \\ \cos\phi\sin\theta \\ \sin\phi \end{bmatrix}$.

It follows that any nonzero vector **X** in $\mathbb{R}^3$ may be written:

$$\mathbf{X} = |\mathbf{X}| \begin{bmatrix} \cos\phi\cos\theta \\ \cos\phi\sin\theta \\ \sin\phi \end{bmatrix},$$

for some choice of angles ϕ, θ.

Transformations of 3-Space

As in the planar case, we define a transformation of 3-space to be a rule T which assigns to every vector $\mathbf{X}$ of 3-space some vector $T(\mathbf{X})$ of 3-space. The vector $T(\mathbf{X})$ is called the *image* of $\mathbf{X}$ under T, and the collection of all vectors which are images of vectors under the transformation T is called the *range* of T. We denote transformations by capital letters, such as A, B, R, S, T, etc.

EXAMPLE 1. Let P denote the transformation which assigns to each vector $\mathbf{X} = \begin{bmatrix} x_1 \\ x_2 \\ x_3 \end{bmatrix}$ the projection to the line along $\mathbf{U} = \begin{bmatrix} 1 \\ 1 \\ 1 \end{bmatrix}$. By formula (1) of Chapter 3.0, we have

$$P(\mathbf{X}) = \left(\frac{\mathbf{X} \cdot \mathbf{U}}{\mathbf{U} \cdot \mathbf{U}} \right) \mathbf{U} = \left(\frac{x_1 + x_2 + x_3}{3} \right) \begin{bmatrix} 1 \\ 1 \\ 1 \end{bmatrix} = \begin{bmatrix} \frac{1}{3}x_1 + \frac{1}{3}x_2 + \frac{1}{3}x_3 \\ \frac{1}{3}x_1 + \frac{1}{3}x_2 + \frac{1}{3}x_3 \\ \frac{1}{3}x_1 + \frac{1}{3}x_2 + \frac{1}{3}x_3 \end{bmatrix}.$$

EXAMPLE 2. Let S denote the transformation which assigns to each vector $\mathbf{X}$ the *reflection of* $\mathbf{X}$ *through the line* along $\mathbf{U} = \begin{bmatrix} 1 \\ 1 \\ 1 \end{bmatrix}$. As in the planar case, $S(\mathbf{X})$ is defined by the condition that the midpoint of the segment between $\mathbf{X}$ and $S(\mathbf{X})$ is the projection of $\mathbf{X}$ to the line along $\mathbf{U}$. Thus

$$S(\mathbf{X}) = 2P(\mathbf{X}) - \mathbf{X}$$

or

$$S\begin{bmatrix} x_1 \\ x_2 \\ x_2 \end{bmatrix} = 2\begin{bmatrix} \frac{1}{3}x_1 + \frac{1}{3}x_2 + \frac{1}{3}x_3 \\ \frac{1}{3}x_1 + \frac{1}{3}x_2 + \frac{1}{3}x_3 \\ \frac{1}{3}x_1 + \frac{1}{3}x_2 + \frac{1}{3}x_3 \end{bmatrix} - \begin{bmatrix} x_1 \\ x_2 \\ x_3 \end{bmatrix} = \begin{bmatrix} -\frac{1}{3}x_1 + \frac{2}{3}x_2 + \frac{2}{3}x_3 \\ \frac{2}{3}x_1 - \frac{1}{3}x_2 + \frac{2}{3}x_3 \\ \frac{2}{3}x_1 + \frac{2}{3}x_2 - \frac{1}{3}x_3 \end{bmatrix}.$$

Exercise 1. In each of the following problems, let P denote projection to the line along $\mathbf{U}$. Find a formula for the coordinates of the image $P\begin{pmatrix} x_1 \\ x_2 \\ x_3 \end{pmatrix}$.

(a) $\mathbf{U} = \begin{pmatrix} 1 \\ 0 \\ 0 \end{pmatrix}$ (so we have projection to the first coordinate axis);

(b) $\mathbf{U} = \begin{pmatrix} 1 \\ 1 \\ 0 \end{pmatrix}$;

(c) $\mathbf{U} = \begin{pmatrix} 1 \\ 1 \\ -1 \end{pmatrix}$;

(d) $\mathbf{U} = \begin{pmatrix} 1 \\ 0 \\ 3 \end{pmatrix}$.

Exercise 2. For each of the vectors in the preceding exercise, find a formula for the reflection $S\begin{pmatrix} x_1 \\ x_2 \\ x_3 \end{pmatrix}$ of the vector $\begin{pmatrix} x_1 \\ x_2 \\ x_2 \end{pmatrix}$ through the line along $\mathbf{U}$.

EXAMPLE 3. Let $\overline{Q}$ denote projection to the x_2x_3-plane and let $\overline{P}$ denote projection to the x_1-axis. Then

$$\overline{Q}\begin{bmatrix} x_1 \\ x_2 \\ x_3 \end{bmatrix} = \begin{bmatrix} 0 \\ x_2 \\ x_3 \end{bmatrix} \quad \text{and} \quad \overline{P}\begin{bmatrix} x_1 \\ x_2 \\ x_3 \end{bmatrix} = \begin{bmatrix} x_1 \\ 0 \\ 0 \end{bmatrix}.$$

Note that $\overline{Q}(\mathbf{X}) + \overline{P}(\mathbf{X}) = \mathbf{X}$ for each $\mathbf{X}$.

EXAMPLE 4. Let Q denote projection to the plane through 0 perpendicular to $\mathbf{U} = \begin{bmatrix} 1 \\ 1 \\ 1 \end{bmatrix}$ and let P denote projection to the line along $\mathbf{U} = \begin{bmatrix} 1 \\ 1 \\ 1 \end{bmatrix}$. Then, as in Example 3, we have $Q(\mathbf{X}) + P(\mathbf{X}) = \mathbf{X}$, so by the formula in Example 1, we have

$$Q\begin{bmatrix} x_1 \\ x_2 \\ x_3 \end{bmatrix} = \begin{bmatrix} x_1 \\ x_2 \\ x_3 \end{bmatrix} - \begin{bmatrix} \frac{1}{3}x_1 + \frac{1}{3}x_2 + \frac{1}{3}x_3 \\ \frac{1}{3}x_1 + \frac{1}{3}x_2 + \frac{1}{3}x_3 \\ \frac{1}{3}x_1 + \frac{1}{3}x_2 + \frac{1}{3}x_3 \end{bmatrix} = \begin{bmatrix} \frac{2}{3}x_1 - \frac{1}{3}x_2 - \frac{1}{3}x_3 \\ -\frac{1}{3}x_1 + \frac{2}{3}x_2 - \frac{1}{3}x_3 \\ -\frac{1}{3}x_1 - \frac{1}{3}x_2 + \frac{2}{3}x_3 \end{bmatrix}.$$

Exercise 3. For each of the vectors **U** of Exercise 1, let Q denote projection to the plane perpendicular to **U**. Find the formula for $Q\begin{pmatrix} x_1 \\ x_2 \\ x_3 \end{pmatrix}$ in terms of the coordinates of $\mathbf{X} = \begin{pmatrix} x_1 \\ x_2 \\ x_3 \end{pmatrix}$.

EXAMPLE 5. Let Π denote the plane through the origin perpendicular to $\mathbf{U} = \begin{bmatrix} 1 \\ 1 \\ 1 \end{bmatrix}$. Let R denote reflection through the plane Π. For any vector **X**, the midpoint of the segment joining **X** to $R(\mathbf{X})$ is the projection $Q(\mathbf{X})$ of **X** to the plane Π. Therefore,

$$R(\mathbf{X}) = 2Q(\mathbf{X}) - \mathbf{X}$$

where Q is the projection in Example 4. Therefore,

$$R\begin{bmatrix} x_1 \\ x_2 \\ x_3 \end{bmatrix} = 2\begin{bmatrix} \frac{2}{3}x_1 - \frac{1}{3}x_2 - \frac{1}{3}x_3 \\ -\frac{1}{3}x_1 + \frac{2}{3}x_2 - \frac{1}{3}x_3 \\ -\frac{1}{3}x_1 - \frac{1}{3}x_2 + \frac{2}{3}x_3 \end{bmatrix} - \begin{bmatrix} x_1 \\ x_2 \\ x_3 \end{bmatrix} = \begin{bmatrix} \frac{1}{3}x_1 - \frac{2}{3}x_2 - \frac{2}{3}x_3 \\ -\frac{2}{3}x_1 + \frac{1}{3}x_2 - \frac{2}{3}x_3 \\ -\frac{2}{3}x_1 - \frac{2}{3}x_2 + \frac{1}{3}x_3 \end{bmatrix}.$$

Exercise 4. For each of the vectors **U** in Exercise 1, let R denote reflection through the plane through the origin perpendicular to **U**. Find the formula for $R\begin{pmatrix} x_1 \\ x_2 \\ x_3 \end{pmatrix}$ in terms of the coordinates of $\mathbf{X} = \begin{pmatrix} x_1 \\ x_2 \\ x_3 \end{pmatrix}$.

EXAMPLE 6. Let D_t denote the transformation which sends any vector into t times itself, where t is some fixed scalar number. Then

$$D_t(\mathbf{X}) = t\mathbf{X},$$

so

$$D_t\begin{bmatrix} x_1 \\ x_2 \\ x_3 \end{bmatrix} = t\begin{bmatrix} x_1 \\ x_2 \\ x_3 \end{bmatrix} = \begin{bmatrix} tx_1 \\ tx_2 \\ tx_3 \end{bmatrix}.$$

As in the planar case, we call D_t *the stretching by* t.

If $t = 0$, then $D_0(\mathbf{X}) = 0 \cdot \mathbf{X} = \mathbf{0}$ for all **X** so D_0 is the *zero transformation*, denoted by 0. If $t = 1$, then $D_1(\mathbf{X}) = 1 \cdot \mathbf{X} = \mathbf{X}$ for all **X**, so D_1 is the *identity transformation* denoted by I.

EXAMPLE 7. For a fixed scalar θ with $0 \leqslant \theta < 2\pi$, we define a rotation R_θ^1 of θ radians about the x_1-axis. This rotation leaves the x_1 component fixed and rotates in the x_2x_3-plane according to the rule for rotating in 2-dimensional space, i.e.,

$$R_\theta^1 \begin{bmatrix} x_1 \\ x_2 \\ x_3 \end{bmatrix} = \begin{bmatrix} x_1 \\ (\cos\theta)x_2 - (\sin\theta)x_3 \\ (\sin\theta)x_2 + (\cos\theta)x_3 \end{bmatrix}.$$

For example,

$$R_{\pi/2}^1 \begin{bmatrix} x_1 \\ x_2 \\ x_3 \end{bmatrix} = \begin{bmatrix} x_1 \\ -x_3 \\ x_2 \end{bmatrix} = R_{-3\pi/2}^1 \begin{bmatrix} x_1 \\ x_2 \\ x_3 \end{bmatrix}$$

and

$$R_{-\pi/2}^1 \begin{bmatrix} x_1 \\ x_2 \\ x_3 \end{bmatrix} = \begin{bmatrix} x_1 \\ x_3 \\ -x_2 \end{bmatrix} = R_{3\pi/2}^1 \begin{bmatrix} x_1 \\ x_2 \\ x_3 \end{bmatrix}.$$

Exercise 5. In terms of the coordinates of $\mathbf{X} = \begin{pmatrix} x_1 \\ x_2 \\ x_3 \end{pmatrix}$, calculate the images of

(a) $R_\pi^1(\mathbf{X})$,
(b) $R_{\pi/4}^1(\mathbf{X})$, and
(c) $R_{-\pi/2}^1(\mathbf{X})$.

EXAMPLE 8. In a similar way, we may define rotations R_θ^2 and R_θ^3 by θ radians about the x_2-axis and the x_3-axis. We have the formulas

$$R_\theta^2 \begin{bmatrix} x_1 \\ x_2 \\ x_3 \end{bmatrix} = \begin{bmatrix} (\cos\theta)x_1 + (\sin\theta)x_3 \\ x_2 \\ (-\sin\theta)x_1 + (\cos\theta)x_3 \end{bmatrix}$$

and

$$R_\theta^3 \begin{bmatrix} x_1 \\ x_2 \\ x_3 \end{bmatrix} = \begin{bmatrix} (\cos\theta)x_1 - (\sin\theta)x_2 \\ (\sin\theta)x_1 + (\cos\theta)x_2 \\ x_3 \end{bmatrix}.$$

Note that the algebraic signs for R_θ^2 are different from those of R_θ^1 and R_θ^3.

Exercise 6. Calculate the images

(a) $R_\pi^2(\mathbf{X})$,
(b) $R_{\pi/4}^3(\mathbf{X})$,
(c) $R_{\pi/2}^1(R_{\pi/2}^2(\mathbf{X}))$,
(d) $R_{\pi/2}^2(R_{\pi/2}^1(\mathbf{X}))$.

Linear Transformations and Matrices

In Chapter 3.1 we examined a number of transformations T of 3-space, all of which have the property that, in terms of the coordinates of $\mathbf{X} = \begin{bmatrix} x_1 \\ x_2 \\ x_3 \end{bmatrix}$, the coordinates of $T(\mathbf{X})$ are given by *linear functions* of these coordinates. In each case the formulae are of the following type:

$$T\begin{bmatrix} x_1 \\ x_2 \\ x_3 \end{bmatrix} = \begin{bmatrix} a_1 x_1 + a_2 x_2 + a_3 x_3 \\ b_1 x_1 + b_2 x_2 + b_3 x_3 \\ c_1 x_1 + c_2 x_2 + c_3 x_3 \end{bmatrix}.$$

Any transformation of this form is called a *linear transformation of 3-space*. The expression

$$\begin{bmatrix} a_1 & a_2 & a_3 \\ b_1 & b_2 & b_3 \\ c_1 & c_2 & c_3 \end{bmatrix}$$

is called the *matrix* of the transformation T and is denoted by $m(T)$.

We can now list the matrices of the linear transformations in Examples 1–8 in Chapter 3.1.

$$m(P) = \begin{bmatrix} \frac{1}{3} & \frac{1}{3} & \frac{1}{3} \\ \frac{1}{3} & \frac{1}{3} & \frac{1}{3} \\ \frac{1}{3} & \frac{1}{3} & \frac{1}{3} \end{bmatrix}, \tag{1}$$

$$m(S) = \begin{bmatrix} -\frac{1}{3} & \frac{2}{3} & \frac{2}{3} \\ \frac{2}{3} & -\frac{1}{3} & \frac{2}{3} \\ \frac{2}{3} & \frac{2}{3} & -\frac{1}{3} \end{bmatrix}, \tag{2}$$

$$m(\overline{Q}) = \begin{bmatrix} 0 & 0 & 0 \\ 0 & 1 & 0 \\ 0 & 0 & 1 \end{bmatrix}, \tag{3}$$

$$m(Q) = \begin{bmatrix} \frac{2}{3} & -\frac{1}{3} & -\frac{1}{3} \\ -\frac{1}{3} & \frac{2}{3} & -\frac{1}{3} \\ -\frac{1}{3} & -\frac{1}{3} & \frac{2}{3} \end{bmatrix}, \tag{4}$$

$$m(R) = \begin{bmatrix} \frac{1}{3} & -\frac{2}{3} & -\frac{2}{3} \\ -\frac{2}{3} & \frac{1}{3} & -\frac{2}{3} \\ -\frac{2}{3} & -\frac{2}{3} & \frac{1}{3} \end{bmatrix}, \tag{5}$$

$$m(D_t) = \begin{bmatrix} t & 0 & 0 \\ 0 & t & 0 \\ 0 & 0 & t \end{bmatrix}, \tag{6}$$

$$m(R_\theta^1) = \begin{bmatrix} 1 & 0 & 0 \\ 0 & \cos\theta & -\sin\theta \\ 0 & \sin\theta & \cos\theta \end{bmatrix}, \tag{7}$$

$$m(R_\theta^2) = \begin{bmatrix} \cos\theta & 0 & \sin\theta \\ & 1 & \\ -\sin\theta & 0 & \cos\theta \end{bmatrix}, \tag{8}$$

$$m(R_\theta^3) = \begin{bmatrix} \cos\theta & -\sin\theta & 0 \\ \sin\theta & \cos\theta & 0 \\ 0 & 0 & 1 \end{bmatrix}.$$

We will denote by id, (read *identity*), the matrix $m(I) = \begin{bmatrix} 1 & 0 & 0 \\ 0 & 1 & 0 \\ 0 & 0 & 1 \end{bmatrix}$.

As in the plane, if T is a linear transformation with matrix $m(T)$ $= \begin{bmatrix} a_1 & a_2 & a_3 \\ b_1 & b_2 & b_3 \\ c_1 & c_2 & c_3 \end{bmatrix}$, then we write

$$\begin{bmatrix} a_1 & a_2 & a_3 \\ b_1 & b_2 & b_3 \\ c_1 & c_2 & c_3 \end{bmatrix} \begin{bmatrix} x_1 \\ x_2 \\ x_3 \end{bmatrix} = \begin{bmatrix} a_1x_1 + a_2x_2 + a_3x_3 \\ b_1x_1 + b_2x_2 + b_3x_3 \\ c_1x_1 + c_2x_2 + c_3x_3 \end{bmatrix}, \tag{9}$$

and we say that the matrix $m(T)$ *acts on the vector* $\begin{bmatrix} x_1 \\ x_2 \\ x_3 \end{bmatrix}$ to yield the vector

$$\begin{bmatrix} a_1x_1 + a_2x_2 + a_3x_3 \\ b_1x_1 + b_2x_2 + b_3x_3 \\ c_1x_1 + c_2x_2 + c_3x_3 \end{bmatrix}.$$

EXAMPLE 1.

$$\begin{bmatrix} 1 & 2 & 3 \\ 4 & 5 & 6 \\ 7 & 8 & 9 \end{bmatrix}\begin{bmatrix} 1 \\ 1 \\ 1 \end{bmatrix} = \begin{bmatrix} 1+2+3 \\ 4+5+6 \\ 7+8+9 \end{bmatrix} = \begin{bmatrix} 6 \\ 15 \\ 24 \end{bmatrix},$$

$$\begin{bmatrix} 1 & 2 & 3 \\ 4 & 5 & 6 \\ 7 & 8 & 9 \end{bmatrix}\begin{bmatrix} 0 \\ -1 \\ 1 \end{bmatrix} = \begin{bmatrix} 0-2+3 \\ 0-5+6 \\ 0-8+9 \end{bmatrix} = \begin{bmatrix} 1 \\ 1 \\ 1 \end{bmatrix},$$

$$\begin{bmatrix} 1 & 2 & 3 \\ 4 & 5 & 6 \\ 7 & 8 & 9 \end{bmatrix}\begin{bmatrix} x_1 \\ x_2 \\ x_3 \end{bmatrix} = \begin{bmatrix} x_1 + 2x_2 + 3x_3 \\ 4x_1 + 5x_2 + 6x_3 \\ 7x_1 + 8x_2 + 9x_3 \end{bmatrix}.$$

EXAMPLE 2.

$$\begin{bmatrix} 0 & 1 & 0 \\ 1 & 0 & 0 \\ 0 & 0 & 1 \end{bmatrix}\begin{bmatrix} x_1 \\ x_2 \\ x_3 \end{bmatrix} = \begin{bmatrix} x_2 \\ x_1 \\ x_3 \end{bmatrix}.$$

We now prove two crucial properties of linear transformations which show how they act on sums and scalar products of vectors. If $\mathbf{X} = \begin{bmatrix} x_1 \\ x_2 \\ x_3 \end{bmatrix}$ and $\mathbf{Y} = \begin{bmatrix} y_1 \\ y_2 \\ y_3 \end{bmatrix}$, then

$$\begin{bmatrix} a_1 & a_2 & a_3 \\ b_1 & b_2 & b_3 \\ c_1 & c_2 & c_3 \end{bmatrix}\left(\begin{bmatrix} x_1 \\ x_2 \\ x_3 \end{bmatrix} + \begin{bmatrix} y_1 \\ y_2 \\ y_3 \end{bmatrix}\right) = \begin{bmatrix} a_1 & a_2 & a_3 \\ b_1 & b_2 & b_3 \\ c_1 & c_2 & c_3 \end{bmatrix}\begin{bmatrix} x_1 + y_1 \\ x_2 + y_2 \\ x_3 + y_3 \end{bmatrix}$$

$$= \begin{bmatrix} a_1(x_1 + y_1) + a_2(x_2 + y_2) + a_3(x_3 + y_3) \\ b_1(x_1 + y_1) + b_2(x_2 + y_2) + b_3(x_3 + y_3) \\ c_1(x_1 + y_1) + c_2(x_2 + y_2) + c_3(x_3 + y_3) \end{bmatrix}$$

$$= \begin{bmatrix} a_1x_1 + a_2x_2 + a_3x_3 \\ b_1x_1 + b_2x_2 + b_3x_3 \\ c_1x_1 + c_2x_2 + c_3x_3 \end{bmatrix} + \begin{bmatrix} a_1y_1 + a_2y_2 + a_3y_3 \\ b_1y_1 + b_2y_2 + b_3y_3 \\ c_1y_1 + c_2y_2 + c_3y_3 \end{bmatrix}$$

$$= \begin{bmatrix} a_1 & a_2 & a_3 \\ b_1 & b_2 & b_3 \\ c_1 & c_2 & c_3 \end{bmatrix}\begin{bmatrix} x_1 \\ x_2 \\ x_3 \end{bmatrix} + \begin{bmatrix} a_1 & a_2 & a_3 \\ b_1 & b_2 & b_3 \\ c_1 & c_2 & c_3 \end{bmatrix}\begin{bmatrix} y_1 \\ y_2 \\ y_3 \end{bmatrix}.$$

Thus, for the associated transformation T with $m(T) = \begin{bmatrix} a_1 & a_2 & a_3 \\ b_1 & b_2 & b_3 \\ c_1 & c_2 & c_3 \end{bmatrix}$, we have

$$T(\mathbf{X} + \mathbf{Y}) = T(\mathbf{X}) + T(\mathbf{Y}). \tag{10a}$$

Similarly, we may show that

$$T(r\mathbf{X}) = rT(\mathbf{X}) \tag{10b}$$

for any scalar r.

Exercise 1. Prove the statement (10b).

Conversely, if T is a transformation such that $T(\mathbf{X} + \mathbf{Y}) = T(\mathbf{X}) + T(\mathbf{Y})$ and $T(r\mathbf{X}) = rT(\mathbf{X})$ for all vectors $\mathbf{X}$, $\mathbf{Y}$ and scalars r, then we may show that the coordinates of $T\begin{bmatrix} x_1 \\ x_2 \\ x_3 \end{bmatrix}$ are given by a set of linear equations in the coordinates of $\mathbf{X} = \begin{bmatrix} x_1 \\ x_2 \\ x_3 \end{bmatrix}$. Specifically,

$$T\begin{bmatrix} x_1 \\ x_2 \\ x_3 \end{bmatrix} = T\left(\begin{bmatrix} x_1 \\ 0 \\ 0 \end{bmatrix} + \begin{bmatrix} 0 \\ x_2 \\ 0 \end{bmatrix} + \begin{bmatrix} 0 \\ 0 \\ x_3 \end{bmatrix}\right)$$

$$= T\left(x_1\begin{bmatrix} 1 \\ 0 \\ 0 \end{bmatrix} + x_2\begin{bmatrix} 0 \\ 1 \\ 0 \end{bmatrix} + x_3\begin{bmatrix} 0 \\ 0 \\ 1 \end{bmatrix}\right)$$

$$= x_1 T\begin{bmatrix} 1 \\ 0 \\ 0 \end{bmatrix} + x_2 T\begin{bmatrix} 0 \\ 1 \\ 0 \end{bmatrix} + x_3 T\begin{bmatrix} 0 \\ 0 \\ 1 \end{bmatrix}.$$

Let $T\begin{bmatrix} 1 \\ 0 \\ 0 \end{bmatrix} = \begin{bmatrix} a_1 \\ b_1 \\ c_1 \end{bmatrix}$, $T\begin{bmatrix} 0 \\ 1 \\ 0 \end{bmatrix} = \begin{bmatrix} a_2 \\ b_2 \\ c_2 \end{bmatrix}$, $T\begin{bmatrix} 0 \\ 0 \\ 1 \end{bmatrix} = \begin{bmatrix} a_3 \\ b_3 \\ c_3 \end{bmatrix}$. Then

$$T\begin{bmatrix} x_1 \\ x_2 \\ x_3 \end{bmatrix} = x_1\begin{bmatrix} a_1 \\ b_1 \\ c_1 \end{bmatrix} + x_2\begin{bmatrix} a_2 \\ b_2 \\ c_2 \end{bmatrix} + x_3\begin{bmatrix} a_3 \\ b_3 \\ c_3 \end{bmatrix}$$

$$= \begin{bmatrix} a_1 x_1 \\ b_1 x_1 \\ c_1 x_1 \end{bmatrix} + \begin{bmatrix} a_2 x_2 \\ b_2 x_2 \\ c_2 x_2 \end{bmatrix} + \begin{bmatrix} a_3 x_3 \\ b_3 x_3 \\ c_3 x_3 \end{bmatrix}$$

$$= \begin{bmatrix} a_1 x_1 + a_2 x_2 + a_3 x_3 \\ b_1 x_1 + b_2 x_2 + b_3 x_3 \\ c_1 x_1 + c_2 x_2 + c_3 x_3 \end{bmatrix}$$

as predicted. In summary, we have:

Theorem 3.1. *Let T be a transformation of 3-space. Then T is a linear transformation if and only if T satisfies the conditions*

$$T(\mathbf{X} + \mathbf{Y}) = T(\mathbf{X}) + T(\mathbf{Y})$$

and

$$T(r\mathbf{X}) = rT(\mathbf{X})$$

for all vectors $\mathbf{X}$ and $\mathbf{Y}$ and all scalars r.

Let T be a linear transformation of $\mathbb{R}^3$. Let L be a straight line in $\mathbb{R}^3$. By the image of L under T, we mean the collection of vectors $T(\mathbf{X})$ for all vectors $\mathbf{X}$ with endpoint lying on L. If Π is a plane, we define the image of Π under T in a similar way.

Theorem 3.2. *Let T be a linear transformation of $\mathbb{R}^3$. The image of a straight line L under T is either a straight line or a point. The image of a plane Π under T is either a plane, a straight line, or a point. The image of $\mathbb{R}^3$ under T is either $\mathbb{R}^3$, a plane, a straight line, or a point.*

The proof proceeds in exact analogy with the proof in dimension 2 (proof of Theorem 2.2 in Chapter 2.2).

Exercise 2. Describe the images of a line $L = \{\mathbf{A} + t\mathbf{U} \mid t \text{ real}\}$ under a linear transformation T.

Exercise 3. Let T be a linear transformation. Let Π be the plane $\{\mathbf{C} + t\mathbf{U} + s\mathbf{V} \mid t, s$ real$\}$, where $\mathbf{U}$ and $\mathbf{V}$ are linearly independent. Show that the image of Π under T is the collection of vectors $T(\mathbf{C}) + tT(\mathbf{U}) + sT(\mathbf{V})$. Under what conditions will the image be a single point? When will the image be a line?

Exercise 4. The image under T of $\mathbb{R}^3 = \{x_1\mathbf{E}_1 + x_2\mathbf{E}_2 + x_3\mathbf{E}_3 \mid x_1, x_2, x_3$ real$\}$ is $\{x_1 T(\mathbf{E}_1) + x_2 T(\mathbf{E}_2) + x_3 T(\mathbf{E}_3)\}$. Under what conditions on $T(\mathbf{E}_1)$, $T(\mathbf{E}_2)$, and $T(\mathbf{E}_3)$ is this all of $\mathbb{R}^3$?

Sums and Products of Linear Transformations

If T and S are linear transformations, then we may define a new transformation $T + S$ by the condition

$$(T + S)(\mathbf{X}) = T(\mathbf{X}) + S(\mathbf{X}) \qquad \text{for every vector } \mathbf{X}.$$

Then by definition, $(T + S)(\mathbf{X} + \mathbf{Y}) = T(\mathbf{X} + \mathbf{Y}) + S(\mathbf{X} + \mathbf{Y})$, and since T and S are linear transformations, this equals $T(\mathbf{X}) + T(\mathbf{Y}) + S(\mathbf{X}) + S(\mathbf{Y}) = T(\mathbf{X}) + S(\mathbf{X}) + T(\mathbf{Y}) + S(\mathbf{Y}) = (T + S)(\mathbf{X}) + (T + S)(\mathbf{Y})$. Thus for every pair $\mathbf{X}, \mathbf{Y}$, we have

$$(T + S)(\mathbf{X} + \mathbf{Y}) = (T + S)(\mathbf{X}) + (T + S)(\mathbf{Y}).$$

Similarly, we may show

$$(T + S)(t\mathbf{X}) = t(T + S)(\mathbf{X}).$$

Therefore, by Theorem 3.1, $T + S$ is a linear transformation. It is called the *sum* of the transformations T and S.

If the matrix of T is $m(T) = \begin{bmatrix} a_{11} & a_{12} & a_{13} \\ a_{21} & a_{22} & a_{23} \\ a_{31} & a_{32} & a_{33} \end{bmatrix}$ and the matrix of S is

$m(S) = \begin{bmatrix} b_{11} & b_{12} & b_{13} \\ b_{21} & b_{22} & b_{23} \\ b_{31} & b_{32} & b_{33} \end{bmatrix}$, then we may calculate the matrix $m(T + S)$ of

$T + S$ as follows:

$$(T + S)\begin{bmatrix} x_1 \\ x_2 \\ x_3 \end{bmatrix} = T\begin{bmatrix} x_1 \\ x_2 \\ x_3 \end{bmatrix} + S\begin{bmatrix} x_1 \\ x_2 \\ x_3 \end{bmatrix}$$

$$= \begin{bmatrix} a_{11} & a_{12} & a_{13} \\ a_{21} & a_{22} & a_{23} \\ a_{31} & a_{32} & a_{33} \end{bmatrix}\begin{bmatrix} x_1 \\ x_2 \\ x_3 \end{bmatrix} + \begin{bmatrix} b_{11} & b_{12} & b_{13} \\ b_{21} & b_{22} & b_{23} \\ b_{31} & b_{32} & b_{33} \end{bmatrix}\begin{bmatrix} x_1 \\ x_2 \\ x_3 \end{bmatrix}$$

$$= \begin{bmatrix} a_{11}x_1 + a_{12}x_2 + a_{13}x_3 \\ a_{21}x_1 + a_{22}x_2 + a_{23}x_3 \\ a_{31}x_1 + a_{32}x_2 + a_{33}x_3 \end{bmatrix} + \begin{bmatrix} b_{11}x_1 + b_{12}x_2 + b_{13}x_3 \\ b_{21}x_1 + b_{22}x_2 + b_{23}x_3 \\ b_{31}x_1 + b_{32}x_2 + b_{33}x_3 \end{bmatrix}$$

$$= \begin{bmatrix} (a_{11} + b_{11})x_1 + (a_{12} + b_{12})x_2 + (a_{13} + b_{13})x_3 \\ (a_{21} + b_{21})x_1 + (a_{22} + b_{22})x_2 + (a_{23} + b_{23})x_3 \\ (a_{31} + b_{31})x_1 + (a_{32} + b_{32})x_2 + (a_{33} + b_{33})x_3 \end{bmatrix}$$

$$= \begin{bmatrix} a_{11} + b_{11} & a_{12} + b_{12} & a_{13} + b_{13} \\ a_{21} + b_{21} & a_{22} + b_{22} & a_{23} + b_{23} \\ a_{31} + b_{31} & a_{32} + b_{32} & a_{33} + b_{33} \end{bmatrix}\begin{bmatrix} x_1 \\ x_2 \\ x_3 \end{bmatrix}.$$

Thus the matrix for the sum of two linear transformations is just the matrix formed by adding the corresponding entries in the matrices of the two linear transformations.

We define matrix addition componentwise by the formula:

$$\begin{bmatrix} a_{11} & a_{12} & a_{13} \\ a_{21} & a_{22} & a_{23} \\ a_{31} & a_{32} & a_{33} \end{bmatrix} + \begin{bmatrix} b_{11} & b_{12} & b_{13} \\ b_{21} & b_{22} & b_{23} \\ b_{31} & b_{32} & b_{33} \end{bmatrix} = \begin{bmatrix} a_{11} + b_{11} & a_{12} + b_{12} & a_{13} + b_{13} \\ a_{21} + b_{21} & a_{22} + b_{22} & a_{23} + b_{23} \\ a_{31} + b_{31} & a_{32} + b_{32} & a_{33} + b_{33} \end{bmatrix}.$$

$$(1)$$

Therefore we may write the matrix of $T + S$ as the sum of the matrices of T and S:

$$m(T + S) = m(T) + m(S). \tag{2}$$

Exercise 1. For any linear transformation T and any scalar t, we define a new transformation tT by the condition: $(tT)(X) = tT(X)$ for all X. Show that tT is a linear transformation by showing that $(tT)(X + Y) = (tT)(X) + (tT)(Y)$ and $(tT)(sX) = s(tT)(X)$ for any vectors X, Y and any scalar s. For any matrix, we define the product of the matrix by the scalar t to be the matrix whose entries are all multiplied by t, i.e.,

$$t \begin{bmatrix} a_{11} & a_{12} & a_{13} \\ a_{21} & a_{22} & a_{23} \\ a_{31} & a_{32} & a_{33} \end{bmatrix} = \begin{bmatrix} ta_{11} & ta_{12} & ta_{13} \\ ta_{21} & ta_{22} & ta_{23} \\ ta_{31} & ta_{32} & ta_{33} \end{bmatrix}. \tag{3}$$

Show that $m(tT) = tm(T)$.

EXAMPLE 1. We may use the above ideas to calculate the matrices of some of the transformations we used in Chapter 3.2, e.g., if S is reflection in the x_3-axis and T is projection to the x_3-axis, then $S = 2T - I$, so

$$m(S) = m(2T - I) = m(2T) - m(I) = 2m(T) - m(I)$$

$$= 2 \begin{bmatrix} 0 & 0 & 0 \\ 0 & 0 & 0 \\ 0 & 0 & 1 \end{bmatrix} - \begin{bmatrix} 1 & 0 & 0 \\ 0 & 1 & 0 \\ 0 & 0 & 1 \end{bmatrix} = \begin{bmatrix} -1 & 0 & 0 \\ 0 & -1 & 0 \\ 0 & 0 & 1 \end{bmatrix}.$$

As in the case of the plane, we may define the product of two linear transformations T and S to be the transformation R given by $R(\mathbf{X}) = S(T(\mathbf{X}))$.

We write $R = ST$, and we call R *the product of S and T*.

EXAMPLE 2. Let K be reflection in the $x_1 x_2$-plane and J be reflection in the $x_2 x_3$-plane. Then

$$J \begin{bmatrix} x_1 \\ x_2 \\ x_3 \end{bmatrix} = \begin{bmatrix} -x_1 \\ x_2 \\ x_3 \end{bmatrix} \quad \text{and} \quad K(J(\mathbf{X})) = K \begin{bmatrix} -x_1 \\ x_2 \\ x_3 \end{bmatrix} = \begin{bmatrix} -x_1 \\ x_2 \\ -x_3 \end{bmatrix}.$$

Also, $J(K(\mathbf{X})) = J \begin{bmatrix} x_1 \\ x_2 \\ -x_3 \end{bmatrix} = \begin{bmatrix} -x_1 \\ x_2 \\ -x_3 \end{bmatrix}$, so $KJ = JK$.

Exercise 2. Show that if S and T are any linear transformations, then ST and TS are linear transformation, using Theorem 3.1.

EXAMPLE 3. Let P be projection to the $x_1 x_2$-plane and let Q be projection to the $x_2 x_3$-plane. Describe PQ and QP. We find $PQ \begin{bmatrix} x_1 \\ x_2 \\ x_3 \end{bmatrix} =$

$$P \begin{bmatrix} Q \begin{bmatrix} x_1 \\ x_2 \\ x_3 \end{bmatrix} \end{bmatrix} = P \begin{bmatrix} 0 \\ x_2 \\ x_3 \end{bmatrix} = \begin{bmatrix} 0 \\ x_2 \\ 0 \end{bmatrix}. \text{ Also, } QP \begin{bmatrix} x_1 \\ x_2 \\ x_3 \end{bmatrix} = \begin{bmatrix} 0 \\ x_2 \\ 0 \end{bmatrix}.$$

Exercise 3. Let R denote projection to the x_3-axis and let P denote projection to the $x_1 x_2$-plane. Find RP and PR.

EXAMPLE 4. Find PP. Since $PP \begin{bmatrix} x_1 \\ x_2 \\ x_3 \end{bmatrix} = P \left(P \begin{bmatrix} x_1 \\ x_2 \\ x_3 \end{bmatrix} \right) = P \begin{bmatrix} x_1 \\ x_2 \\ 0 \end{bmatrix} = \begin{bmatrix} x_1 \\ x_2 \\ 0 \end{bmatrix}$. There-

fore, $PP(\mathbf{X}) = P(\mathbf{X})$ for all $\mathbf{X}$, so we have $PP = P$.

Exercise 4. Show that $RR = R$ and $QQ = Q$, where R and Q are as above.

EXAMPLE 5. Find KQ and QK, when K and Q are the transformations defined in Examples 2 and 3. We have

$$KQ \begin{bmatrix} x_1 \\ x_2 \\ x_3 \end{bmatrix} = K \begin{bmatrix} 0 \\ x_2 \\ x_3 \end{bmatrix} = \begin{bmatrix} 0 \\ x_2 \\ -x_3 \end{bmatrix}, \qquad QK \begin{bmatrix} x_1 \\ x_2 \\ x_3 \end{bmatrix} = Q \begin{bmatrix} x_1 \\ x_2 \\ -x_3 \end{bmatrix} = \begin{bmatrix} 0 \\ x_2 \\ -x_3 \end{bmatrix}.$$

Thus $KQ = QK$.

Exercise 5. Find KP and PK, and RK and KR, where K, P, and R are the transformations in Examples 2, 3, and 4.

EXAMPLE 6. Let S be a linear transformation and let I denote the identity transformation. Find SI and IS. For each $\mathbf{X}$, we have $SI(\mathbf{X}) = S(\mathbf{X})$ and $IS(\mathbf{X}) = I(S(\mathbf{X})) = S(\mathbf{X})$. Thus $SI = S = IS$.

If T and S are linear transformations with matrices

$$m(T) = \begin{bmatrix} a_{11} & a_{12} & a_{13} \\ a_{21} & a_{22} & a_{23} \\ a_{31} & a_{32} & a_{33} \end{bmatrix} \quad \text{and} \quad m(S) = \begin{bmatrix} b_{11} & b_{12} & b_{13} \\ b_{21} & b_{22} & b_{23} \\ b_{31} & b_{32} & b_{33} \end{bmatrix},$$

then we know that TS is also a linear transformation, so we may calculate its matrix as follows. To find the first column of $m(TS)$, we must find the image of $\mathbf{E}_1$, i.e., $TS(\mathbf{E}_1) = T(S(\mathbf{E}_1))$, where $S(\mathbf{E}_1) = \begin{bmatrix} b_{11} \\ b_{21} \\ b_{31} \end{bmatrix}$. Thus

$$T(S(\mathbf{E}_1)) = \begin{bmatrix} a_{11} & a_{12} & a_{13} \\ a_{21} & a_{22} & a_{23} \\ a_{31} & a_{32} & a_{33} \end{bmatrix} \begin{bmatrix} b_{11} \\ b_{21} \\ b_{31} \end{bmatrix} = \begin{bmatrix} a_{11}b_{11} + a_{12}b_{21} + a_{13}b_{31} \\ a_{21}b_{11} + a_{22}b_{21} + a_{23}b_{31} \\ a_{31}b_{11} + a_{32}b_{21} + a_{33}b_{31} \end{bmatrix}.$$

Similarly, we may find the second column of $m(TS)$ by computing $T(S(\mathbf{E}_2))$ and the third column of $m(TS)$ by computing $T(S(\mathbf{E}_3))$.

We define the product of the two matrices $m(T)$ and $m(S)$ to be the matrix $m(TS)$ of the product transformation TS. Thus $m(T)m(S) =$

$m(TS)$ or

$$\begin{bmatrix} a_{11} & a_{12} & a_{13} \\ a_{21} & a_{22} & a_{23} \\ a_{31} & a_{32} & a_{33} \end{bmatrix} \begin{bmatrix} b_{11} & b_{12} & b_{13} \\ b_{21} & b_{22} & b_{23} \\ b_{31} & b_{32} & b_{33} \end{bmatrix} = m(TS)$$

$$= \begin{bmatrix} a_{11}b_{11} + a_{12}b_{21} + a_{13}b_{31} & a_{11}b_{12} + a_{12}b_{22} + a_{13}b_{32} & a_{11}b_{13} + a_{12}b_{23} + a_{13}b_{33} \\ a_{21}b_{11} + a_{22}b_{21} + a_{23}b_{31} & a_{21}b_{12} + a_{22}b_{22} + a_{23}b_{32} & a_{21}b_{13} + a_{22}b_{23} + a_{23}b_{33} \\ a_{31}b_{11} + a_{32}b_{21} + a_{33}b_{31} & a_{31}b_{12} + a_{32}b_{22} + a_{33}b_{32} & a_{31}b_{13} + a_{32}b_{23} + a_{33}b_{33} \end{bmatrix}.$$

$$(4)$$

EXAMPLE 7. If $m(T) = \begin{bmatrix} 2 & 1 & 2 \\ 1 & 3 & 1 \\ 3 & 1 & 4 \end{bmatrix}$ and $m(S) = \begin{bmatrix} 1 & 1 & 2 \\ 2 & 3 & 1 \\ 1 & 3 & 1 \end{bmatrix}$, then

$m(T)m(S)$

$$= \begin{bmatrix} 2\cdot 1 + 1\cdot 2 + 2\cdot 1 & 2\cdot 1 + 1\cdot 3 + 2\cdot 3 & 2\cdot 2 + 1\cdot 1 + 2\cdot 1 \\ 1\cdot 1 + 3\cdot 2 + 1\cdot 1 & 1\cdot 1 + 3\cdot 3 + 1\cdot 3 & 1\cdot 2 + 3\cdot 1 + 1\cdot 1 \\ 3\cdot 1 + 1\cdot 2 + 4\cdot 1 & 3\cdot 1 + 1\cdot 3 + 4\cdot 3 & 3\cdot 2 + 1\cdot 1 + 4\cdot 1 \end{bmatrix}$$

$$= \begin{bmatrix} 6 & 11 & 7 \\ 8 & 13 & 6 \\ 9 & 18 & 11 \end{bmatrix}$$

and

$$m(S)m(T) = \begin{bmatrix} 1 & 1 & 2 \\ 2 & 3 & 1 \\ 1 & 3 & 1 \end{bmatrix} \begin{bmatrix} 2 & 1 & 2 \\ 1 & 3 & 1 \\ 3 & 1 & 4 \end{bmatrix} = \begin{bmatrix} 9 & 6 & 11 \\ 10 & 12 & 11 \\ 8 & 11 & 9 \end{bmatrix}.$$

Note that $m(TS) \neq m(ST)$ in this case.

When computing the product of matrices in the 3-dimensional case, we find that each entry is the dot product of a *row* of the first matrix with a *column* of the second. The entry in the second row, third column, of $m(T)m(S)$ is given by the dot product of the second row of $m(T)$ with the third column of $m(S)$. We indicate this entry in the examples above.

Exercise 6. Calculate the products of the following matrices:

$$\begin{pmatrix} 2 & 1 & 0 \\ 0 & 3 & 3 \\ 0 & 2 & 1 \end{pmatrix} \begin{pmatrix} 7 & 0 & 1 \\ 1 & 1 & 1 \\ 2 & 1 & 2 \end{pmatrix}, \quad \begin{pmatrix} 3 & 0 & 0 \\ 0 & 1 & 0 \\ 0 & 0 & 3 \end{pmatrix} \begin{pmatrix} 2 & 0 & 6 \\ 1 & 1 & 1 \\ 1 & 3 & 2 \end{pmatrix}, \quad \begin{pmatrix} 0 & 1 & 0 \\ 1 & 0 & 0 \\ 0 & 0 & 1 \end{pmatrix} \begin{pmatrix} 1 & 2 & 3 \\ 4 & 5 & 6 \\ 7 & 8 & 9 \end{pmatrix}.$$

Exercise 7. Each of the transformations J, K, P, Q, R of Examples 1, 2, and 3 has the property that the transformation equals its own square, i.e., $JJ = J$. Verify that

$m(J) \cdot m(J) = m(J)$, and verify that each of the matrices of the other linear transformations equals its own square.

Exercise 8. Since $PR = 0$ as linear transformations, it follows that $m(PR)$
$= m(0) = \begin{pmatrix} 0 & 0 & 0 \\ 0 & 0 & 0 \\ 0 & 0 & 0 \end{pmatrix}$. Verify that $m(P)m(R) = 0 = m(R)m(P)$.

Exercise 9. Show that $\begin{pmatrix} 0 & 0 & 1 \\ 1 & 0 & 0 \\ 0 & 1 & 0 \end{pmatrix}$ has cube = id but a square $\neq$ id. Describe the linear transformation with this matrix.

As in the 2-dimensional case, the distributive law of matrix multiplication over matrix addition is a consequence of this law for linear transformations. The same is true for the associative laws of matrix multiplication and matrix addition. Thus

$$(m(T)m(S))m(R) = m(TS)m(R) = m((TS)R) = m(T(SR))$$

$$= m(T)m(SR) = m(T)(m(S)m(R)), \qquad (5)$$

$$m(T)m(S) + m(T)m(R) = m(TS) + m(TR) = m(TS + TR)$$

$$= m(T(S + R)) = m(T)m(S + R)$$

$$= m(T)(m(S) + m(R)). \qquad (6)$$

Exercise 10. Let **U** be a unit vector with coordinates $\begin{pmatrix} u_1 \\ u_2 \\ u_3 \end{pmatrix}$. Show that the projec-

tion P to the line along **U** has the matrix $\begin{bmatrix} u_1^2 & u_2 u_1 & u_3 u_1 \\ u_1 u_2 & u_2^2 & u_3 u_2 \\ u_1 u_3 & u_2 u_3 & u_3^2 \end{bmatrix}$.

Exercise 11. Show that the matrix for the reflection R through the line along **U** is

$$m(R) = \begin{bmatrix} 2u_1^2 - 1 & 2u_2 u_1 & 2u_3 u_1 \\ 2u_1 u_2 & 2u_2^2 - 1 & 2u_3 u_2 \\ 2u_1 u_3 & 2u_2 u_3 & 2u_3^2 - 1 \end{bmatrix}.$$

EXAMPLE 8. If $\mathbf{U} = \begin{bmatrix} 1/\sqrt{3} \\ 1/\sqrt{3} \\ 1/\sqrt{3} \end{bmatrix}$, then, defining P and R as in Exercises 10 and 11,

$$m(P) = \begin{pmatrix} \frac{1}{3} & \frac{1}{3} & \frac{1}{3} \\ \frac{1}{3} & \frac{1}{3} & \frac{1}{3} \\ \frac{1}{3} & \frac{1}{3} & \frac{1}{3} \end{pmatrix} = \frac{1}{3} \begin{pmatrix} 1 & 1 & 1 \\ 1 & 1 & 1 \\ 1 & 1 & 1 \end{pmatrix},$$

while

$$m(R) = 2m(P) - m(I) = \frac{1}{3} \begin{bmatrix} 1 & 2 & 2 \\ 2 & 1 & 2 \\ 2 & 2 & 2 \end{bmatrix}.$$

Exercise 12. Find the matrices of projection to the line along $U = \frac{1}{3} \begin{pmatrix} 2 \\ 1 \\ 2 \end{pmatrix}$ and of

reflection through this line.

§1. Elementary Matrices and Diagonal Matrices

As in the 2×2 case, there are certain simple 3×3 matrices from which we can build up an arbitrary matrix. We consider matrices of three types: shear matrices, permutation matrices, and diagonal matrices.

(1) *Shear Matrices*: For $i \neq j$, let e_{ij}^s = matrix with all 1's on the diagonal and s in the i, j position and 0 otherwise. For example,

$$e_{12}^s = \begin{pmatrix} 1 & s & 0 \\ 0 & 1 & 0 \\ 0 & 0 & 1 \end{pmatrix}, \qquad e_{23}^s = \begin{pmatrix} 1 & 0 & 0 \\ 0 & 1 & s \\ 0 & 0 & 1 \end{pmatrix}.$$

(2) *Permutation Matrices*: p_{ij} = matrix which is obtained from the identity by interchanging the i'th and j'th rows and leaving the remaining row unchanged. For example,

$$P_{13} = \begin{bmatrix} 0 & 0 & 1 \\ 0 & 1 & 0 \\ 1 & 0 & 0 \end{bmatrix}, \qquad P_{23} = \begin{bmatrix} 1 & 0 & 0 \\ 0 & 0 & 1 \\ 0 & 1 & 0 \end{bmatrix}.$$

A matrix of one of these types is called an *elementary matrix*. If e is an elementary matrix and m is any given matrix, it is easy to describe the matrices em and me.

In the following discussion, we set

$$m = \begin{bmatrix} a_1 & a_2 & a_3 \\ b_1 & b_2 & b_3 \\ c_1 & c_2 & c_3 \end{bmatrix}.$$

EXAMPLE 9.

$$e_{21}^s m = \begin{pmatrix} 1 & 0 & 0 \\ s & 1 & 0 \\ 0 & 0 & 1 \end{pmatrix} \begin{bmatrix} a_1 & a_2 & a_3 \\ b_1 & b_2 & b_3 \\ c_1 & c_2 & c_3 \end{bmatrix} = \begin{bmatrix} a_1 & a_2 & a_3 \\ sa_1 + b_1 & sa_2 + b_2 & sa_3 + b_3 \\ c_1 & c_2 & c_3 \end{bmatrix},$$

We see that $e_{21}^s m$ is the matrix we get, starting with m, by adding s times the first row of m to the second row of m and leaving the other rows alone.

Exercise 13. Show that $e_{12}^s m$ is the matrix obtained from m by adding s times the second row to the first row and leaving the other rows alone.

EXAMPLE 10.

$$me_{21}^s = \begin{bmatrix} a_1 & a_2 & a_3 \\ b_1 & b_2 & b_3 \\ c_1 & c_2 & c_3 \end{bmatrix} \begin{pmatrix} 1 & 0 & 0 \\ s & 1 & 0 \\ 0 & 0 & 1 \end{pmatrix} = \begin{bmatrix} a_1 + sa_2 & a_2 & a_3 \\ b_1 + sb_2 & b_2 & b_3 \\ c_1 + sc_2 & c_2 & c_3 \end{bmatrix},$$

which is the matrix obtained from m by adding s times the second column to the first column and leaving the other columns alone.

Exercise 14. Describe the matrix me_{13}^s.

The preceding calculations work in all cases, and so we have:

Proposition 1. *For each* i, j *with* $i \neq j$, $e_{ij}^s m$ *is the matrix obtained from* m *by adding* s *times the* j'*th row to the* i'*th row of* m *and leaving the other rows alone. Also,* me_{ij}^s *is the matrix obtained from* m *by adding* s *times the* i'*th column to the* j'*th column.*

Note: Recall that $id = \begin{bmatrix} 1 & 0 & 0 \\ 0 & 1 & 0 \\ 0 & 0 & 1 \end{bmatrix}$, so $e_{ij}^s id = e_{ij}^s$.

Thus we can instantly remember what multiplication by e_{ij}^s does to an arbitrary matrix by thinking of e_{ij}^s itself as the result of multiplying the identity matrix by e_{ij}^s. For instance, $e_{31}^s = \begin{bmatrix} 1 & 0 & 0 \\ 0 & 1 & 0 \\ s & 0 & 1 \end{bmatrix}$, so the product $e_{31}^s id$ = the matrix obtained by adding s times the first row of the identity to its last row. Hence, this is what multiplication by e_{31}^s does to *any* matrix: it adds s times the first row to the last row.

EXAMPLE 11.

$$p_{13} m = \begin{bmatrix} 0 & 0 & 1 \\ 0 & 1 & 0 \\ 1 & 0 & 0 \end{bmatrix} \begin{bmatrix} a_1 & a_2 & a_3 \\ b_1 & b_2 & b_3 \\ c_1 & c_2 & c_3 \end{bmatrix} = \begin{bmatrix} c_1 & c_2 & c_3 \\ b_1 & b_2 & b_3 \\ a_1 & a_2 & a_3 \end{bmatrix}.$$

Thus $p_{13} m$ is the matrix obtained from m by interchanging the first and last rows and leaving the second row alone.

Exercise 15. Show that mp_{13} is the matrix obtained by interchanging the first and third columns of m and leaving the second column alone.

For every permutation matrix p_{ij}, the situation is similar to that we have just found; so we have:

Proposition 2. *For each i, j with $i \neq j$, $p_{ij}m$ is the matrix obtained from m by interchanging the i'th and j'th rows, and mp_{ij} is the matrix obtained from m by interchanging the i'th and j'th columns.*

Again, to remember how p_{ij} acts on an arbitrary matrix m, we need only look at p_{ij} and remember that $p_{ij} = p_{ij}\text{id}$.

We call a matrix $\begin{bmatrix} t_1 & 0 & 0 \\ 0 & t_2 & 0 \\ 0 & 0 & t_3 \end{bmatrix}$ a *diagonal* matrix.

EXAMPLE 12. Let $d = \begin{bmatrix} t_1 & 0 & 0 \\ 0 & t_2 & 0 \\ 0 & 0 & t_3 \end{bmatrix}$.

$$dm = \begin{bmatrix} t_1 & 0 & 0 \\ 0 & t_2 & 0 \\ 0 & 0 & t_3 \end{bmatrix} \begin{bmatrix} a_1 & a_2 & a_3 \\ b_1 & b_2 & b_3 \\ c_1 & c_2 & c_3 \end{bmatrix} = \begin{bmatrix} t_1 a_1 & t_1 a_2 & t_1 a_3 \\ t_2 b_1 & t_2 b_2 & t_2 b_3 \\ t_3 c_1 & t_3 c_2 & t_3 c_3 \end{bmatrix}.$$

Exercise 16. With d as in the preceding example, calculate md.

By the same calculation as in Example 12 and Exercise 16, we find:

Proposition 3. *If d is any diagonal matrix, dm is the matrix obtained from m by multiplying the i'th row of m by t_i for each i, where t_i is the entry in d in the (i,i)-position. Also, md is obtained in a similar way, the i'th column of m being multiplied by t_i.*

Exercise 17.

(a) Show that $(p_{12})^2 = \text{id}$ and similarly for p_{13} and p_{23}.
(b) Show that $p_{12}p_{23} \neq p_{23}p_{12}$.

Exercise 18. Show that if

$$d_t = \begin{bmatrix} t_1 & 0 & 0 \\ 0 & t_2 & 0 \\ 0 & 0 & t_3 \end{bmatrix} \quad \text{and} \quad d_s = \begin{bmatrix} s_1 & 0 & 0 \\ 0 & s_2 & 0 \\ 0 & 0 & s_3 \end{bmatrix},$$

then $d_s d_t = d_t d_s$.

Exercise 19. Calculate the following matrices:

(a) $m = \begin{pmatrix} 1 & 0 & 0 \\ s & 1 & 0 \\ 0 & 0 & 1 \end{pmatrix} \begin{pmatrix} 4 & 0 & 0 \\ 0 & 5 & 0 \\ 0 & 0 & 6 \end{pmatrix} \begin{pmatrix} 1 & 0 & 0 \\ 0 & 1 & 0 \\ 0 & t & 0 \end{pmatrix}$,

(b) $m = \begin{pmatrix} 1 & 0 & 0 \\ 0 & 0 & 1 \\ 0 & 1 & 0 \end{pmatrix} \begin{pmatrix} 2 & 0 & 0 \\ 0 & -2 & 0 \\ 0 & 0 & 0 \end{pmatrix} \begin{pmatrix} 0 & 0 & 1 \\ 0 & 1 & 0 \\ 1 & 0 & 0 \end{pmatrix}$.

By multiplying elementary matrices together, several at a time, we can build up *all* matrices.

Theorem 3.3. *Let m be a 3 × 3 matrix. We can find a diagonal matrix d and elementary matrices $e_1 e_2, \ldots, e_n$ and $f_1, \ldots, f_m$ so that*

$$m = e_1 e_2 \ldots e_n d f_1 f_2 \ldots f_m .$$

We shall give the proof of this theorem in the next chapter.

Figures 3.11a–l indicate the effects of elementary transformations, diagonal matrices, and projections. Figure 3.11a shows the identity transformation I; Figure 3.11b shows the matrix with diagonal entries 2, 1, and 1; Figure 3.11c shows the matrix with diagonal entries 1, 2, and 1; Figure 3.11d shows the matrix with diagonal entries 2, 2, and 1; Figure 3.11e shows the shear transformation m_{13}^2; Figure 3.11f shows the matrix with diagonal entries 1, 1, and 2; Figure 3.11g shows the matrix with diagonal entries -1, -1, and 1; Figure 3.11h shows the shear transformation m_{13}^{-1}; Figure 3.11i shows the matrix with diagonal entries -1, -1, and -1; Figure 3.11j shows the matrix with diagonal entries 1, -1, and -1; Figure 3.11k shows the shear m_{12}^1; and Figure 3.11l shows the projection given by the diagonal matrix with entries 1, 1, and 0.

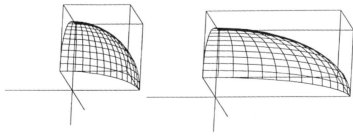

(a)   (b)

Figure 3.11

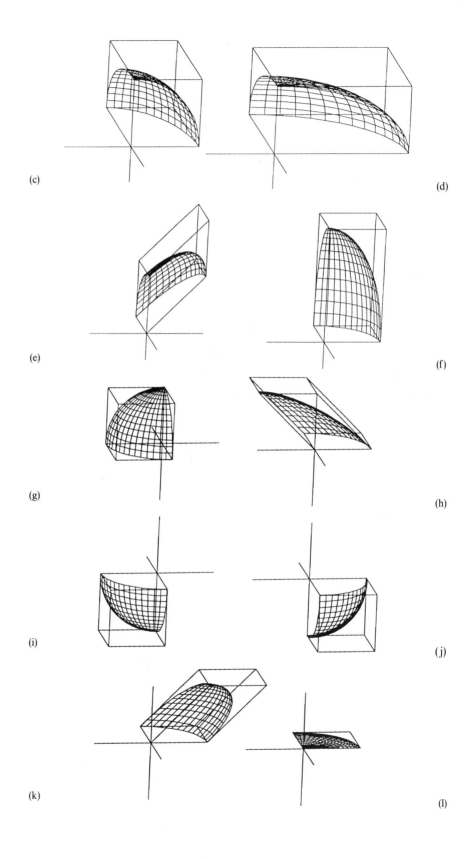

(c)

(d)

(e)

(f)

(g)

(h)

(i)

(j)

(k)

(l)

Inverses and Systems of Equations

Let T be a linear transformation of $\mathbb{R}^3$. As in the case of two dimensions, we say that the linear transformation S is an inverse of T if

$$ST = I \quad \text{and} \quad TS = I. \tag{1}$$

Can a linear transformation have more than one inverse? The answer is no and the proof is the same as in $\mathbb{R}^2$. (See Chapter 2.4, page 52.)

EXAMPLE 1. If $t \neq 0$, then $D_{1/t}$ is the inverse of D_t.

EXAMPLE 2. The inverse of R_θ^1, rotation by θ radians around the x_1-axis, then, is $R_{-\theta}^1 = R_{2\pi - \theta}^1$.

EXAMPLE 3. Let A be the transformation with matrix $\begin{bmatrix} 1 & 0 & 0 \\ 0 & 2 & 0 \\ 0 & 0 & 3 \end{bmatrix}$. The trans-

formation B with matrix $\begin{bmatrix} 1 & 0 & 0 \\ 0 & \frac{1}{2} & 0 \\ 0 & 0 & \frac{1}{3} \end{bmatrix}$ is the inverse of A, because

$$AB \begin{bmatrix} x_1 \\ x_2 \\ x_3 \end{bmatrix} = A \left[\begin{bmatrix} 1 & 0 & 0 \\ 0 & \frac{1}{2} & 0 \\ 0 & 0 & \frac{1}{3} \end{bmatrix} \begin{bmatrix} x_1 \\ x_2 \\ x_3 \end{bmatrix} \right] = A \left[\begin{bmatrix} x_1 \\ \frac{1}{2} x_2 \\ \frac{1}{3} x_3 \end{bmatrix} \right] = \begin{bmatrix} 1 & 0 & 0 \\ 0 & 2 & 0 \\ 0 & 0 & 3 \end{bmatrix} \begin{bmatrix} x_1 \\ \frac{1}{2} x_2 \\ \frac{1}{3} x_3 \end{bmatrix} = \begin{bmatrix} x_1 \\ x_2 \\ x_3 \end{bmatrix},$$

so $AB = I$, and, similarly, $BA = I$.

EXAMPLE 4. Let π be a plane and let P be projection on the plane π. We claim P does not have an inverse, and we shall give two proofs for this statement.

Suppose Q is a transformation satisfying $PQ = QP = I$.

(a) Choose a vector $\mathbf{X}$ which is not in the plane π. Then $\mathbf{X} = I(\mathbf{X}) = PQ(\mathbf{X}) = P(Q(\mathbf{X}))$.

But $\mathbf{X}$ is not in π, while $P(Q(\mathbf{X}))$ is in π, so we have a contradiction. Thus, no such Q exists.

(b) Choose a vector $\mathbf{X} \neq \mathbf{0}$ with $\mathbf{X} \perp \pi$. Then $P(\mathbf{X}) = \mathbf{0}$, and so $Q(P(\mathbf{X})) = \mathbf{0}$. Thus $\mathbf{X} = I(\mathbf{X}) = Q(P(\mathbf{X})) = \mathbf{0}$. This is a contradiction. Therefore, no such Q exists.

EXAMPLE 5. Let T have matrix

$$\begin{bmatrix} 1 & -1 & 0 \\ 0 & 1 & -1 \\ -1 & 0 & 1 \end{bmatrix}.$$

We claim T has no inverse. Suppose S satisfies $ST = TS = I$. Let $\mathbf{X} = \begin{bmatrix} x_1 \\ x_2 \\ x_3 \end{bmatrix}$ be a vector in $\mathbb{R}^3$. Set $S(\mathbf{X}) = \begin{bmatrix} y_1 \\ y_2 \\ y_2 \end{bmatrix}$. Then

$$\mathbf{X} = I(\mathbf{X}) = T(S(\mathbf{X})) = T\begin{bmatrix} y_1 \\ y_2 \\ y_3 \end{bmatrix}$$

$$= \begin{bmatrix} 1 & -1 & 0 \\ 0 & 1 & -1 \\ -1 & 0 & 1 \end{bmatrix}\begin{bmatrix} y_1 \\ y_2 \\ y_3 \end{bmatrix} = \begin{bmatrix} y_1 - y_2 \\ y_2 - y_3 \\ -y_1 + y_3 \end{bmatrix}.$$

Since $(y_1 - y_2) + (y_2 - y_3) + (-y_1 + y_3) = 0$, $\mathbf{X}$ lies on the plane: $x_1 + x_2 + x_3 = 0$. Thus every vector in $\mathbb{R}^3$ lies on this plane. This is false, so S does not exist, and the claim is proved.

Exercise 1. Find an inverse for the transformation T which reflects each vector in the plane $x_3 = 0$.

Exercise 2. Find conditions on the numbers a, b, c so that the transformation T with matrix

$$\begin{pmatrix} a & 0 & 0 \\ 0 & b & 0 \\ 0 & 0 & c \end{pmatrix}$$

has an inverse S. Calculate S when it exists.

Exercise 3. Find an inverse S for the transformation T whose matrix is

$$\begin{pmatrix} 0 & 1 & 0 \\ 1 & 0 & 0 \\ 0 & 0 & 1 \end{pmatrix}.$$

Hint: $T\begin{pmatrix} x_1 \\ x_2 \\ x_3 \end{pmatrix} = \begin{pmatrix} x_2 \\ x_1 \\ x_3 \end{pmatrix}$ for every $\begin{pmatrix} x_1 \\ x_2 \\ x_3 \end{pmatrix}$.

Exercise 4. Find an inverse S for the transformation with matrix $\begin{pmatrix} 0 & 1 & 0 \\ 0 & 0 & 1 \\ 1 & 0 & 0 \end{pmatrix}$.

Exercise 5. Show that the transformation T with matrix $\begin{pmatrix} a & b & c \\ 2a & 2b & 2c \\ 3a & 3b & 3c \end{pmatrix}$ has no inverse for any values of a, b, c.

Exercise 6. Show that the transformation T with matrix $\begin{bmatrix} a & b & c \\ d & e & f \\ a+d & b+e & c+f \end{bmatrix}$ has no inverse for any values of a, b, c, d, e, f.

Exercise 7. Let T be the transformation with matrix $\begin{bmatrix} a_1 & a_2 & a_3 \\ b_1 & b_2 & b_3 \\ c_1 & c_2 & c_3 \end{bmatrix}$. Suppose that there exist scalars t_1, t_2, t_3, not all 0, such that

$$t_1 \begin{bmatrix} a_1 \\ a_2 \\ a_3 \end{bmatrix} + t_2 \begin{bmatrix} b_1 \\ b_2 \\ b_3 \end{bmatrix} + t_3 \begin{bmatrix} c_1 \\ c_2 \\ c_3 \end{bmatrix} = 0.$$

(a) Show that there is a plane π such that for every vector $\mathbf{X}$, $T(\mathbf{X})$ lies in π.
(b) Conclude that T has no inverse.

Let T be a linear transformation of $\mathbb{R}^3$. For a given vector $\mathbf{Y}$, we may try to solve the equation

$$T(\mathbf{X}) = \mathbf{Y} \tag{1}$$

by some vector $\mathbf{X}$. Suppose S is an inverse of T. Then

$$T(S(\mathbf{Y})) = TS(\mathbf{Y}) = I(\mathbf{Y}) = \mathbf{Y},$$

so $\mathbf{X} = S(\mathbf{Y})$ solves (2).
Now let $\mathbf{X}$ be a solution of (2). Then

$$S(\mathbf{Y}) = S(T(\mathbf{X})) = ST(\mathbf{X}) = I(\mathbf{X}) = \mathbf{X}.$$

So $S(\mathbf{Y})$ is the only solution of (2).
We have shown: if T has an inverse, then Eq. (2) has exactly one solution $\mathbf{X}$ for each given $\mathbf{Y}$.
Conversely, let T be a linear transformation which has the property that Eq. (2) possesses exactly one solution for each $\mathbf{Y}$. We shall show that it follows that T has an inverse.
Denote by $S(\mathbf{Y})$ the solution of (2). Then $T(S(\mathbf{Y})) = \mathbf{Y}$, and if $T(\mathbf{X}) = \mathbf{Y}$, then $\mathbf{X} = S(\mathbf{Y})$. We have thus defined a transformation S which sends each vector $\mathbf{Y}$ into $S(\mathbf{Y})$. By the definition of S, $TS(\mathbf{Y}) = T(S(\mathbf{Y})) = \mathbf{Y}$, for each $\mathbf{Y}$ in $\mathbb{R}^3$. So $TS = I$. Also, if $\mathbf{X}$ is any vector, set $\mathbf{Y} = T(\mathbf{X})$. Then $\mathbf{X} = S(\mathbf{Y})$, by the definition of S, and so $\mathbf{X} = S(T(\mathbf{X})) = ST(\mathbf{X})$. Hence $I = ST$. Thus we have shown that S is an inverse of T.
Our conscience should bother us about one point. S is a transformation

of $\mathbb{R}^3$; but is it a *linear* transformation? To answer this question, choose two vectors, **X** and **Y**.

$$T(S(\mathbf{X}) + S(\mathbf{Y})) = T(S(\mathbf{X})) + T(S(\mathbf{Y})) \qquad \text{(since } T \text{ is linear)}$$
$$= TS(\mathbf{X}) + TS(\mathbf{Y}) = I(\mathbf{X}) + I(\mathbf{Y}) = \mathbf{X} + \mathbf{Y},$$

so

$$S(\mathbf{X}) + S(\mathbf{Y})$$

solves (2) with $\mathbf{X} + \mathbf{Y}$ as the right-hand side. By the definition of S, it follows that $S(\mathbf{X}) + S(\mathbf{Y}) = S(\mathbf{X} + \mathbf{Y})$. Similarly, $S(t\mathbf{X}) = tS(\mathbf{X})$ whenever t is a scalar.

Exercise 8. Prove the last statement.

We have proved:

Proposition 1. *Let T be a linear transformation of $\mathbb{R}^3$. Then T has an inverse if and only if the equation*

$$T(\mathbf{X}) = \mathbf{Y} \tag{2}$$

has, for each $\mathbf{Y}$, one and only one solution $\mathbf{X}$.

Corollary. *If T has an inverse, then*

$$T(\mathbf{X}) = \mathbf{0} \qquad \text{implies that} \quad \mathbf{X} = \mathbf{0}. \tag{3}$$

PROOF. Set $\mathbf{Y} = \mathbf{0}$ in Proposition 1.

Suppose, conversely, that T is a linear transformation for which (3) holds, i.e., $\mathbf{0}$ is the only vector which T sends into $\mathbf{0}$. It follows that the equation $T(\mathbf{X}) = \mathbf{Y}$ has *at most* one solution for each $\mathbf{Y}$. To see this, suppose $T(\mathbf{U}) = \mathbf{Y}$ and $T(\mathbf{V}) = \mathbf{Y}$. Then

$$T(\mathbf{U} - \mathbf{V}) = T(\mathbf{U}) - T(\mathbf{V}) = \mathbf{Y} - \mathbf{Y} = \mathbf{0},$$

so by (3), $\mathbf{U} - \mathbf{V} = \mathbf{0}$, or $\mathbf{U} = \mathbf{V}$.

We proceed to show that (3) also implies the *existence* of a solution of $T(\mathbf{X}) = \mathbf{Y}$ for each $\mathbf{Y}$. We saw, in Chapter 3.2, that the image of $\mathbb{R}^3$ under any linear transformation is either a plane through $\mathbf{0}$, a line through $\mathbf{0}$, $\mathbb{R}^3$, or the origin. Suppose the image of $\mathbb{R}^3$ under T is a plane π through $\mathbf{0}$. The vectors $T(\mathbf{E}_1)$, $T(\mathbf{E}_2)$, $T(\mathbf{E}_3)$ all then lie in π. Three vectors in a plane are linearly dependent. Thus we can find scalars t_1, t_2, t_3, not all 0, such that $t_1 T(\mathbf{E}_1) + t_2 T(\mathbf{E}_2) + t_3 T(\mathbf{E}_3) = \mathbf{0}$. Therefore,

$$T(t_1\mathbf{E}_1 + t_2\mathbf{E}_2 + t_3\mathbf{E}_3) = t_1 T(\mathbf{E}_1) + t_2 T(\mathbf{E}_2) + t_3 T(\mathbf{E}_3) = \mathbf{0},$$

while $t_1\mathbf{E}_1 + t_2\mathbf{E}_2 + t_3\mathbf{E}_3 = \begin{bmatrix} t_1 \\ t_2 \\ t_3 \end{bmatrix} \neq \mathbf{0}$. This contradicts (3). Hence, the image of $\mathbb{R}^3$ under T cannot be a plane. In the same way, we conclude that the image cannot be a line or the origin, and so the image must be all of $\mathbb{R}^3$. This means that for each $\mathbf{Y}$, there is some $\mathbf{X}$ with $T(\mathbf{X}) = \mathbf{Y}$. So the existence of a solution is established. We have just proved: If T is a linear transformation of $\mathbb{R}^3$ such that (3) holds, then Eq. (2) has one and only one solution for each $\mathbf{Y}$.

By Proposition 1, it follows that T has an inverse. Thus we have:

Proposition 2. *Let T be a linear transformation of $\mathbb{R}^3$. If (3) holds, i.e., if $T(\mathbf{X}) = \mathbf{0}$ implies that $\mathbf{X} = \mathbf{0}$, then T has an inverse. Conversely, if T has an inverse, then (3) holds.*

We now need a practical test to decide whether or not (3) holds, given the matrix of some linear transformation T. In two dimensions, the test was this: If $\begin{pmatrix} a & b \\ c & d \end{pmatrix}$ is the matrix of a transformation A, then A has an inverse if and only if $ad - bc \neq 0$. We seek a similar test in $\mathbb{R}^3$.

Let T be the linear transformation whose matrix is

$$\begin{bmatrix} a_1 & a_2 & a_3 \\ b_1 & b_2 & b_3 \\ c_1 & c_2 & c_3 \end{bmatrix}.$$

We call the vectors

$$\mathbf{A} = \begin{bmatrix} a_1 \\ a_2 \\ a_3 \end{bmatrix}, \qquad \mathbf{B} = \begin{bmatrix} b_1 \\ b_2 \\ b_3 \end{bmatrix}, \qquad \mathbf{C} = \begin{bmatrix} c_1 \\ c_2 \\ c_3 \end{bmatrix}$$

the *row vectors* of this matrix. The vector $\mathbf{X} = \begin{bmatrix} x_1 \\ x_2 \\ x_3 \end{bmatrix}$ satisfies $T(\mathbf{X}) = \mathbf{0}$ if and only if

$$\begin{cases} a_1 x_1 + a_2 x_2 + a_3 x_3 = 0, \\ b_1 x_1 + b_2 x_2 + b_3 x_3 = 0, \\ c_1 x_1 + c_2 x_2 + c_3 x_3 = 0. \end{cases} \tag{4}$$

On p. 124, Chapter 3.0, we showed that (4) has a nonzero solution $\mathbf{X} = \begin{bmatrix} x_1 \\ x_2 \\ x_3 \end{bmatrix}$ if and only if

$$\mathbf{A} \cdot (\mathbf{B} \times \mathbf{C}) = \mathbf{0}. \tag{5}$$

Combining this last result with Proposition 2, we obtain:

Theorem 3.4. *Let T be a linear transformation of $\mathbb{R}^3$ and denote by* **A**, **B**, **C** *the row vectors of the matrix of T. Then T has an inverse if and only if* $\mathbf{A} \cdot (\mathbf{B} \times \mathbf{C}) \neq 0$.

Let T be a linear transformation of $\mathbb{R}^3$ and assume that $\mathbf{A} \cdot (\mathbf{B} \times \mathbf{C}) \neq 0$. We wish to calculate the matrix of T^{-1}. Set

$$m(T^{-1}) = \begin{bmatrix} u_1 & v_1 & w_1 \\ u_2 & v_2 & w_2 \\ u_3 & v_3 & w_3 \end{bmatrix},$$

where the u_i, v_i, w_i are to be found. Then $T^{-1}\begin{bmatrix} 1 \\ 0 \\ 0 \end{bmatrix} = \begin{bmatrix} u_1 \\ u_2 \\ u_3 \end{bmatrix}$, so $\begin{bmatrix} 1 \\ 0 \\ 0 \end{bmatrix} = T\begin{bmatrix} u_1 \\ u_2 \\ u_3 \end{bmatrix}$,

which we can write

$$\begin{aligned} a_1 u_1 + a_2 u_2 + a_3 u_3 &= 1, \\ b_1 u_1 + b_2 u_2 + b_3 u_3 &= 0, \\ c_1 u_1 + c_2 u_2 + c_3 u_3 &= 0. \end{aligned} \tag{6}$$

Set $\Delta = \mathbf{A} \cdot (\mathbf{B} \times \mathbf{C})$. Then we have

$$\mathbf{A} \cdot (\mathbf{B} \times \mathbf{C}) = \Delta,$$
$$\mathbf{B} \cdot (\mathbf{B} \times \mathbf{C}) = 0,$$
$$\mathbf{C} \cdot (\mathbf{B} \times \mathbf{C}) = 0,$$

and hence, using the hypothesis $\Delta \neq 0$,

$$\mathbf{A} \cdot \frac{\mathbf{B} \times \mathbf{C}}{\Delta} = 1,$$
$$\mathbf{B} \cdot \frac{\mathbf{B} \times \mathbf{C}}{\Delta} = 0,$$
$$\mathbf{C} \cdot \frac{\mathbf{B} \times \mathbf{C}}{\Delta} = 0.$$

Thus $\begin{bmatrix} u_1 \\ u_2 \\ u_3 \end{bmatrix} = \mathbf{B} \times \mathbf{C}/\Delta$ satisfies (6), so

$$T\left(\frac{\mathbf{B} \times \mathbf{C}}{\Delta}\right) = \begin{bmatrix} 1 \\ 0 \\ 0 \end{bmatrix},$$

and so

$$\frac{\mathbf{B} \times \mathbf{C}}{\Delta} = T^{-1}\begin{bmatrix} 1 \\ 0 \\ 0 \end{bmatrix}. \tag{7}$$

Now recall that by Exercise 6, Chapter 3.0,

$$\mathbf{A} \cdot (\mathbf{B} \times \mathbf{C}) = \mathbf{B} \cdot (\mathbf{C} \times \mathbf{A}) = \mathbf{C} \cdot (\mathbf{A} \times \mathbf{B}).$$

By a calculation like the one which led to (7), we get

$$\frac{\mathbf{C} \times \mathbf{A}}{\Delta} = T^{-1} \begin{pmatrix} 0 \\ 1 \\ 0 \end{pmatrix}$$

and

$$\frac{\mathbf{A} \times \mathbf{B}}{\Delta} = T^{-1} \begin{pmatrix} 0 \\ 0 \\ 1 \end{pmatrix}.$$

By the definition of the cross product,

$$\mathbf{B} \times \mathbf{C} = \begin{pmatrix} \begin{vmatrix} b_2 & b_3 \\ c_2 & c_3 \end{vmatrix} \\ \begin{vmatrix} b_3 & b_1 \\ c_3 & c_1 \end{vmatrix} \\ \begin{vmatrix} b_1 & b_2 \\ c_1 & c_2 \end{vmatrix} \end{pmatrix} \quad \text{and}$$

$$\mathbf{C} \times \mathbf{A} = \begin{pmatrix} \begin{vmatrix} c_2 & c_3 \\ a_2 & a_3 \end{vmatrix} \\ \begin{vmatrix} c_3 & c_1 \\ a_3 & a_1 \end{vmatrix} \\ \begin{vmatrix} c_1 & c_2 \\ a_1 & a_2 \end{vmatrix} \end{pmatrix} \quad \text{and}$$

$$\mathbf{A} \times \mathbf{B} = \begin{pmatrix} \begin{vmatrix} a_2 & a_3 \\ b_2 & b_3 \end{vmatrix} \\ \begin{vmatrix} a_3 & a_1 \\ b_3 & b_1 \end{vmatrix} \\ \begin{vmatrix} a_1 & a_2 \\ b_1 & b_2 \end{vmatrix} \end{pmatrix}.$$

We conclude that

$$m(T^{-1}) = \frac{1}{\Delta} \begin{pmatrix} \begin{vmatrix} b_2 & b_3 \\ c_2 & c_3 \end{vmatrix} & \begin{vmatrix} c_2 & c_3 \\ a_2 & a_3 \end{vmatrix} & \begin{vmatrix} a_2 & a_3 \\ b_2 & b_3 \end{vmatrix} \\ \begin{vmatrix} b_3 & b_1 \\ c_3 & c_1 \end{vmatrix} & \begin{vmatrix} c_3 & c_1 \\ a_3 & a_1 \end{vmatrix} & \begin{vmatrix} a_3 & a_1 \\ b_3 & b_1 \end{vmatrix} \\ \begin{vmatrix} b_1 & b_2 \\ c_1 & c_2 \end{vmatrix} & \begin{vmatrix} c_1 & c_2 \\ a_1 & a_2 \end{vmatrix} & \begin{vmatrix} a_1 & a_2 \\ b_1 & b_2 \end{vmatrix} \end{pmatrix}. \tag{8}$$

Note that

$$\Delta = \mathbf{A} \cdot (\mathbf{B} \times \mathbf{C}) = a_1 \begin{vmatrix} b_2 & b_3 \\ c_2 & c_3 \end{vmatrix} + a_2 \begin{vmatrix} b_3 & b_1 \\ c_3 & c_1 \end{vmatrix} + a_3 \begin{vmatrix} b_1 & b_2 \\ c_1 & c_2 \end{vmatrix}.$$

EXAMPLE 6. If $m(T) = \begin{bmatrix} 1 & 2 & 3 \\ 1 & 0 & -1 \\ 0 & 1 & 1 \end{bmatrix}$, find $m(T^{-1})$.

$$\Delta = 1 \begin{vmatrix} 0 & -1 \\ 1 & 1 \end{vmatrix} + 2 \begin{vmatrix} -1 & 1 \\ 1 & 0 \end{vmatrix} + 3 \begin{vmatrix} 1 & 0 \\ 0 & 1 \end{vmatrix}$$

$$= 1 \cdot 1 + 2(-1) + 3 \cdot 1 = 2.$$

By (8),

$$m(T^{-1}) = \tfrac{1}{2} \begin{bmatrix} 1 & -(-1) & -2 \\ -1 & -(-1) & 4 \\ 1 & -(1) & -2 \end{bmatrix} = \begin{bmatrix} \tfrac{1}{2} & \tfrac{1}{2} & -1 \\ -\tfrac{1}{2} & \tfrac{1}{2} & 2 \\ \tfrac{1}{2} & -\tfrac{1}{2} & -1 \end{bmatrix}.$$

Exercise 9. Verify that the matrix just obtained for $m(T^{-1})$ satisfies $m(T)m(T^{-1})$ $= m(I)$ and $m(T^{-1})m(T) = m(I)$.

Exercise 10. Use (8) to find inverses for

(a) $\begin{pmatrix} 0 & 1 & 0 \\ 1 & 0 & 0 \\ 0 & 0 & 1 \end{pmatrix}$,

(b) $\begin{pmatrix} 1 & 0 & 0 \\ 0 & 0 & -1 \\ 0 & 1 & 0 \end{pmatrix}$,

(c) $\begin{pmatrix} 5 & 5 & 1 \\ 1 & 0 & 0 \\ 0 & 0 & 1 \end{pmatrix}$.

Let m be a matrix which possesses an inverse matrix m^{-1}, i.e.,

$$mm^{-1} = m^{-1}m = \text{identity matrix} = \text{id}.$$

Let $m = \begin{bmatrix} a_1 & a_2 & a_3 \\ b_1 & b_2 & b_3 \\ c_1 & c_2 & c_3 \end{bmatrix}$. We shall give a new approach to the problem of finding the entries in the matrix m^{-1}.

Given a vector $\begin{bmatrix} y_1 \\ y_2 \\ y_3 \end{bmatrix}$, consider the following system of equations for the unknowns x_1, x_2, x_3:

$$a_1 x_1 + a_2 x_2 + a_3 x_3 = y_1,$$
$$b_1 x_1 + b_2 x_2 + b_3 x_3 = y_2, \qquad (9)$$
$$c_1 x_1 + c_2 x_2 + c_3 x_3 = y_3.$$

Suppose we can solve the system (9) be setting

$$x_1 = d_1 y_1 + d_2 y_2 + d_3 y_3,$$
$$x_2 = e_1 y_1 + e_2 y_2 + e_3 y_3, \qquad (10)$$
$$x_3 = f_1 y_1 + f_2 y_2 + f_3 y_3,$$

where d_i, e_i, f_i are certain numbers which do not depend on y_1, y_2, y_3. We define a matrix n by

$$n = \begin{bmatrix} d_1 & d_2 & d_3 \\ e_1 & e_2 & e_3 \\ f_1 & f_2 & f_3 \end{bmatrix}. \tag{11}$$

Then (9) and (10) state that, setting

$$\mathbf{X} = \begin{bmatrix} x_1 \\ x_2 \\ x_3 \end{bmatrix} \quad \text{and} \quad \mathbf{Y} = \begin{bmatrix} y_1 \\ y_2 \\ y_3 \end{bmatrix},$$

$m(\mathbf{X}) = \mathbf{Y}$ and $n(\mathbf{Y}) = \mathbf{X}$. So

$$nm(\mathbf{X}) = n(m(\mathbf{X})) = n(\mathbf{Y}) = \mathbf{X}.$$

This holds for each vector $\mathbf{X}$. Hence

$$nm = \text{id}.$$

It follows that

$$n = (nm)(m^{-1}) = \text{id}(m^{-1}) = m^{-1}.$$

We thus have the following result:

The matrix n defined by (11) is the inverse of the matrix m.

EXAMPLE 7. $m = \begin{bmatrix} 1 & 2 & 3 \\ 1 & 0 & -1 \\ 0 & 1 & 1 \end{bmatrix}$. Find m^{-1} by the preceding method.

The system (9) here is

$$\begin{aligned} x_1 + 2x_2 + 3x_3 &= y_1, \\ x_1 \quad\quad - x_3 &= y_2, \\ x_2 + x_3 &= y_3. \end{aligned}$$

We can solve this by eliminating x_3 from the last equation by setting

$$x_3 = y_3 - x_2.$$

Inserting this expression in the first two equations yields

$$\begin{aligned} x_1 + 2x_2 + 3(y_3 - x_2) &= y_1, \\ x_1 - (y_3 - x_2) &= y_2, \end{aligned}$$

or

$$\begin{aligned} x_1 - x_2 &= y_1 - 3y_3, \\ x_1 + x_2 &= y_2 + y_3. \end{aligned}$$

Solving this system for x_1 and x_2, we find

$$2x_1 = y_1 + y_2 - 2y_3,$$

$$2x_2 = (y_2 + y_3) - (y_1 - 3y_3) = -y_1 + y_2 + 4y_3,$$

so

$$x_1 = \tfrac{1}{2} y_1 + \tfrac{1}{2} y_2 - y_3,$$
$$x_2 = -\tfrac{1}{2} y_1 + \tfrac{1}{2} y_2 + 2 y_3,$$

and

$$x_3 = \tfrac{1}{2} y_1 - \tfrac{1}{2} y_2 - y_3,$$

since

$$x_3 = y_3 - x_2 = \tfrac{1}{2} y_1 - \tfrac{1}{2} y_2 - y_3.$$

Thus, here,

$$m^{-1} = \begin{pmatrix} d_1 & d_2 & d_3 \\ e_1 & e_2 & e_3 \\ f_1 & f_2 & f_3 \end{pmatrix} = \begin{pmatrix} \tfrac{1}{2} & \tfrac{1}{2} & -1 \\ -\tfrac{1}{2} & \tfrac{1}{2} & 2 \\ \tfrac{1}{2} & -\tfrac{1}{2} & -1 \end{pmatrix}.$$

To our great relief, the answer we found agrees with our earlier answer to Example 6.

Exercise 11. Using the method just described, find inverses for the following matrices:

(a) $\begin{pmatrix} 0 & 1 & 0 \\ 1 & 0 & 0 \\ 0 & 0 & 1 \end{pmatrix}$,

(b) $\begin{pmatrix} 5 & 5 & 1 \\ 1 & 0 & 0 \\ 0 & 0 & 1 \end{pmatrix}$,

(c) $\begin{pmatrix} 1 & -1 & 2 \\ 6 & 0 & 1 \\ 3 & 2 & 1 \end{pmatrix}$.

Exercise 12. Using the methods just described, find an inverse for

$$\begin{pmatrix} 1 & a & b \\ 0 & 1 & c \\ 0 & 0 & 1 \end{pmatrix}.$$

Exercise 13. By computing $\mathbf{A} \cdot (\mathbf{B} \times \mathbf{C})$, show that $\begin{bmatrix} 1 & s & s^2 \\ 1 & t & t^2 \\ 1 & u & u^2 \end{bmatrix}$ has an inverse provided s, t, u are all distinct. *Hint:* Simplify the expression you get for $\mathbf{A} \cdot (\mathbf{B} \times \mathbf{C})$, writing it as a product.

Exercise 14. S and T are linear transformations of $\mathbb{R}^3$ which have inverses. Show that ST has an inverse and that $(ST)^{-1} = T^{-1}S^{-1}$.

Exercise 15. S is an invertible linear transformation of $\mathbb{R}^3$, D is a linear transformation of $\mathbb{R}^3$, and $T = S^{-1}DS$. Show that

$$T^2 = S^{-1}D^2S, \qquad T^3 = S^{-1}D^3S.$$

Exercise 16. n is a 3×3 matrix such that $n^3 = 0$. Show that

$$(\mathrm{id} + n)^{-1} = \mathrm{id} - n + n^2.$$

Exercise 17. Set $n = \begin{pmatrix} 0 & a & b \\ 0 & 0 & c \\ 0 & 0 & 0 \end{pmatrix}$. Show that $n^3 = 0$. Using Exercise 16, find the

inverse of $\text{id} + n = \begin{pmatrix} 1 & a & b \\ 0 & 1 & c \\ 0 & 0 & 1 \end{pmatrix}$. Compare your result with the answer to Exer-

cise 12.

Exercise 18. For what values of a, b, c, d does $\begin{pmatrix} 1 & 0 & 0 \\ 0 & a & b \\ 0 & c & d \end{pmatrix}$ have an inverse?
Calculate the inverse when it exists.

We defined the inverse S of a linear transformation T by the two conditions:

(i) $ST = I$

and

(ii) $TS = I$.

Suppose that only one of the two conditions is satisfied, say, (i) holds. Does (ii) follow?

Proposition 3. *Let S, T be linear transformations of $\mathbb{R}^3$ such that $ST = I$. Then $TS = I$.*

PROOF. Choose a vector $\mathbf{X}$ such that $T(\mathbf{X}) = \mathbf{0}$. Then $\mathbf{X} = I(\mathbf{X}) = ST(\mathbf{X}) = S(\mathbf{0}) = \mathbf{0}$. Then, by Proposition 2, T has an inverse, T^{-1}, with $T^{-1}T = TT^{-1} = I$. Since $ST = I$, $S = SI = S(TT^{-1}) = (ST)T^{-1} = IT^{-1} = IT^{-1} = T^{-1}$. Hence $TS = T(T^{-1}) = I$.

Exercise 19. Let S, T be two linear transformations. Show that if the product ST has an inverse, then S has an inverse and T has an inverse.

§1. Inverses of Elementary Matrices and Diagonal Matrices

Recall the elementary matrices e_{ij}^s and p_{ij} that we studied in Chapter 3.3. Let us find the inverses of these matrices.

EXAMPLE 8. $e_{31}^s e_{31}^t = \begin{bmatrix} 1 & 0 & 0 \\ 0 & 1 & 0 \\ s & 0 & 1 \end{bmatrix} \begin{bmatrix} 1 & 0 & 0 \\ 0 & 1 & 0 \\ t & 0 & 1 \end{bmatrix} = \begin{bmatrix} 1 & 0 & 0 \\ 0 & 1 & 0 \\ s+t & 0 & 1 \end{bmatrix}$. Choosing
$t = -s$, this gives $e_{31}^s e_{31}^{-s} = \text{id}$, and replacing s by $-s$ and t by s, we get
$e_{31}^{-s} e_{31}^s = \text{id}$. So $(e_{31}^s)^{-1} = e_{31}^{-s}$.

Exercise 20. Show that $(e_{32}^s)^{-1} = e_{32}^{-s}$.

Exercise 21. Show that $(e_{12}^s)^{-1} = e_{12}^{-s}$.

In the general case, for all i, j, and s, a similar calculation gives us:

Proposition 4. $(e_{ij}^s)^{-1} = e_{ij}^{-s}$.

EXAMPLE 9. Recall that $p_{31} = \begin{bmatrix} 0 & 0 & 1 \\ 0 & 1 & 0 \\ 1 & 0 & 0 \end{bmatrix}$. We know that for each matrix m, $p_{31}m$ is obtained from m by interchanging the first and third rows and leaving the second row alone. Hence,

$$p_{31}p_{31} = \text{id},$$

so $(p_{31})^{-1} = p_{31}$.

Exercise 22. Find $(p_{12})^{-1}$.

In the general case, we have:

Proposition 5. *For every* i, j, $i \neq j$,

$$(p_{ij})^{-1} = p_{ij}.$$

Next, consider the diagonal matrix

$$d = \begin{bmatrix} t_1 & 0 & 0 \\ 0 & t_2 & 0 \\ 0 & 0 & t_3 \end{bmatrix}.$$

If any of the numbers t_1, t_2, t_3 is 0, then by Theorem 3.4, d has no inverse. If all $t_i \neq 0$, we have

$$\begin{bmatrix} t_1 & 0 & 0 \\ 0 & t_2 & 0 \\ 0 & 0 & t_3 \end{bmatrix}\begin{bmatrix} 1/t_1 & 0 & 0 \\ 0 & 1/t_2 & 0 \\ 0 & 0 & 1/t_3 \end{bmatrix} = \text{id},$$

so $d^{-1} = \begin{bmatrix} 1/t_1 & 0 & 0 \\ 0 & 1/t_2 & 0 \\ 0 & 0 & 1/t_3 \end{bmatrix}$. Thus we have:

Proposition 6. *A diagonal matrix* d *has an inverse if and only if its diagonal entries are all* $\neq 0$, *and in that case* d^{-1} *is the diagonal matrix whose diagonal entries are the reciprocals of those for* d.

We shall now show that, by multiplying a given matrix m by suitable elementary matrices, we can convert m into a diagonal matrix.

Let $m = \begin{bmatrix} a_1 & a_2 & a_3 \\ b_1 & b_2 & b_3 \\ c_1 & c_2 & c_3 \end{bmatrix}$. If $a_1 \neq 0$, we can choose t so that $b_1 + ta_1 = 0$.

Then

$$e_{21}^t m = \begin{bmatrix} a_1 & a_2 & a_3 \\ 0 & b_2 + ta_2 & b_3 + ta_3 \\ c_1 & c_2 & c_3 \end{bmatrix}.$$

Similarly, we can choose s so that

$$e_{31}^s e_{21}^t m = \begin{bmatrix} a_1 & a_2 & a_3 \\ 0 & b_2 + ta_2 & b_3 + ta_3 \\ 0 & c_2 + sa_2 & c_3 + sa_3 \end{bmatrix}.$$

Thus $e_{31}^s e_{21}^t m = \begin{bmatrix} a_1 & a_2 & a_3 \\ 0 & x & y \\ 0 & z & w \end{bmatrix}$. Similarly, we can find r, q so that

$$e_{31}^s e_{21}^t m e_{12}^r e_{13}^q = \begin{bmatrix} a_1 & 0 & 0 \\ 0 & x & y \\ 0 & z & w \end{bmatrix}. \qquad (12)$$

If $a_1 = 0$, either m is the zero matrix or some entry of m is $\neq 0$, say, $b_3 \neq 0$.
Then

$$mp_{13} = \begin{bmatrix} a_3 & a_2 & a_1 \\ b_3 & b_2 & b_1 \\ c_3 & c_2 & c_1 \end{bmatrix} \quad \text{and} \quad p_{12}mp_{13} = \begin{bmatrix} b_3 & b_2 & b_1 \\ a_3 & a_2 & a_1 \\ c_3 & c_2 & c_1 \end{bmatrix}.$$

Since $b_3 \neq 0$, we can apply formula (12) to the matrix $p_{12}mp_{13}$ and get

$$e_{31}^s e_{21}^t p_{12} m p_{13} e_{12}^r e_{13}^q = \begin{bmatrix} b_3 & 0 & 0 \\ 0 & x & y \\ 0 & z & w \end{bmatrix}. \qquad (13)$$

If $x \neq 0$, we can proceed as earlier and find scalars i, j so that (2) yields

$$e_{32}^i e_{31}^s e_{21}^t p_{12} m p_{13} e_{12}^r e_{13}^q e_{23}^j = \begin{bmatrix} b_3 & 0 & 0 \\ 0 & x & 0 \\ 0 & 0 & u \end{bmatrix}.$$

The right-hand side is a diagonal matrix. If $x = 0$, we consider two
possibilities. If y, z, w in (13) are all 0, then the right-hand side of (13) is a
diagonal matrix. If at least one of y, z, $w \neq 0$, say $w \neq 0$, we can multiply
the left-hand side in (13) on the left by p_{23} and on the right by p_{23} and
obtain

$$p_{23} e_{31}^s e_{21}^t p_{12} m p_{13} e_{12}^r e_{13}^q p_{23} = \begin{bmatrix} b_3 & 0 & 0 \\ 0 & w & z \\ 0 & y & x \end{bmatrix},$$

where now $w \neq 0$. Proceeding as we did earlier when we had $x \neq 0$, we find that

$$e_{32}^i p_{23} e_{31}^s e_{21}^t p_{12} m p_{13} e_{12}^r e_{13}^q p_{23} e_{23}^j = \begin{bmatrix} b_3 & 0 & 0 \\ 0 & w & 0 \\ 0 & 0 & u \end{bmatrix},$$

which is again a diagonal matrix. So in every case, we can find elementary matrices $e_1, \ldots, e_k, f_1, \ldots, f_l$ so that

$$e_1 e_2 \ldots e_k m f_1 f_2 \ldots f_l = d,$$

where d is a diagonal matrix. It follows, by multiplying the last equation by e_1^{-1}, that

$$e_2 \ldots e_k m f_1 \ldots f_l = e_1^{-1} d.$$

Continuing, we obtain

$$m f_1 \ldots f_l = e_k^{-1} \ldots e_2^{-1} e_1^{-1} d,$$

and

$$m = e_k^{-1} \ldots e_2^{-1} e_1^{-1} d f_l^{-1} f_{l-1}^{-1} \ldots f_1^{-1}.$$

Note that the inverse of an elementary matrix is again an elementary matrix. We have thus proved:

Theorem 3.5. *Let m be a 3×3 matrix. Then we can find elementary matrices $g_1, \ldots, g_k, h_1, \ldots, h_l$ and a diagonal matrix d so that*

$$m = g_1 g_2 \ldots g_k d h_1 h_2 \ldots h_l. \tag{14}$$

EXAMPLE 10. Let us express the matrix

$$m = \begin{bmatrix} 1 & 1 & 0 \\ 3 & 1 & 2 \\ 5 & 2 & 4 \end{bmatrix}$$

in the form (14),

$$e_{21}^{-3} m = \begin{bmatrix} 1 & 1 & 0 \\ 0 & -2 & 2 \\ 5 & 2 & 4 \end{bmatrix},$$

$$e_{31}^{-5} e_{21}^{-3} m = \begin{bmatrix} 1 & 1 & 0 \\ 0 & -2 & 2 \\ 0 & -3 & 4 \end{bmatrix},$$

$$e_{32}^{-3/2} e_{31}^{-5} e_{21}^{-3} m = \begin{bmatrix} 1 & 1 & 0 \\ 0 & -2 & 2 \\ 0 & 0 & 1 \end{bmatrix}.$$

Also,

$$\begin{bmatrix} 1 & 1 & 0 \\ 0 & -2 & 2 \\ 0 & 0 & 1 \end{bmatrix} e_{12}^{-1} = \begin{bmatrix} 1 & 0 & 0 \\ 0 & -2 & 2 \\ 0 & 0 & 1 \end{bmatrix}$$

and

$$\begin{bmatrix} 1 & 0 & 0 \\ 0 & -2 & 2 \\ 0 & 0 & 1 \end{bmatrix} e_{12}^{-1} e_{23}^{1} = \begin{bmatrix} 1 & 0 & 0 \\ 0 & -2 & 0 \\ 0 & 0 & 1 \end{bmatrix}.$$

Setting $d = \begin{bmatrix} 1 & 0 & 0 \\ 0 & -2 & 0 \\ 0 & 0 & 1 \end{bmatrix}$, we thus have $e_{32}^{-3/2} e_{31}^{-5} e_{21}^{-3} m e_{12}^{-1} e_{23}^{1} = d$, so m

$= e_{21}^{3} e_{31}^{5} e_{32}^{3/2} d e_{23}^{-1} e_{12}^{1}$ or

$$\begin{bmatrix} 1 & 1 & 0 \\ 3 & 1 & 2 \\ 5 & 2 & 4 \end{bmatrix} = \begin{bmatrix} 1 & 0 & 0 \\ 3 & 1 & 0 \\ 0 & 0 & 1 \end{bmatrix} \begin{bmatrix} 1 & 0 & 0 \\ 0 & 1 & 0 \\ 5 & 0 & 1 \end{bmatrix} \begin{bmatrix} 1 & 0 & 0 \\ 0 & 1 & 0 \\ 0 & 3/2 & 1 \end{bmatrix}$$

$$\times \begin{bmatrix} 1 & 0 & 0 \\ 0 & -2 & 0 \\ 0 & 0 & 1 \end{bmatrix} \begin{bmatrix} 1 & 0 & 0 \\ 0 & 1 & -1 \\ 0 & 0 & 1 \end{bmatrix} \begin{bmatrix} 1 & 1 & 0 \\ 0 & 1 & 0 \\ 0 & 0 & 1 \end{bmatrix}.$$

Exercise 23. Express the following matrices in the form of (14):

(a) $\begin{pmatrix} 0 & 0 & 1 \\ 1 & 0 & 0 \\ 0 & 1 & 0 \end{pmatrix}$,

(b) $\begin{pmatrix} 1 & 3 & 0 \\ 2 & 0 & 0 \\ 0 & 0 & 0 \end{pmatrix}$,

(c) $\begin{pmatrix} 1 & 1 & 1 \\ 0 & 2 & 2 \\ 0 & 0 & 3 \end{pmatrix}$.

§2. Systems of Three Linear Equations in Three Unknowns

We consider the following system of three equations in three unknowns:

$$\begin{aligned} a_1 x_1 + a_2 x_2 + a_3 x_3 &= u_1, \\ b_1 x_1 + b_2 x_2 + b_3 x_3 &= u_2, \\ c_1 x_1 + c_2 x_2 + c_3 x_3 &= u_3. \end{aligned} \tag{15}$$

For each choice of numbers u_1, u_2, u_3, we may ask: Does the system (15) have a solution x_1, x_2, x_3? And if (15) has a solution, is this solution unique?

We may write the above expression in matrix form by introducing the linear transformation T with matrix

$$m = \begin{bmatrix} a_1 & a_2 & a_3 \\ b_1 & b_2 & b_3 \\ c_1 & c_2 & c_3 \end{bmatrix}.$$

Then the system (15) may be written

$$T(\mathbf{X}) = \mathbf{U}, \tag{16}$$

where $\mathbf{X}$ is the vector $\begin{bmatrix} x_1 \\ x_2 \\ x_3 \end{bmatrix}$ and $\mathbf{U} = \begin{bmatrix} u_1 \\ u_2 \\ u_3 \end{bmatrix}$. We call (16) the *nonhomogeneous system with matrix m.*

If the matrix m has an inverse, then the linear transformation T has an inverse T^{-1}. We then have

$$T(T^{-1}(\mathbf{U})) = (TT^{-1})(\mathbf{U}) = \mathbf{U};$$

so $\mathbf{X} = T^{-1}(\mathbf{U})$ is a solution of (16), and, conversely, if $T(\mathbf{X}) = \mathbf{U}$, then $\mathbf{X} = T^{-1}(T(\mathbf{X})) = T^{-1}(\mathbf{U})$.

So there is a unique solution vector $\mathbf{X}$ for each choice of $\mathbf{U}$.

In particular, if $\mathbf{U} = \mathbf{0}$, we find that $\mathbf{X} = \begin{bmatrix} 0 \\ 0 \\ 0 \end{bmatrix} = T^{-1} \begin{bmatrix} 0 \\ 0 \\ 0 \end{bmatrix}$ is the unique solution of the system

$$T(\mathbf{X}) = \mathbf{0}. \tag{17}$$

This system, with the zero vector on the right-hand side, is called the *homogeneous system* associated with the system (15).

No matter what the matrix m is, the homogeneous system has at least one solution, namely, the solution $\mathbf{X} = \begin{bmatrix} 0 \\ 0 \\ 0 \end{bmatrix}$. This is called the *trivial solution* of the homogeneous system, and we have seen above that if m has an inverse, then the trivial solution is the only solution of the homogeneous system.

But what if m does not possess an inverse? Will the system (17) then have a nontrivial solution? Proposition 2 of this chapter tells us that the answer is yes.

As in the 2-dimensional case, we obtain the following three general results.

Proposition 7. *The system* (16) *has a unique solution* $\mathbf{X}$ *for every* $\mathbf{U}$ *if and only if the transformation T has an inverse.*

Proposition 8. *The homogeneous system* (17) *has a non-trivial solution if and only if T fails to have an inverse.*

In the case that T fails to have an inverse, the general solution of (16) is described as follows.

Proposition 9. *If $\overline{\mathbf{X}}$ is a particular solution of* (16), *so that $T(\overline{\mathbf{X}}) = \mathbf{U}$, we may express every solution of* (16) *in the form $\overline{\mathbf{X}} + \mathbf{X}^h$, where $\mathbf{X}^h$ is a solution of the homogeneous system* (17).

EXAMPLE 11. Find all solutions of the nonhomogeneous system

$$\begin{cases} x_1 + x_2 + x_3 = 1, \\ 2x_1 - x_2 = 5, \\ 5x_1 + 2x_2 + 3x_3 = 8. \end{cases}$$

If x_1, x_2, x_3 solves the corresponding homogeneous system

$$\begin{cases} x_1 + x_2 + x_3 = 0 \\ 2x_1 - x_2 = 0, \\ 5x_1 + 2x_2 + 3x_3 = 0, \end{cases}$$

then $2x_1 = x_2$ and $x_1 + 2x_1 + x_3 = 0$, so $x_3 = -3x_1$. Hence, the most general solution $\mathbf{X}^h = \begin{pmatrix} x_1 \\ x_2 \\ x_3 \end{pmatrix}$ of the homogeneous system is

$$\mathbf{X}^h = \begin{pmatrix} x_1 \\ 2x_1 \\ -3x_1 \end{pmatrix}.$$

A particular solution $\overline{\mathbf{X}}$ of the nonhomogeneous system is

$$\overline{\mathbf{X}} = \begin{pmatrix} 2 \\ -1 \\ 0 \end{pmatrix}.$$

By Proposition 9, the general solution of the nonhomogeneous system is then

$$\mathbf{X} = \begin{pmatrix} 2 \\ -1 \\ 0 \end{pmatrix} + \begin{pmatrix} x_1 \\ 2x_1 \\ -3x_1 \end{pmatrix} = \begin{pmatrix} 2 + x_1 \\ -1 + 2x_1 \\ -3x_1 \end{pmatrix},$$

where x_1 is an arbitrary real number.

§3. Two Equations in Three Unknowns

In the system (15) under consideration, the number of unknowns equals the number of equations. Suppose we are given a system of two equations in three unknowns:

$$\begin{aligned} a_1x_1 + a_2x_2 + a_3x_3 &= u_1, \\ b_1x_1 + b_2x_2 + b_3x_3 &= u_2. \end{aligned} \tag{18}$$

The homogeneous system corresponding to (18) is

$$\begin{aligned} a_1x_1 + a_2x_2 + a_3x_3 &= 0, \\ b_1x_1 + b_2x_2 + b_3x_3 &= 0. \end{aligned} \tag{19}$$

Setting $\mathbf{X} = \begin{bmatrix} x_1 \\ x_2 \\ x_3 \end{bmatrix}$, $\mathbf{A} = \begin{bmatrix} a_1 \\ a_2 \\ a_3 \end{bmatrix}$, $\mathbf{B} = \begin{bmatrix} b_1 \\ b_2 \\ b_3 \end{bmatrix}$, we may write the system (18) as

$$\mathbf{A} \cdot \mathbf{X} = u_1, \qquad \mathbf{B} \cdot \mathbf{X} = u_2 \tag{18'}$$

and the system (19) as

$$\mathbf{A} \cdot \mathbf{X} = 0, \qquad \mathbf{B} \cdot \mathbf{X} = 0. \tag{19'}$$

Proposition 10. *If* $\overline{\mathbf{X}}$ *is one solution of* (18), *then any solution of* (18) *can be written as* $\overline{\mathbf{X}} + \mathbf{X}^h$, *where* $\mathbf{X}^h$ *is a solution of the homogeneous system* (19).

PROOF. If $\mathbf{A} \cdot \overline{\mathbf{X}} = \mathbf{A} \cdot \mathbf{X} = u_1$ and $\mathbf{B} \cdot \overline{\mathbf{X}} = \mathbf{B} \cdot \mathbf{X} = u_2$, then $\mathbf{A} \cdot (\mathbf{X} - \overline{\mathbf{X}}) = 0$ and $\mathbf{B} \cdot (\mathbf{X} - \overline{\mathbf{X}}) = 0$, so $\mathbf{X} - \overline{\mathbf{X}}$ is a solution $\mathbf{X}^h$ of (19).

If $\mathbf{A}$ and $\mathbf{B}$ are linearly independent, then the solution space of (19) is just the line perpendicular to the plane spanned by $\mathbf{A}$ and $\mathbf{B}$, i.e., the line along the nonzero vector $\mathbf{A} \times \mathbf{B}$.

If $\mathbf{A}$ and $\mathbf{B}$ are linearly dependent, but not both $\mathbf{0}$, then the solution space of (19) will be the plane through the origin perpendicular to the line containing $\mathbf{A}$ and $\mathbf{B}$.

If $\mathbf{A} = \mathbf{B} = \mathbf{0}$, then the solution of (19) is all of $\mathbb{R}^3$.

Determinants

Let m be the matrix

$$\begin{bmatrix} a_1 & a_2 & a_3 \\ b_1 & b_2 & b_3 \\ c_1 & c_2 & c_3 \end{bmatrix}.$$

Let $\mathbf{A}$, $\mathbf{B}$, $\mathbf{C}$ denote the row vectors of this matrix. The quantity

$$\mathbf{A} \cdot (\mathbf{B} \times \mathbf{C}) = \mathbf{B} \cdot (\mathbf{C} \times \mathbf{A}) = \mathbf{C} \cdot (\mathbf{A} \times \mathbf{B})$$

is called the *determinant of m* and is denoted $\det(m)$ or

$$\begin{vmatrix} a_1 & a_2 & a_3 \\ b_1 & b_2 & b_3 \\ c_1 & c_2 & c_3 \end{vmatrix}.$$

Expressed in these terms, Theorem 3.4 of Chapter 3.4 says that m has an inverse if and only if $\det(m) \neq 0$. In Exercise 6 in Chapter 3.0, we saw that

$$\mathbf{A} \cdot (\mathbf{B} \times \mathbf{C}) = \mathbf{B} \cdot (\mathbf{C} \times \mathbf{A}) = \mathbf{C} \cdot (\mathbf{A} \times \mathbf{B}).$$

Also $\mathbf{A} \times \mathbf{B} = -\mathbf{B} \times \mathbf{A}$.

In what follows, we shall frequently make use of these relations.

(i) *If two rows are interchanged, the determinant changes sign.*

PROOF.

$$\begin{bmatrix} b_1 & b_2 & b_3 \\ a_1 & a_2 & a_3 \\ c_1 & c_2 & c_3 \end{bmatrix} = \mathbf{B} \cdot (\mathbf{A} \times \mathbf{C}) = \mathbf{B} \cdot (-\mathbf{C} \times \mathbf{A}) = -\mathbf{B} \cdot (\mathbf{C} \times \mathbf{A})$$

$$= -\mathbf{A} \cdot (\mathbf{B} \times \mathbf{C}) = - \begin{vmatrix} a_1 & a_2 & a_3 \\ b_1 & b_2 & b_3 \\ c_1 & c_2 & c_3 \end{vmatrix}.$$

Interchanging the last two rows, we get

$$\begin{vmatrix} a_1 & a_2 & a_3 \\ c_1 & c_2 & c_3 \\ b_1 & b_2 & b_3 \end{vmatrix} = \mathbf{A} \cdot (\mathbf{C} \times \mathbf{B}) = -\mathbf{A} \cdot (\mathbf{B} \times \mathbf{C}) = - \begin{vmatrix} a_1 & a_2 & a_3 \\ b_1 & b_2 & b_3 \\ c_1 & c_2 & c_3 \end{vmatrix}.$$

Finally, interchanging the first and third rows,

$$\begin{vmatrix} c_1 & c_2 & c_3 \\ b_1 & b_2 & b_3 \\ a_1 & a_2 & a_3 \end{vmatrix} = \mathbf{C} \cdot (\mathbf{B} \times \mathbf{A}) = -\mathbf{C} \cdot (\mathbf{A} \times \mathbf{B}) = - \begin{vmatrix} a_1 & a_2 & a_3 \\ b_1 & b_2 & b_3 \\ c_1 & c_2 & c_3 \end{vmatrix}.$$

Thus (i) is proved.

(ii) *If a row is* 0, *then* $\det(m) = 0$.

PROOF. If $\mathbf{A} = \mathbf{0}$, $\det(m) = \mathbf{0} \cdot (\mathbf{B} \times \mathbf{C}) = 0$, and if $\mathbf{B}$ or $\mathbf{C} = \mathbf{0}$, $\det(m) = \mathbf{A} \cdot (\mathbf{0} \times \mathbf{C})$ or $\mathbf{A} \cdot (\mathbf{B} \times \mathbf{0})$, and so $\det(m) = 0$.

(iii) *If two rows are equal, then* $\det(m) = 0$.

PROOF. If $\mathbf{A} = \mathbf{B}$, $\det(m) = \mathbf{A} \cdot (\mathbf{A} \times \mathbf{C}) = 0$, by Chapter 3.0., p. 121. Similarly, if $\mathbf{A} = \mathbf{C}$, $\det(m) = 0$. If $\mathbf{B} = \mathbf{C}$, $\det(m) = \mathbf{A} \cdot (\mathbf{B} \times \mathbf{B}) = \mathbf{A} \cdot \mathbf{0} = 0$. So (iii) is proved.

(iv) *Suppose the three rows* $\mathbf{A}$, $\mathbf{B}$, $\mathbf{C}$ *are linearly dependent. Then* $\det(m) = 0$.

PROOF. If $\mathbf{B}$ and $\mathbf{C}$ are linearly dependent, then $\mathbf{B} \times \mathbf{C} = \mathbf{0}$, so

$$\det(m) = \mathbf{A} \cdot (\mathbf{B} \times \mathbf{C}) = \mathbf{A} \cdot \mathbf{0} = 0.$$

If $\mathbf{B}$ and $\mathbf{C}$ are linearly independent, then $\mathbf{A} = c_1\mathbf{B} + c_2\mathbf{C}$, and so

$$\det(m) = \mathbf{A} \cdot (\mathbf{B} \times \mathbf{C}) = (c_1\mathbf{B} + c_2\mathbf{C}) \cdot \mathbf{B} \times \mathbf{C}$$
$$= c_1\mathbf{B} \cdot (\mathbf{B} \times \mathbf{C}) + c_2\mathbf{C} \cdot (\mathbf{B} \times \mathbf{C}) = c_1 0 + c_2 0 = 0,$$

so (iv) is proved.

(v) *Suppose the three rows* $\mathbf{A}$, $\mathbf{B}$, $\mathbf{C}$ *are linearly independent. Then* $\det(m) \neq 0$.

PROOF. If $\mathbf{X} = \begin{bmatrix} x_1 \\ x_2 \\ x_3 \end{bmatrix}$ is a vector that m sends into $\mathbf{0}$, then

$$\begin{bmatrix} a_1 & a_2 & a_3 \\ b_1 & b_2 & b_3 \\ c_1 & c_2 & c_3 \end{bmatrix} \begin{bmatrix} x_1 \\ x_2 \\ x_3 \end{bmatrix} = \begin{bmatrix} 0 \\ 0 \\ 0 \end{bmatrix},$$

so $\mathbf{A} \cdot \mathbf{X} = 0$, $\mathbf{B} \cdot \mathbf{X} = 0$, $\mathbf{C} \cdot \mathbf{X} = 0$. By Proposition 1, Chapter 3.0, we conclude that $\mathbf{X} \cdot \mathbf{X} = 0$ and so $\mathbf{X} = \mathbf{0}$.

So m sends only $\mathbf{0}$ into $\mathbf{0}$. By Proposition 2 of Chapter 3.4, it follows that m has an inverse, and so by Theorem 3.4 of Chapter 3.4, $\det(m) \neq 0$.

Putting (iv) and (v) together, we have:

(vi) $\det(m) \neq 0$ *if and only if the rows of m are linearly independent.*

Exercise 1. Show that (iii) and (ii) are consequences of (iv).

(vii) *If a scalar multiple of one row of a matrix is added to another row, the determinant is unchanged.*

PROOF.

$$\begin{vmatrix} a_1 & a_2 & a_3 \\ b_1 + ta_1 & b_2 + ta_2 & b_3 + ta_3 \\ c_1 & c_2 & c_3 \end{vmatrix} = \mathbf{A} \cdot \left[(\mathbf{B} + t\mathbf{A}) \times \mathbf{C} \right]$$

$$= \mathbf{A} \cdot (\mathbf{B} \times \mathbf{C} + t\mathbf{A} \times \mathbf{C})$$

$$= \mathbf{A} \cdot (\mathbf{B} \times \mathbf{C}) + t\mathbf{A} \cdot (\mathbf{A} \times \mathbf{C})$$

$$= \mathbf{A} \cdot (\mathbf{B} \times \mathbf{C}) = \begin{vmatrix} a_1 & a_2 & a_3 \\ b_1 & b_2 & b_3 \\ c_1 & c_2 & c_3 \end{vmatrix}.$$

Similar reasoning gives the result in the other cases.

Exercise 2. Verify (vii) for the case when t times the last row is added to the first row.

§1. The Transpose of a Matrix

Let $m = \begin{bmatrix} a_1 & a_2 & a_3 \\ b_1 & b_2 & b_3 \\ c_1 & c_2 & c_3 \end{bmatrix}$. We call the line through a_1, b_2, c_3 the *diagonal* of m.

The following pairs of entries lie symmetrically placed with respect to the diagonal:

$$(a_2, b_1), \qquad (a_3, c_1), \qquad (b_3, c_2).$$

Let us interchange the elements in each pair, but leave the elements on the diagonal alone, and write down the matrix this gives:

$$\begin{bmatrix} a_1 & b_1 & c_1 \\ a_2 & b_2 & c_2 \\ a_3 & b_3 & c_3 \end{bmatrix}.$$

We call this new matrix the *transpose of m* and denote it by m^*. Note that the columns of m^* are the rows of m and the rows of m^* are the columns of m.

(viii) $\det(m^*) = \det(m)$.

PROOF.

$$\det(m^*) = a_1 \begin{vmatrix} b_2 & c_2 \\ b_3 & c_3 \end{vmatrix} - b_1 \begin{vmatrix} a_2 & c_2 \\ a_3 & c_3 \end{vmatrix} + c_1 \begin{vmatrix} a_2 & b_2 \\ a_3 & b_3 \end{vmatrix}$$

$$= a_1 b_2 c_3 - a_1 c_2 b_3 - b_1 a_2 c_3 + b_1 c_2 a_3 + c_1 a_2 b_3 - c_1 a_3 b_2$$

$$\det(m) = a_1 \begin{vmatrix} b_2 & b_3 \\ c_2 & c_3 \end{vmatrix} - a_2 \begin{vmatrix} b_1 & b_3 \\ c_1 & c_3 \end{vmatrix} + a_3 \begin{vmatrix} b_1 & b_2 \\ c_1 & c_2 \end{vmatrix}$$

$$= a_1 b_2 c_3 - a_1 b_3 c_2 - a_2 b_1 c_3 + a_2 b_2 c_1 + a_3 b_1 c_2 - a_3 b_2 c_1 .$$

Thus $\det(m) = \det(m^*)$, as asserted.

We can use this result to give another characterization of matrices with determinant $\neq 0$.

(ix) $\det(m) \neq 0$ *if and only if the columns of m are linearly independent.*

PROOF. If $\det(m) \neq 0$, then $\det(m^*) \neq 0$ by (viii). Hence, the rows of m^* are linearly independent by (vi). But the rows of m^* are the columns of m, so the columns of m are independent.

Conversely, if the columns of m are independent, then the rows of m^* are independent, so $\det(m^*) \neq 0$ and $\det(m) \neq 0$. The statement is proved.

Is the analogue of (vii) true when columns are used instead of rows?

EXAMPLE 1. Fix a matrix

$$m = \begin{bmatrix} a_1 & a_2 & a_3 \\ b_1 & b_2 & b_3 \\ c_1 & c_2 & c_3 \end{bmatrix};$$

let t be a scalar, and set

$$m_t = \begin{vmatrix} a_1 & a_2 + ta_1 & a_3 \\ b_1 & b_2 + tb_1 & b_3 \\ c_1 & c_2 + tc_1 & c_3 \end{vmatrix}.$$

Then

$$m_t^* = \begin{bmatrix} a_1 & b_1 & c_1 \\ a_2 + ta_1 & b_2 + tb_1 & c_2 + tc_1 \\ a_3 & b_3 & c_3 \end{bmatrix}.$$

By (viii),

$$\det(m_t) = \det(m_t^*).$$

By (vii),

$$\det(m_t^*) = \begin{vmatrix} a_1 & b_1 & c_1 \\ a_2 & b_2 & c_2 \\ a_3 & b_3 & b_3 \end{vmatrix} = \det(m^*).$$

Using (viii), we again get $\det(m^*) = \det(m)$, so $\det(m_t) = \det(m)$. Thus

$$\begin{vmatrix} a_1 & a_2 + ta_1 & a_3 \\ b_1 & b_2 + tb_1 & b_2 \\ c_1 & c_2 + tc_1 & c_3 \end{vmatrix} = \begin{vmatrix} a_1 & a_2 & a_3 \\ b_1 & b_2 & b_3 \\ c_1 & c_2 & c_3 \end{vmatrix},$$

and so the analogue of (vii) holds in this case.

Exercise 3. Show that

$$\begin{vmatrix} a_1 + ta_3 & a_2 & a_3 \\ b_1 + tb_3 & b_2 & b_3 \\ c_1 + tc_3 & c_2 & c_3 \end{vmatrix} = \begin{vmatrix} a_1 & a_2 & a_3 \\ b_1 & b_2 & b_3 \\ c_1 & c_2 & c_3 \end{vmatrix}.$$

Reasoning as in Example 1 and Exercise 2, we find that:

(x) *If a scalar multiple of one column of a matrix is added to another column, the determinant is unchanged.*

§2. Elementary Matrices

Recall the elementary matrices e_{ij}^t and p_{ij} we studied in earlier chapters. Let us find their determinants.

EXAMPLE 2.

$$\det(e_{12}^t) = \begin{vmatrix} 1 & t & 0 \\ 0 & 1 & 0 \\ 0 & 0 & 1 \end{vmatrix} = 1 \begin{vmatrix} 1 & 0 \\ 0 & 1 \end{vmatrix} - t \begin{vmatrix} 0 & 0 \\ 0 & 1 \end{vmatrix} + 0 = 1.$$

(xi) *For every i, j, t, $\det(e_{ij}^t) = 1$, and for every $1, j$, $\det(p_{ij}) = -1$.*

Exercise 4. Prove (xi).

EXAMPLE 3. Let $m = \begin{bmatrix} a_1 & a_2 & a_3 \\ b_1 & b_2 & b_3 \\ c_1 & c_2 & c_3 \end{bmatrix}$. Then

$$e_{12}^t m = \begin{bmatrix} a_1 + tb_1 & a_2 + tb_2 & a_3 + tb_3 \\ b_1 & b_2 & b_3 \\ c_1 & c_2 & c_3 \end{bmatrix}.$$

Hence, by (vii), $\det(e_{12}^t m) = \det(m)$.

EXAMPLE 4. Let m be as before. Then

$$me_{12}^t = \begin{bmatrix} a_1 & a_2 + ta_1 & a_3 \\ b_1 & b_2 + tb_1 & b_3 \\ c_1 & c_2 + tc_1 & c_3 \end{bmatrix}.$$

Hence, by (x), $\det(me_{12}^t) = \det(m)$.

EXAMPLE 5. Let m be as before.

$$p_{13}m = \begin{bmatrix} c_1 & c_2 & c_3 \\ b_1 & b_2 & b_3 \\ a_1 & a_2 & a_3 \end{bmatrix}.$$

Then, by (i), $\det(p_{13}m) = -\det m$. By (xi), $\det(p_{13}) = -1$. So we have

$$\det(p_{13}m) = (\det(p_{13}))\det(m).$$

Exercise 5. Show that for each matrix m,

$$\det(e_{13}^t m) = \det(e_{13}^t)\det(m)$$

and

$$\det(me_{13}^t) = \det(m)\det(e_{13}^t).$$

Exercise 6. Show that for each matrix m,

$$\det(p_{12}m) = \det(p_{12})\det(m).$$

Reasoning as in the preceding examples and exercises, we find:

(xii) *For every i, j, t and every matrix m,*

$$\det(e_{ij}^t m) = \det(e_{ij}^t)\det(m)$$

and

$$\det(me_{ij}^t) = \det(m)\det(e_{ij}^t).$$

Also:

(xiii) *For every i, j,*

$$\det(p_{ij}m) = \det(mp_{ij}) = (\det m)(\det p_{ij}).$$

Let $d = \begin{bmatrix} t_1 & 0 & 0 \\ 0 & t_2 & 0 \\ 0 & 0 & t_3 \end{bmatrix}$ and $m = \begin{bmatrix} a_1 & a_2 & a_3 \\ b_1 & b_2 & b_3 \\ c_1 & c_2 & c_3 \end{bmatrix}$. Then

$$dm = \begin{bmatrix} t_1 a_1 & t_1 a_2 & t_1 a_3 \\ t_2 b_2 & t_2 b_2 & t_2 b_3 \\ t_3 c_1 & t_3 c_2 & t_3 c_3 \end{bmatrix}.$$

If $\mathbf{A}, \mathbf{B}, \mathbf{C}$ are the rows of m, then $t_1\mathbf{A}, t_2\mathbf{B}, t_3\mathbf{C}$ are the rows of dm. Hence

$$\det(dm) = t_1\mathbf{A} \cdot (t_2\mathbf{B} \times t_3\mathbf{C})$$

$$= t_1 t_2 t_3 \mathbf{A} \cdot \mathbf{B} \times \mathbf{C} = t_1 t_2 t_3 \det(m).$$

Also, $\det(d) = t_1 t_2 t_3$. Thus we have proved:

(xiv) *If d is a diagonal matrix and m is any matrix, then* $\det(dm) = \det(d)$
$\det(m)$. *Similarly, we get* $\det(md) = \det(m)\det(d)$.

In Theorem 2.7 of Chapter 2.5, we showed that if a, b are two 2×2
matrices, then $\det(ab) = (\det a)(\det b)$. We now proceed to prove the corresponding relation for 3×3 matrices.

Theorem 3.6. *Let a, b be two 3×3 matrices. Then* $\det(ab) = (\det a)(\det b)$.

(*Note:* Theorems 3.5 and 3.6 appear below.)
In the 2×2 case, we proved the corresponding result by direct computation. Although it would be possible to do the same with 3×3 case, we
prefer to give a proof based on the properties of elementary matrices.

PROOF. By Theorem 3.3 of Chapter 3.4, there exists a diagonal matrix d and
elementary matrices e_i, f_j such that

$$a = e_1 \ldots e_k df_1 \ldots f_l$$

Using relations (xii), (xiii), and (xiv), we see that

$$\det a = (\det e_1) \ldots (\det e_k)(\det d)(\det f_1) \ldots (\det f_l).$$

Similarly, there exists a diagonal matrix d' and elementary matrices g_i, h_j
so that

$$b = g_1 \ldots g_r d' h_1 \ldots h_s$$

and

$$\det b = (\det g_1) \ldots (\det g_r)(\det d')(\det h_1) \ldots (\det h_s).$$

Hence

$$ab = e_1 \ldots e_k df_1 \ldots f_l g_1 \ldots g_r d' h_1 \ldots h_s$$

and

$$\det(ab) = (\det(e_1)) \ldots (\det e_k)(\det d)(\det f_1) \ldots (\det h_s).$$

So

$$\det(ab) = \det(a)\det(b).$$

Note: Even though, in general, $ab \neq ba$, we now see that $\det(ab) = \det(ba)$, because both are equal to $(\det a)(\det b) = (\det b)(\det a)$.

§3. Geometric Meaning of 3×3 Determinants

Next we proceed to extend to determinants of 3×3 matrices the results we found in Chapter 2.5 concerning the relations between determinants and *orientation* and between determinants and *area*.

Consider a triplet of vectors X_1, X_2, X_3 regarded as an ordered triplet with X_1 first, X_2 second, and X_3 third. Suppose X_1 and X_2 are linearly independent. Then $X_1 \times X_2$ is perpendicular to the plane of X_1 and X_2, and it is so chosen that the rotation about $X_1 \times X_2$ which sends X_1 to X_2 is through a positive angle α. The *upper half-space* determined by the ordered pair X_1, X_2 is the set of all vectors X such that $(X_1 \times X_2) \cdot X > 0$.

The triplet X_1, X_2, X_3 is said to be *positively oriented* if X_3 lies in the upper half-space determined by the ordered pair X_1, X_2 i.e., if $(X_1 \times X_2) \cdot X_3 > 0$. If $(X_1 \times X_2) \cdot X_3 < 0$, the triplet is said to be *negatively oriented*.

EXAMPLE 6.

(a) The triplet E_1, E_2, E_3 is positively oriented since $(E_1 \times E_2) \cdot E_3 = E_3 \cdot E_3 = 1 > 0$.
(b) The triplet E_1, E_2, $-E_3$ is negatively oriented, since $(E_1 \times E_2) \cdot (-E_3) = -1$.
(c) The triplet E_2, E_1, E_3 is negatively oriented, since $(E_2 \times E_1) \cdot E_3 = (-E_3) \cdot E_3 = -1$ (see Fig. 3.12).

Let

$$X_1 = \begin{bmatrix} x_{11} \\ x_{21} \\ x_{31} \end{bmatrix}, \quad X_2 = \begin{bmatrix} x_{12} \\ x_{22} \\ x_{32} \end{bmatrix}, \quad X_3 = \begin{bmatrix} x_{13} \\ x_{23} \\ x_{33} \end{bmatrix}$$

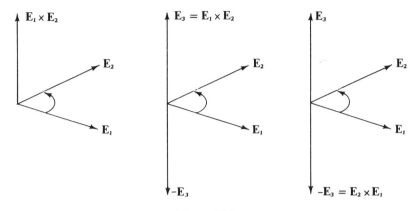

Figure 3.12

be an oriented triplet of vectors. Then

$$(\mathbf{X}_1 \times \mathbf{X}_2) \cdot \mathbf{X}_3 = \begin{vmatrix} x_{11} & x_{21} & x_{31} \\ x_{12} & x_{22} & x_{32} \\ x_{13} & x_{23} & x_{33} \end{vmatrix}.$$

By (viii), the right-hand term equals

$$\begin{vmatrix} x_{11} & x_{12} & x_{13} \\ x_{21} & x_{22} & x_{23} \\ x_{31} & x_{32} & x_{33} \end{vmatrix},$$

which is the determinant of the matrix whose columns are the vectors $\mathbf{X}_1$, $\mathbf{X}_2$, $\mathbf{X}_3$. Let us denote this matrix by $(\mathbf{X}_1 | \mathbf{X}_2 | \mathbf{X}_3)$. Thus

$$(\mathbf{X}_1 \times \mathbf{X}_2) \cdot \mathbf{X}_3 = \det(\mathbf{X}_1 | \mathbf{X}_2 | \mathbf{X}_3),$$

and so we obtain:

Proposition 1. *The triplet* $\mathbf{X}_1$, $\mathbf{X}_2$, $\mathbf{X}_3$ *is positively oriented if and only if* $\det(\mathbf{X}_1 | \mathbf{X}_2 | \mathbf{X}_3) > 0$.

Next let A be a linear transformation which has an inverse. We say that A *preserves orientation* if whenever $\mathbf{X}_1$, $\mathbf{X}_2$, $\mathbf{X}_3$ is a positively oriented triplet, then the triplet $A(\mathbf{X}_1)$, $A(\mathbf{X}_2)$, $A(\mathbf{X}_3)$ of image vectors is also positively oriented.

In Chapter 2.5, Theorem 2.5, we showed that a linear transformation of $\mathbb{R}^2$ preserves orientation if and only if the determinant of its matrix > 0. Is the analogous result true for $\mathbb{R}^3$?

Let A be a linear transformation of $\mathbb{R}^3$ with $m(A) = \begin{bmatrix} a_{11} & a_{12} & a_{13} \\ a_{21} & a_{22} & a_{23} \\ a_{31} & a_{32} & a_{33} \end{bmatrix}$.

Suppose A has an inverse and preserves orientation. Since the triplet $\mathbf{E}_1$, $\mathbf{E}_2$, $\mathbf{E}_3$ is positively oriented, it follows that the triplet $A(\mathbf{E}_1)$, $A(\mathbf{E}_2)$, $A(\mathbf{E}_3)$ is positively oriented. Hence, by Theorem 3.5, the determinant of the matrix

$(A(\mathbf{E}_1)|A(\mathbf{E}_2)|A(\mathbf{E}_3))$ is > 0. But $A(\mathbf{E}_1) = \begin{bmatrix} a_{11} \\ a_{21} \\ a_{31} \end{bmatrix}$, etc. So $(A(\mathbf{E}_1)|A(\mathbf{E}_2)$

$|A(\mathbf{E}_3)) = m(A)$, and so $\det(m(A)) > 0$. Conversely, let A be a linear transformation and suppose $\det(m(A)) > 0$. Let $\mathbf{X}_1, \mathbf{X}_2, \mathbf{X}_3$ be a positively

oriented triplet of vectors with $\mathbf{X}_1 = \begin{bmatrix} x_{11} \\ x_{21} \\ x_{31} \end{bmatrix}$, with $\mathbf{X}_2$ and $\mathbf{X}_3$ expressed

similarly. Then $A(\mathbf{X}_1) = \begin{bmatrix} a_{11}x_{11} + a_{12}x_{21} + a_{13}x_{31} \\ a_{21}x_{11} + a_{22}x_{21} + a_{23}x_{31} \\ a_{31}x_{11} + a_{32}x_{21} + a_{33}x_{31} \end{bmatrix}$, and we have similar

expressions for $A(\mathbf{X}_2), A(\mathbf{X}_3)$.

The matrix

$$(A(\mathbf{X}_1)|A(\mathbf{X}_2)|A(\mathbf{X}_3))$$

$$= \begin{bmatrix} a_{11}x_{11} + a_{12}x_{21} + a_{13}x_{31} & a_{11}x_{12} + a_{12}x_{22} + a_{13}x_{32} & a_{11}x_{13} + a_{12}x_{23} + a_{13}x_{33} \\ a_{21}x_{11} + a_{22}x_{21} + a_{23}x_{31} & a_{21}x_{12} + a_{22}x_{22} + a_{23}x_{32} & a_{21}x_{13} + a_{22}x_{23} + a_{23}x_{33} \\ a_{31}x_{11} + a_{32}x_{21} + a_{33}x_{31} & a_{31}x_{12} + a_{32}x_{22} + a_{33}x_{32} & a_{31}x_{13} + a_{32}x_{23} + a_{33}x_{33} \end{bmatrix}$$

$$= \begin{bmatrix} a_{11} & a_{12} & a_{13} \\ a_{21} & a_{22} & a_{23} \\ a_{31} & a_{32} & a_{33} \end{bmatrix} \begin{bmatrix} x_{11} & x_{12} & x_{13} \\ x_{21} & x_{22} & x_{23} \\ x_{31} & x_{32} & x_{33} \end{bmatrix}.$$

Then, by Theorem 3.7,

$$\det(A(\mathbf{X}_1)|A(\mathbf{X}_2)|A(\mathbf{X}_3)) = \begin{vmatrix} a_{11} & a_{12} & a_{13} \\ a_{21} & a_{22} & a_{23} \\ a_{31} & a_{32} & a_{33} \end{vmatrix} \begin{vmatrix} x_{11} & x_{12} & x_{13} \\ x_{21} & x_{22} & x_{23} \\ x_{31} & x_{32} & x_{33} \end{vmatrix}.$$

Since the triplet $\mathbf{X}_1, \mathbf{X}_2, \mathbf{X}_3$ is positively oriented, the second determinant on the right-hand side is > 0, and since by hypothesis $\det(m(A)) > 0$, the first determinant on the right-hand side is also > 0. Hence $A(\mathbf{X}_1), A(\mathbf{X}_2), A(\mathbf{X}_3)$ is a positively oriented triplet. Thus A preserves orientation. We have proved:

Theorem 3.7. *A linear transformation A on $\mathbb{R}^3$ preserves orientation if and only if $\det(m(A)) > 0$.*

We now proceed to describe the relation between 3×3 determinants and volume. Let

$$\mathbf{X}_1 = \begin{bmatrix} x_{11} \\ x_{21} \\ x_{31} \end{bmatrix}, \qquad \mathbf{X}_2 = \begin{bmatrix} x_{12} \\ x_{22} \\ x_{32} \end{bmatrix}, \qquad \mathbf{X}_3 = \begin{bmatrix} x_{13} \\ x_{23} \\ x_{33} \end{bmatrix}$$

be a positively oriented triplet of vectors. Denote by Π the parallelepiped with edges along these vectors, i.e., Π consists of all vectors

$$\mathbf{X} = t_1\mathbf{X}_1 + t_2\mathbf{X}_2 + t_3\mathbf{X}_3,$$

where t_1, t_2, t_3 are scalars between 0 and 1. By (ix) of Chapter 3.0,

$$\text{volume }(\Pi) = \mathbf{X}_3 \cdot (\mathbf{X}_1 \times \mathbf{X}_2) = \begin{vmatrix} x_{13} & x_{23} & x_{33} \\ x_{11} & x_{21} & x_{31} \\ x_{12} & x_{22} & x_{32} \end{vmatrix} = \begin{vmatrix} x_{11} & x_{21} & x_{31} \\ x_{12} & x_{22} & x_{32} \\ x_{13} & x_{23} & x_{33} \end{vmatrix}.$$

By (viii), the right-hand side equals

$$\begin{vmatrix} x_{11} & x_{12} & x_{13} \\ x_{21} & x_{22} & x_{23} \\ x_{31} & x_{32} & x_{33} \end{vmatrix} = \det(\mathbf{X}_1 \,|\, \mathbf{X}_2 \,|\, \mathbf{X}_3).$$

So

$$\text{volume }(\Pi) = \det(\mathbf{X}_1 \,|\, \mathbf{X}_2 \,|\, \mathbf{X}_3). \tag{1}$$

Now let T be a linear transformation having an inverse. Denote by $T(\Pi)$ the image of Π under T. $T(\Pi)$ is the parallelepiped determined by the vectors $T(\mathbf{X}_1)$, $T(\mathbf{X}_2)$, $T(\mathbf{X}_3)$. Hence, by formula (1) (see Fig. 3.13), we have

$$\text{volume}(T(\Pi)) = \det(T(\mathbf{X}_1), T(\mathbf{X}_2), T(\mathbf{X}_3)). \tag{2}$$

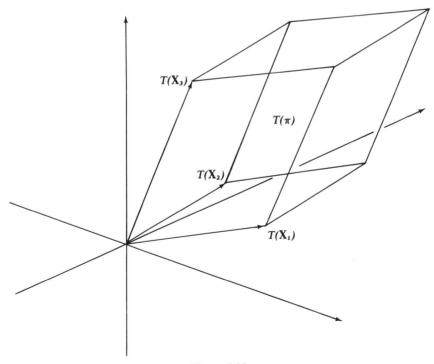

Figure 3.13

On the other hand, the calculation that led us to Theorem 3.5 gives

$$\det(T(\mathbf{X}_1) \mid T(\mathbf{X}_2) \mid T(\mathbf{X}_3)) = (\det(m(T)))\det(\mathbf{X}_1 \mid \mathbf{X}_2 \mid \mathbf{X}_3)$$
$$= \det(m(T)) \cdot \text{volume}(\Pi).$$

By (2), this gives

$$\text{volume}(T(\Pi)) = (\det(m(T)))(\text{volume}(\Pi)). \tag{3}$$

We have thus found the following counterpart to Theorem 2.6 of Chapter 2.5.

Theorem 3.8. *Let T be an orientation-preserving linear transformation of $\mathbb{R}^3$. If Π is any parallelepiped, then*

$$\text{volume}(T(\Pi)) = (\det(m(T)))(\text{volume}(\Pi)). \tag{4}$$

Note: If T reverses orientation, the same argument yields formula (4) with a minus sign on the right-hand side, so for every invertible transformation T, we have

$$\text{volume}(T(\Pi)) = |\det(m(T))|\text{volume}(\Pi). \tag{5}$$

If T has *no inverse*, $T(\Pi)$ degenerates into a figure lying in a plane, so the left-hand side is 0 while the right-hand side is 0, since $\det(m(T)) = 0$ in this case. Thus, formula (5) is valid for every linear transformation of $\mathbb{R}^3$.

Eigenvalues

EXAMPLE 1. Let π be a plane through the origin and let S be the transformation which reflects each vector through π. If $\mathbf{Y}$ is a vector on π, then $S(\mathbf{Y}) = \mathbf{Y}$, and if $\mathbf{U}$ is a vector perpendicular to π, then $S(\mathbf{U}) = -\mathbf{U}$. Thus for $t = 1$ and $t = -1$, there exist nonzero vectors $\mathbf{X}$ satisfying $S(\mathbf{X}) = t\mathbf{X}$. If $\mathbf{X}$ is any vector which is neither on π nor perpendicular to π, then $S(\mathbf{X})$ is not a multiple of $\mathbf{X}$.

Let T be a linear transformation of $\mathbb{R}^3$ and let t be a real number. We say that t is an *eigenvalue* of T if there is some nonzero vector $\mathbf{X}$ such that

$$T(\mathbf{X}) = t\mathbf{X} \quad \text{and} \quad \mathbf{X} \neq \mathbf{0}.$$

If t is an eigenvalue of T, then we call a vector $\mathbf{Y}$ an *eigenvector of T corresponding to t* if $T(\mathbf{Y}) = t\mathbf{Y}$.

For example, the eigenvalues of S are 1 and -1. The eigenvectors of S corresponding to 1 are all the vectors in π and the eigenvectors of S corresponding to -1 are all the vectors perpendicular to π.

EXAMPLE 2. Fix λ in $\mathbb{R}$. Let D_λ be stretching by λ. Then for every vector $\mathbf{X}$, $D_\lambda(\mathbf{X}) = \lambda\mathbf{X}$. Hence λ is an eigenvalue of D_λ. Every vector $\mathbf{X}$ in $\mathbb{R}^3$ is an eigenvector of D_λ corresponding to the eigenvalue λ.

EXAMPLE 3. Let D be the linear transformation with the diagonal matrix

$$\begin{bmatrix} \lambda_1 & 0 & 0 \\ 0 & \lambda_2 & 0 \\ 0 & 0 & \lambda_3 \end{bmatrix}.$$

If $\mathbf{X} = \begin{bmatrix} x_1 \\ x_2 \\ x_3 \end{bmatrix}$, then $D(\mathbf{X}) = \begin{bmatrix} \lambda_1 & 0 & 0 \\ 0 & \lambda_2 & 0 \\ 0 & 0 & \lambda_3 \end{bmatrix} \begin{bmatrix} x_1 \\ x_2 \\ x_3 \end{bmatrix} = \begin{bmatrix} \lambda_1 x_1 \\ \lambda_2 x_2 \\ \lambda_3 x_3 \end{bmatrix}$. It follows that setting

$\mathbf{E}_1 = \begin{bmatrix} 1 \\ 0 \\ 0 \end{bmatrix}$, $\mathbf{E}_2 = \begin{bmatrix} 0 \\ 1 \\ 0 \end{bmatrix}$, $\mathbf{E}_3 = \begin{bmatrix} 0 \\ 0 \\ 1 \end{bmatrix}$, we have

$$D(\mathbf{E}_1) = \lambda_1 \mathbf{E}_1, \qquad D(\mathbf{E}_2) = \lambda_2 \mathbf{E}_2, \qquad D(\mathbf{E}_3) = \lambda_3 \mathbf{E}_3.$$

Thus $\lambda_1, \lambda_2, \lambda_3$ are eigenvalues of D and $\mathbf{E}_1, \mathbf{E}_2, \mathbf{E}_3$ are eigenvectors. Does D have any other eigenvalues? Suppose $D(\mathbf{X}) = t\mathbf{X}$, where $\mathbf{X} = \begin{bmatrix} x_1 \\ x_2 \\ x_3 \end{bmatrix}$, $\mathbf{X} \neq 0$,

and t is in $\mathbb{R}$. Then $\begin{bmatrix} tx_1 \\ tx_2 \\ tx_3 \end{bmatrix} = \begin{bmatrix} \lambda_1 x_1 \\ \lambda_2 x_2 \\ \lambda_3 x_3 \end{bmatrix}$, so $tx_i = \lambda_i x_i$ for $i = 1, 2, 3$. Since $x_i \neq 0$

for some i, $t = \lambda_i$. Therefore, $\lambda_1, \lambda_2, \lambda_3$ are all the eigenvalues of D.

Exercise 1. With D as in Example 3, find all the eigenvectors of D.

Exercise 2. Let T be a linear transformation of $\mathbb{R}^3$ satisfying $T^2 = I$. Let t be an eigenvalue of T and $\mathbf{X}$ an eigenvector corresponding to t with $\mathbf{X} \neq \mathbf{0}$.

(a) Show that $T^2(\mathbf{X}) = t^2 \mathbf{X}$.
(b) Show that $t = 1$ or $t = -1$.
(c) Apply what you have found to the reflection S in Example 1.

Exercise 3. Let T be a linear transformation of $\mathbb{R}^3$ such that $T^2 = 0$.

(a) Show that 0 is an eigenvalue of T.
(b) Show that 0 is the only eigenvalue of T.

Exercise 4. Let T be the linear transformation with matrix

$$m(T) = \begin{pmatrix} 0 & 0 & 1 \\ 0 & 0 & 0 \\ 0 & 0 & 0 \end{pmatrix}.$$

(a) Show that $T^2 = 0$.
(b) Apply Exercise 3 to determine the eigenvalues of T.
(c) Find all eigenvectors of T.

§1. Characteristic Equation

Given a transformation T with matrix $\begin{bmatrix} a_1 & a_2 & a_3 \\ b_1 & b_2 & b_3 \\ c_1 & c_2 & c_3 \end{bmatrix}$, how can we determine

the eigenvalues of T? We proceed as we did for the corresponding problem in two dimensions.

Assume t is an eigenvalue of T and $\begin{pmatrix} x_1 \\ x_2 \\ x_3 \end{pmatrix}$ is a corresponding eigenvector

with $\begin{pmatrix} x_1 \\ x_2 \\ x_3 \end{pmatrix} \neq \begin{pmatrix} 0 \\ 0 \\ 0 \end{pmatrix}$. Then $T\begin{pmatrix} x_1 \\ x_2 \\ x_3 \end{pmatrix} = t\begin{pmatrix} x_1 \\ x_2 \\ x_3 \end{pmatrix}$, so

$$\begin{pmatrix} a_1 & a_2 & a_3 \\ b_1 & b_2 & b_3 \\ c_1 & c_2 & c_3 \end{pmatrix}\begin{pmatrix} x_1 \\ x_2 \\ x_3 \end{pmatrix} = t\begin{pmatrix} x_1 \\ x_2 \\ x_3 \end{pmatrix}$$

or

$$\begin{cases} a_1 x_1 + a_2 x_2 + a_3 x_3 = t x_1, \\ b_1 x_1 + b_2 x_2 + b_3 x_3 = t x_2, \\ c_1 x_1 + c_2 x_2 + c_3 x_3 = t x_3. \end{cases} \tag{1}$$

Transposing the right-hand terms, we get

$$\begin{cases} (a_1 - t)x_1 + a_2 x_2 + a_3 x_3 = 0, \\ b_1 x_1 + (b_2 - t)x_2 + b_3 x_3 = 0, \\ c_1 x_1 + c_2 x_2 + (c_3 - t)x_3 = 0. \end{cases} \tag{2}$$

Thus x_1, x_2, x_3 is a nonzero solution of the homogeneous system (2). By Proposition 8 and Theorem 3.4 of Chapter 3.4, it follows that the determinant

$$\begin{vmatrix} a_1 - t & a_2 & a_3 \\ b_1 & b_2 - t & b_3 \\ c_1 & c_2 & c_3 - t \end{vmatrix} = 0. \tag{3}$$

If the left-hand side is expanded, this equation has the form

$$-t^3 + u_1 t^2 + u_2 t + u_3 = 0, \tag{4}$$

where u_1, u_2, u_3 are certain constants.

Equation (3) is called the *characteristic equation* for the transformation T.

We just saw that if t is an eigenvalue of T, then t is a root of the characteristic equation of T. Conversely, if t is a root of the characteristic equation, then (3) holds. Hence, by Proposition 8 of Chapter 3.4, the system (2) has a nonzero solution x_1, x_2, x_3, and so (1) also has this solution. Therefore,

$$T\begin{pmatrix} x_1 \\ x_2 \\ x_3 \end{pmatrix} = t\begin{pmatrix} x_1 \\ x_2 \\ x_3 \end{pmatrix}.$$

Hence t is an eigenvalue of T. We have proved:

Theorem 3.8. *A real number t is an eigenvalue of the transformation T of $\mathbb{R}^3$ if and only if t is a root of the characteristic equation (3).*

Theorem 3.8 appears later.

EXAMPLE 4. Let T have matrix

$$\begin{pmatrix} 1 & 0 & 0 \\ -5 & 2 & 0 \\ 2 & 3 & 7 \end{pmatrix}.$$

The characteristic equation of T is

$$\begin{vmatrix} 1-t & 0 & 0 \\ -5 & 2-t & 0 \\ 2 & 3 & 7-t \end{vmatrix} = 0$$

or

$$(1-t)\begin{vmatrix} 2-t & 0 \\ 3 & 7-t \end{vmatrix} = (1-t)(2-t)(7-t) = 0.$$

The roots of this equation are 1, 2, and 7, and so these are the eigenvalues of T. Let us calculate the eigenvectors corresponding to the eigenvalue 2. If $\begin{pmatrix} x_1 \\ x_2 \\ x_3 \end{pmatrix}$ is such a vector, then

$$\begin{pmatrix} 1 & 0 & 0 \\ -5 & 2 & 0 \\ 2 & 3 & 7 \end{pmatrix}\begin{pmatrix} x_1 \\ x_2 \\ x_3 \end{pmatrix} = 2\begin{pmatrix} x_1 \\ x_2 \\ x_3 \end{pmatrix} = \begin{pmatrix} 2x_1 \\ 2x_2 \\ 2x_3 \end{pmatrix}.$$

Then

$$x_1 = 2x_1,$$
$$-5x_1 + 2x_2 = 2x_2,$$
$$2x_1 + 3x_2 + 7x_3 = 2x_3.$$

The first equation gives $x_1 = 0$. The second equation puts no restriction on x_2. The third equation yields

$$5x_3 = -3x_2 \quad \text{or} \quad x_3 = -\tfrac{3}{5}x_2.$$

Thus an eigenvector of T corresponding to the eigenvalue 2 must have the form $\begin{pmatrix} 0 \\ x_2 \\ -\frac{3}{5}x_2 \end{pmatrix} = \begin{pmatrix} 0 \\ 5y \\ -3y \end{pmatrix}$, if we set $y = \tfrac{1}{5}x_2$. Is every vector of this form an eigenvector?

$$\begin{pmatrix} 1 & 0 & 0 \\ -5 & 2 & 0 \\ 2 & 3 & 7 \end{pmatrix}\begin{pmatrix} 0 \\ 5y \\ -3y \end{pmatrix} = \begin{pmatrix} 0 \\ 10y \\ 15y - 21y \end{pmatrix} = \begin{pmatrix} 0 \\ 10y \\ -6y \end{pmatrix} = 2\begin{pmatrix} 0 \\ 5y \\ -3y \end{pmatrix}.$$

Thus the answer is yes and we have: a vector is an eigenvector of T with eigenvalue 2 if and only if it has the form $\begin{pmatrix} 0 \\ 5y \\ -3y \end{pmatrix}$. Note that the vectors of

this form fill up the line $\left\{ y \begin{bmatrix} 0 \\ 5 \\ -3 \end{bmatrix} \middle| y \text{ in } \mathbb{R} \right\}$, which passes through the origin.

Exercise 5. Find all eigenvectors corresponding to the eigenvalue 1 for the transformation T of Example 4. Show that these vectors fill up a line through the origin.

EXAMPLE 5. Fix θ with $0 \le \theta < 2\pi$. R_θ^3 is the transformation of $\mathbb{R}^3$ which rotates each vector by θ degrees around the positive x_3-axis. Find all eigenvalues of R_θ^3. We have

$$m(R_\theta^3) = \begin{bmatrix} \cos\theta & -\sin\theta & 0 \\ \sin\theta & \cos\theta & 0 \\ 0 & 0 & 1 \end{bmatrix},$$

so the characteristic equation is

$$\begin{vmatrix} \cos\theta - t & -\sin\theta & 0 \\ \sin\theta & \cos\theta - t & 0 \\ 0 & 0 & 1 - t \end{vmatrix} = - \begin{vmatrix} 0 & 0 & 1 - t \\ \sin\theta & \cos\theta - t & 0 \\ \cos\theta - t & -\sin\theta & 0 \end{vmatrix}$$

$$= -(1 - t)\left[-(\sin\theta)^2 - (\cos\theta - t)^2 \right]$$

$$= 0.$$

If t is a real root, then either $1 - t = 0$ or $(\sin\theta)^2 + (\cos\theta - t)^2 = 0$. The second equation implies that $\sin\theta = 0$ and $\cos\theta = t$.

Case 1: $\theta \ne 0, \pi$. In this case, $t = 1$ is the only root of the characteristic equation, and so 1 is the only eigenvalue. If $\mathbf{X}$ lies on the x_3-axis, evidently $T(\mathbf{X}) = \mathbf{X} = 1 \cdot \mathbf{X}$, so the x_3-axis consists of eigenvectors with eigenvalue 1. There are no other eigenvectors.

Case 2: $\theta = 0$. In this case, $R_\theta^3 = I$, so 1 is the only eigenvalue and every vector in $\mathbb{R}^3$ is an eigenvector corresponding to this eigenvalue.

Case 3: $\theta = \pi$. The characteristic equation of R_π^3 is

$$(1 - t)(-1 - t)^2 = 0$$

with roots $t = 1$ and $t = -1$, so the eigenvalues of R_π^3 are 1 and -1.

Exercise 6. Find the eigenvectors of R_π which correspond to the eigenvalue -1. Describe, in geometrical terms, how $R_\pi^3\mathbf{X}$ is obtained from $\mathbf{X}$ if $\mathbf{X}$ is any vector. Then explain, geometrically, why 1 and -1 occur as eigenvalues of R_π^3.

Now let A be a given linear transformation and let

$$-t^3 + at^2 + bt + c = 0$$

be the characteristic equation of A.

Define $f(t) = -t^3 + at^2 + bt + c$. Then f is a function defined for all real t. The equation $f(t) = 0$ must have at least one real root t_0. To see this, note that $f(t) < 0$ when t is a large positive number, while $f(t) > 0$ when t is a negative number with large absolute value. Therefore, at some point t_0, the graph of f must cross the t-axis. Dividing $f(t)$ by $t - t_0$, we get a quadratic polynomial $-t^2 + dt + e$, where d and e are certain constants. Thus

$$f(t) = (t - t_0)(-t^2 + dt + e).$$

The polynomial $g(t) = -t^2 + dt + e$ may be factored

$$g(t) = -(t - t_1)(t - t_2),$$

where t_1, t_2 are the roots of g, which may be real or conjugate complex numbers. We can distinguish three possibilities.

(i) t_1, t_2 are complex numbers, $t_1 = u + iv$, $t_2 = u - iv$, with $v \neq 0$. Then $f(t) = 0$ has exactly one real root, namely, t_0. In this case, the graph of f appears as in Fig. 3.14.

EXAMPLE 6. For $A = R_\theta^3$, we found in Example 5,

$$f(t) = (t - 1)\left[-(\sin\theta)^2 - (\cos\theta - t)^2\right].$$

Here $g(t) = [-(\sin\theta)^2 - (\cos\theta - t)^2] = -t^2 + 2(\cos\theta)t - 1$. If $\theta \neq 0$ or π, then g has no real roots, so possibility (i) occurs.

(ii) t_1 and t_2 are real and $t_1 = t_2$. If $t_0 = t_1 = t_2$, then $f(t) = 0$ has a triple root at t_0. The graph of f now appears as in Fig. 3.15. If $t_0 \neq t_1$, then f has one simple real root, t_0, and one double real root, $t_1 = t_2$. The graph of f appears as in Fig. 3.16.

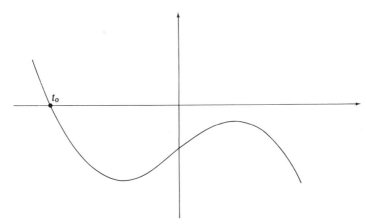

Figure 3.14

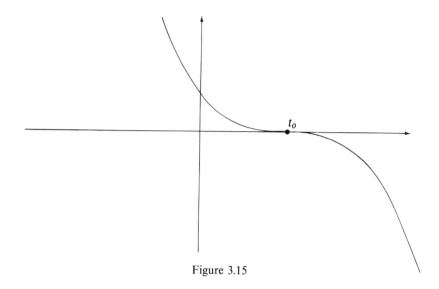

Figure 3.15

EXAMPLE 7. For $A = R_\pi^3$, we found

$$f(t) = (1 - t)(-1 - t)^2,$$

so $t_0 = 1$, $t_1 = t_2 = -1$, and possibility (ii) occurs.
    For $A = D_\lambda$, we have

$$f(t) = \begin{vmatrix} \lambda - t & 0 & 0 \\ 0 & \lambda - t & 0 \\ 0 & 0 & \lambda - t \end{vmatrix} = (\lambda - t)^3,$$

so $t_0 = t_1 = t_2 = \lambda$.

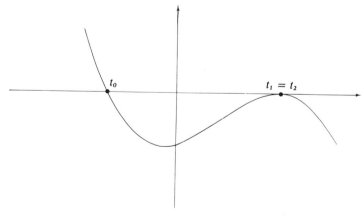

Figure 3.16

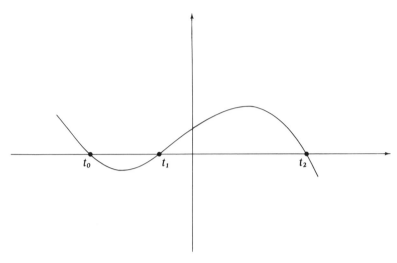

Figure 3.17

(iii) t_1 and t_2 are real and $t_1 \neq t_2$. If $t_1 = t_0$ or $t_2 = t_0$, the situation is as in
    case (ii). If $t_1 \neq t_0$ and $t_2 \neq t_0$, then the equation $f(t) = 0$ has the three
    distinct real roots t_0, t_1, and t_2. The graph of f now appears as in Fig.
    3.17.

EXAMPLE 8. The transformation T of Example 4 had 1, 2, and 7 as the roots
of its characteristic equation and, so, illustrated case (iii).

Let us summarize what we have found. Using Theorem 3.9 we can
conclude:

Proposition 1. *If A is a linear transformation of $\mathbb{R}^3$, then A always has at least
one eigenvalue and may have one, two, or three distinct eigenvalues.*

Exercise 7. Find all eigenvalues and eigenvectors of the transformation whose
matrix is $\begin{pmatrix} 0 & 0 & 2 \\ 0 & 1 & 0 \\ 1 & 0 & 0 \end{pmatrix}$.

Exercise 8. Let T be the transformation with matrix $\begin{pmatrix} c & 0 & b \\ 0 & c & 0 \\ 0 & 0 & c \end{pmatrix}$.

(i) Find the characteristic equation of T.
(ii) Show that c is the only eigenvalue of T.
(iii) Show the eigenvectors of T corresponding to this eigenvalue fill up a plane
    through 0, and give an equation of this plane.

Exercise 9.

(i) Find the characteristic equation for the transformation with matrix
$$\begin{pmatrix} 0 & 1 & 0 \\ 0 & 0 & 1 \\ a_1 & a_2 & a_3 \end{pmatrix}.$$

(ii) Show that given any three numbers a, b, c, there is some linear transformation of $\mathbb{R}^3$ whose characteristic equation is $-t^3 + at^2 + bt + c = 0$.

Let A be a linear transformation of $\mathbb{R}^3$ and let t be an eigenvalue of A. By the *eigenspace* E_t we mean the collection of all eigenvectors of A which correspond to the eigenvalue t.

EXAMPLE 9. Let P be the linear transformation which projects each vector on the plane π through the origin. Then for each vector $\mathbf{X}$ in π, $P(\mathbf{X}) = \mathbf{X}$, while for each $\mathbf{X}$ perpendicular to π, $P(\mathbf{X}) = \mathbf{0}$. P has no other eigenvectors. Hence, here $E_1 = $ the plane π, $E_0 = $ the line through $\mathbf{0}$ perpendicular to π.

Let A be an arbitrary linear transformation of $\mathbb{R}^3$ and t an eigenvalue of A. If $\mathbf{X}$ is a vector belonging to the eigenspace E_t, then $A(\mathbf{X}) = t\mathbf{X}$. Hence, for each scalar c, $A(c\mathbf{X}) = cA(\mathbf{X}) = ct\mathbf{X} = t(c\mathbf{X})$, and so $c\mathbf{X}$ is also in E_t. Thus E_t contains the line along $\mathbf{X}$. If E_t is not equal to this line, then there is some $\mathbf{Y}$ in E_t such that $\mathbf{X}$ and $\mathbf{Y}$ are linearly independent. For each pair of scalars c_1, c_2, $A(c_1\mathbf{X} + c_2\mathbf{Y}) = c_1A(\mathbf{X}) + c_2A(\mathbf{Y}) = c_1t\mathbf{X} + c_2t\mathbf{Y} = t(c_1\mathbf{X} + c_2\mathbf{Y})$. Thus $c_1\mathbf{X} + c_2\mathbf{Y}$ is in E_t. So the entire plane

$$\{(c_1\mathbf{X} + c_2\mathbf{Y}) \,|\, c_1, c_2 \text{ in } \mathbb{R}\}$$

is contained in E_t. If E_t does not coincide with this plane, then there is some vector $\mathbf{Z}$ in E_t such that $\mathbf{X}$, $\mathbf{Y}$, $\mathbf{Z}$ are linearly independent. By Proposition 1 of Chapter 3.0, $\{c_1\mathbf{X} + c_2\mathbf{Y} + c_3\mathbf{z} \,|\, c_1, c_2, c_3 \text{ in } \mathbb{R}\}$ is all of $\mathbb{R}^3$ and this set is contained in E_t. So, in that case, $E_t = \mathbb{R}^3$. We have shown:

Proposition 2. *If A is a linear transformation of $\mathbb{R}^3$, each eigenspace of A is either a line through the origin, or a plane through the origin, or all of $\mathbb{R}^3$.*

§2. Isometries of $\mathbb{R}^3$

In Chapter 2.5 we found all the length-preserving linear transformations of the plane. They turned out to be the rotations and reflections of the plane. Let us try to solve the corresponding problem in 3-space.

A linear transformation T of $\mathbb{R}^3$ which preserves lengths of segments is called an *isometry*. Exactly as in $\mathbb{R}^2$, we find that T is an isometry if and only if

$$|T(\mathbf{X})| = |\mathbf{X}| \qquad \text{for every vector } \mathbf{X}. \tag{5}$$

Proposition 3. *An isometry T preserves the dot product, i.e., for all vectors* **X, Y,**

$$T(\mathbf{X}) \cdot T(\mathbf{Y}) = \mathbf{X} \cdot \mathbf{Y}.$$

PROOF. Since (5) holds for each vector,

$$|T(\mathbf{X} - \mathbf{Y})|^2 = |\mathbf{X} - \mathbf{Y}|^2$$

or

$$(T(\mathbf{X} - \mathbf{Y})) \cdot (T(\mathbf{X} - \mathbf{Y})) = (\mathbf{X} - \mathbf{Y}) \cdot (\mathbf{X} - \mathbf{Y}).$$

So

$$(T(\mathbf{X}) - T(\mathbf{Y})) \cdot (T(\mathbf{X}) - T(\mathbf{Y})) = (\mathbf{X} - \mathbf{Y}) \cdot (\mathbf{X} - \mathbf{Y})$$

or

$$T(\mathbf{X}) \cdot T(\mathbf{X}) - 2T(\mathbf{X}) \cdot T(\mathbf{Y}) + T(\mathbf{Y}) \cdot T(\mathbf{Y}) = \mathbf{X} \cdot \mathbf{X} - 2\mathbf{X} \cdot \mathbf{Y} + \mathbf{Y} \cdot \mathbf{Y}.$$

Again by (5), $T(\mathbf{X}) \cdot T(\mathbf{X}) = \mathbf{X} \cdot \mathbf{X}$ and similarly for **Y**, so cancelling we get

$$-2T(\mathbf{X}) \cdot T(\mathbf{Y}) = -2\mathbf{X} \cdot \mathbf{Y},$$

and so

$$T(\mathbf{X}) \cdot T(\mathbf{Y}) = \mathbf{X} \cdot \mathbf{Y}.$$

Proposition 4. *If T is an isometry of* $\mathbb{R}^3$, *then T has* 1 *or* -1 *as an eigenvalue and has no other eigenvalues.*

PROOF. Every linear transformation T of $\mathbb{R}^3$ has an eigenvalue t, so for some vector $\mathbf{X} \neq \mathbf{0}$, $T(\mathbf{X}) = t\mathbf{X}$. Then

$$|\mathbf{X}| = |T(\mathbf{X})| = |t\mathbf{X}| = |t| \, |\mathbf{X}|, \qquad \text{so} \quad |t| = 1.$$

Hence

$$t = 1 \quad \text{or} \quad t = -1.$$

Proposition 5. *If T is an isometry of* $\mathbb{R}^3$, *then* $\det(T) = 1$ *or* $\det(T) = -1$.

Note: We write $\det(T)$ for $\det(m(T))$, the determinant of the matrix of T.

PROOF. Consider the cube Q with edges $\mathbf{E}_1$, $\mathbf{E}_2$, $\mathbf{E}_3$. The vectors $T(\mathbf{E}_1)$, $T(\mathbf{E}_2)$, $T(\mathbf{E}_3)$ are edges of the image, $T(Q)$, of Q under T. For each i, $|T(\mathbf{E}_i)| = |\mathbf{E}_i| = 1$, and for each i, j with $i \neq j$, $T(\mathbf{E}_i) \cdot T(\mathbf{E}_j) = \mathbf{E}_i \cdot \mathbf{E}_j = 0$. Thus $T(Q)$ is a cube of side 1. Hence, $\mathrm{vol}(T(Q)) = 1 = \mathrm{vol}\, Q$. Also, by Theorem 3.6 of Chapter 3.5, $\mathrm{vol}\, T(Q) = |\det(T)| \cdot \mathrm{vol}\, Q$. Hence $|\det(T)| = 1$. So $\det T = 1$ or $\det T = -1$.

One example of an isometry is a *rotation about an axis*. Fix a vector **F** and denote by π the plane through 0 orthogonal to **F**. Fix a number θ. We denote by R_θ the transformation of the plane π which rotates each vector in

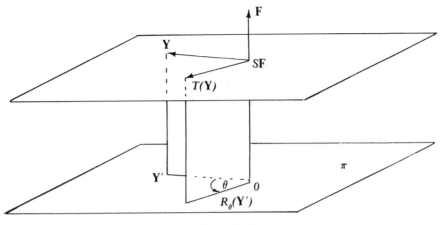

Figure 3.18

π counterclockwise by an angle θ about $\mathbf{F}$. Let $\mathbf{Y}$ be any vector in $\mathbb{R}^3$. We decompose $\mathbf{Y}$ as

$$\mathbf{Y} = \mathbf{Y}' + s\mathbf{F},$$

where $\mathbf{Y}'$ is the projection of $\mathbf{Y}$ on π and s is a scalar (see Fig. 3.18). We now define

$$T(\mathbf{Y}) = R_\theta(\mathbf{Y}') + s\mathbf{F}.$$

Note that $T(\mathbf{Y})$ lies in the plane through $\mathbf{Y}$ perpendicular to $\mathbf{F}$. We call the transformation T *a rotation about the axis* $\mathbf{F}$ *by the angle* θ.

Exercise 10. Prove that T is a linear transformation of $\mathbb{R}^3$ and that $T(\mathbf{F}) = \mathbf{F}$.

For each $\mathbf{Y}$, $|T(\mathbf{Y})|^2 = |R_\theta(\mathbf{Y}') + s\mathbf{F}|^2 = (R_\theta(\mathbf{Y}') + s\mathbf{F}) \cdot (R_\theta(\mathbf{Y}') + s\mathbf{F}) = |R_\theta(\mathbf{Y}')|^2 + s^2|\mathbf{F}|^2$, since $R_\theta(\mathbf{Y}')$ is orthogonal to $\mathbf{F}$. Also, $|\mathbf{Y}|^2 = |\mathbf{Y}'|^2 + s^2|\mathbf{F}|^2$. Since R_θ is a rotation of π, $|R_\theta(\mathbf{Y}')| = |\mathbf{Y}'|$, and so $|T(\mathbf{Y})|^2 = |\mathbf{Y}|^2$. Thus T is an isometry.

Now choose orthogonal unit vectors $\mathbf{X}_1$, $\mathbf{X}_2$ in π such that $\mathbf{X}_1 \times \mathbf{X}_2 = \mathbf{F}$. Then the triplet $\mathbf{X}_1$, $\mathbf{X}_2$, $\mathbf{F}$ is positively oriented. The triplet of vectors $T(\mathbf{X}_1)$, $T(\mathbf{X}_2)$, $\mathbf{F}$ is also positively oriented, and so T preserves orientation. The proof is contained in Exercise 11.

Exercise 11.

(a) Express $T(\mathbf{X}_1)$ and $T(\mathbf{X}_2)$ as linear combinations of $\mathbf{X}_1$ and $\mathbf{X}_2$ with coefficients depending on θ.

(b) Compute $(T(\mathbf{X}_1) \times T(\mathbf{X}_2)) \cdot \mathbf{F}$ and show it equals $(\mathbf{X}_1 \times \mathbf{X}_2) \cdot \mathbf{F}$ and hence is positive. Thus T preserves orientation.

(c) Using Theorem 3.5 of Chapter 3.5, conclude that $\det T > 0$.

Since T is an isometry, $\det T = \pm 1$ and so, since $\det T > 0$, $\det T = 1$. In sum, we have proved:

Proposition 6. *If T is a rotation about an axis, then T is an isometry and* $\det T = 1$.

What about the converse of this statement? Suppose T is an isometry with $\det T = 1$. Let $\mathbf{F}$ be an eigenvector of T with $|\mathbf{F}| = 1$ and $T(\mathbf{F}) = t\mathbf{F}$. By Proposition 4, we know that $t = \pm 1$. We consider the two cases separately. Let us first suppose $t = 1$. Then

$$T(\mathbf{F}) = \mathbf{F}.$$

Let π be the plane orthogonal to $\mathbf{F}$ and passing through the origin. If $\mathbf{X}$ is a vector in π,

$$T(\mathbf{X}) \cdot \mathbf{F} = T(\mathbf{X}) \cdot T(\mathbf{F}) = \mathbf{X} \cdot \mathbf{F} = 0,$$

so $T(\mathbf{X})$ is orthogonal to $\mathbf{F}$. Hence $T(\mathbf{X})$ lies in π. Thus T transforms π into itself (see Fig. 3.19). Let us denote by T_π the resulting transformation of the plane π. T_π is evidently a linear transformation of π and an isometry of π since T has these properties on $\mathbb{R}^3$. In Chapter 2.5 we showed that an isometry of the plane is either a rotation or a reflection. Hence either T_π is a rotation of π through some angle θ or T_π is a reflection of π across a line in π through the origin.

Case 1: T_π is a rotation of π through an angle θ. By the discussion following after Proposition 5, we conclude that T is a rotation of $\mathbb{R}^3$ about the axis $\mathbf{F}$.

Case 2: T_π is a reflection across a line in π. In this case there exist nonzero vectors $\mathbf{X}_1$, $\mathbf{X}_2$ in π with $T_\pi(\mathbf{X}_1) = \mathbf{X}_1$, $T_\pi(\mathbf{X}_2) = -\mathbf{X}_2$, and we can choose

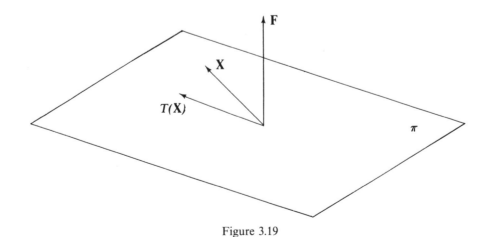

Figure 3.19

these vectors so that the triplet $\mathbf{X}_1$, $\mathbf{X}_2$, $\mathbf{F}$ is positively oriented. Also, $T(\mathbf{X}_1) = T_\pi(\mathbf{X}_1) = \mathbf{X}_1$ and $T(\mathbf{X}_2) = T_\pi(\mathbf{X}_2) = -\mathbf{X}_2$, so the triplet $T(\mathbf{X}_1)$, $T(\mathbf{X}_2)$, $T(\mathbf{F})$ is the triplet $\mathbf{X}_1$, $-\mathbf{X}_2$, $\mathbf{F}$ which is negatively oriented. But $\det T = 1$ by hypothesis, and so we have a contradiction. Thus Case 2 cannot occur. We conclude: If $t = 1$, then T is a rotation about an axis.

Now let us suppose that $t = -1$, so $T(\mathbf{F}) = -\mathbf{F}$. We again form the plane π orthogonal to $\mathbf{F}$ and the transformation T_π of π on itself. As before, T_π is either a rotation of π or reflection in a line of π.

Suppose T_π is a rotation of π by an angle θ. Choose orthogonal unit vectors $\mathbf{X}_1$, $\mathbf{X}_2$ in π such that $\mathbf{X}_1$, $\mathbf{X}_2$, $\mathbf{F}$ is a positively oriented triplet. Then the triplet $T_\pi(\mathbf{X}_1)$, $T_\pi(\mathbf{X}_2)$, $\mathbf{F}$ is again positively oriented.

Exercise 12. Prove this last statement by calculating $(T_\pi(\mathbf{X}_1) \times T_\pi(\mathbf{X}_2)) \cdot \mathbf{F}$ and showing that it is positive.

It follows that the triplet $T_\pi(\mathbf{X}_1)$, $T_\pi(\mathbf{X}_2)$, $-\mathbf{F}$ is negatively oriented. But this is exactly the triplet $T(\mathbf{X}_1)$, $T(\mathbf{X}_2)$, $T(\mathbf{F})$. Since $\mathbf{X}_1$, $\mathbf{X}_2$, $\mathbf{F}$ was a positively oriented triplet, this contradicts the fact that $\det T = 1$. Hence T_π is not a rotation of π so it must be a reflection of π. Therefore we can find orthogonal unit vectors $\mathbf{X}_1$, $\mathbf{X}_2$ in π with $T_\pi(\mathbf{X}_1) = -\mathbf{X}_1$, $T_\pi(\mathbf{X}_2) = \mathbf{X}_2$. Now consider the plane π' determined by the vectors $\mathbf{F}$ and $\mathbf{X}_1$ (see Fig. 3.20). We note that since $T(\mathbf{F}) = -\mathbf{F}$ and $T(\mathbf{X}_1) = T_\pi(\mathbf{X}_1) = -\mathbf{X}_1$, T coincides on the plane π' with minus the identity transformation. Thus T rotates the vectors of π' by $180°$ about the $\mathbf{X}_2$-axis. Also, $T(\mathbf{X}_2) = T_\pi(\mathbf{X}_2) = \mathbf{X}_2$. Hence T acts on $\mathbb{R}^3$ by rotation by $180°$ about the $\mathbf{X}_2$-axis.

In summary, we have proved:

Theorem 3.9. *Let T be an isometry of $\mathbb{R}^3$ and let $\det T = 1$. Then T is rotation about an axis.*

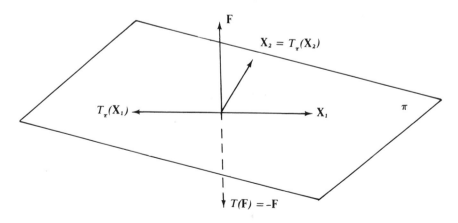

Figure 3.20

Let S and T be two isometries of $\mathbb{R}^3$. What can be said about their product ST?

Exercise 13.

(a) If S and T are isometries, then ST and TS are also isometries.
(b) If S is an isometry, then S^{-1} also is an isometry.

Exercise 14. If S is rotation about an axis and T is rotation about a possibly different axis, then ST is rotation about an axis.

Note: If the axes for S and for T are distinct, our conclusion that ST is again a rotation about some axis, though correct, is by no means evident.

Exercise 15. Let S be rotation by 90° about the x_3-axis and T be rotation by 90° about the x_1-axis. Find the axes for the rotations ST and TS.

Exercise 16. Let T be an isometry with det $T = -1$.

(a) Show that $-T$ is a rotation.
(b) Conclude that T is the result of first performing a rotation and then reflecting every vector about the origin.

§3. Orthogonal Matrices

In Chapter 2.5, we found that a 2×2 matrix m is the matrix of an isometry if and only if it has one of the following forms:

(i)
$$\begin{pmatrix} \cos\theta & -\sin\theta \\ \sin\theta & \cos\theta \end{pmatrix},$$

(ii)
$$\begin{pmatrix} \cos\theta & \sin\theta \\ \sin\theta & -\cos\theta \end{pmatrix}.$$

Note that in each case the columns are mutually orthogonal unit vectors $\mathbb{R}^2$. It turns out that the analogous statement is true in 3 dimensions. Let

$$m = \begin{bmatrix} a_{11} & a_{12} & a_{13} \\ a_{21} & a_{22} & a_{23} \\ a_{31} & a_{32} & a_{33} \end{bmatrix}.$$

Exercise 17. Assume m is the matrix of an isometry T.

(a) Show $T(\mathbf{E}_i) \cdot T(\mathbf{E}_j) = \begin{cases} 1 & \text{if } i = j, \\ 0 & \text{if } i \neq j, \end{cases}$ $1 \leq i, j \leq 3$.

(b) Show that the columns of m are orthogonal unit vectors in $\mathbb{R}^3$.

Exercise 18. Assume m is the matrix of an isometry. Show that the inverse m^{-1} equals the transpose m^*.

Exercise 19. Show that, conversely, if m is a matrix satisfying $m^{-1} = m^*$, then m is the matrix of an isometry.

A matrix m such that

$$m^{-1} = m^*$$ (6)

is called an *orthogonal* matrix. Exercises 18 and 19 together prove the following result:

Proposition 7. *A* 3×3 *matrix* m *is the matrix of an isometry if and only if* m *is an orthogonal matrix.*

Exercise 20. Let R_θ^3 be rotation around the x_3-axis by an angle θ. Let $m = m(R_\theta^3)$. Directly show that m satisfies (6).

Exercise 21. Let T be reflection in the plane $x + y + z = 0$. Let $m = m(T)$. Directly show that m satisfies (6).

Exercise 22. Let S be rotation by $180°$ around the line: $x = t$, $y = t$, $z = t$. Let $m = m(S)$. Directly show that m satisfies (6).

Exercise 23. Let m be an orthogonal matrix.

(a) Show that m^* is an orthogonal matrix.
(b) Show that the rows of m are orthogonal unit vectors in $\mathbb{R}^3$.

Exercise 24. Show that the product of two orthogonal matrices is an orthogonal matrix.

Exercise 25. Consider the system of equation

$$\begin{cases} a_{11}x_1 + a_{12}x_2 + a_{13}x_3 = y_1, \\ a_{21}x_1 + a_{22}x_2 + a_{23}x_3 = y_2, \\ a_{31}x_1 + a_{32}x_2 + a_{33}x_3 = y_3. \end{cases}$$ (7)

Assume the coefficient matrix

$$\begin{pmatrix} a_{11} & a_{12} & a_{13} \\ a_{21} & a_{22} & a_{23} \\ a_{31} & a_{32} & a_{33} \end{pmatrix}$$

is an orthogonal matrix. Show that, given y_1, y_2, y_3, the system (7) is solved by setting

$$\begin{cases} x_1 = a_{11}y_1 + a_{21}y_2 + a_{31}y_3, \\ x_2 = a_{12}y_1 + a_{22}y_2 + a_{32}y_3, \\ x_3 = a_{13}y_1 + a_{23}y_2 + a_{33}y_3. \end{cases}$$ (8)

Exercise 26. Let m be a 3×3 matrix. Assume that the column vectors of m are orthogonal unit vectors. Prove that m is an orthogonal matrix.

CHAPTER 3.7

Symmetric Matrices

In the 2-dimensional case, we saw that a special role is played by matrices $\begin{pmatrix} a & b \\ b & d \end{pmatrix}$ which have both off-diagonal elements equal. The corresponding condition in 3 dimensions is symmetry about the diagonal. We say that a matrix is *symmetric* if the entry in the ith position in the jth column is the same as the entry in the jth position in the ith column, i.e., $a_{ij} = a_{ji}$ for all i, j,

$$\begin{bmatrix} a & b & c \\ b & d & e \\ c & e & f \end{bmatrix} = \begin{bmatrix} a_{11} & a_{12} & a_{13} \\ a_{12} & a_{22} & a_{23} \\ a_{13} & a_{23} & a_{33} \end{bmatrix}. \tag{1}$$

Note that this condition does not place any restriction on the diagonal elements themselves, but as soon as we know the elements on the diagonal and above the diagonal, we can fill in the rest of the entries in a symmetric matrix:

If the following matrix is symmetric,

$$\begin{bmatrix} 1 & 2 & -1 \\ x & 7 & 0 \\ y & z & 3 \end{bmatrix}$$

then $x = 2$, $y = -1$, $z = 0$.

We may express the symmetry condition succinctly by using the notion of transpose. Recall that the *transpose* m^* of a matrix m is the matrix whose columns are the rows of m. Thus a matrix m is symmetric if and only if $m^* = m$.

Exercise 1. For which values of the letters x, y, z, will the following matrices be symmetric?

(a) $\begin{pmatrix} 1 & 2 & 1 \\ x & 2 & 1 \\ y & z & 4 \end{pmatrix}$,

(b) $\begin{pmatrix} 2 & 5 & z \\ x & 1 & y \\ 2 & 4 & 0 \end{pmatrix}$,

(c) $\begin{pmatrix} x & \cdot 5 & 2 \\ 5 & y & 1 \\ 2 & 1 & z \end{pmatrix}$,

(d) $\begin{pmatrix} 1 & x & 3 \\ y & 2 & 4 \\ 3 & z & 6 \end{pmatrix}$,

(e) $\begin{pmatrix} 1 & y & 4 \\ 3 & x & 5 \\ 4 & 6 & z \end{pmatrix}$.

Exercise 2. Show that every diagonal matrix is symmetric.

Exercise 3. Show that if A and B are symmetric, then $A + B$ is symmetric and cA is symmetric for any c.

Exercise 4. True or false? If a and b are symmetric, then ab is symmetric. (Show that $(ab)^*= b^* a^*$).

Exercise 5. Show that if a is any matrix, then the average $\frac{1}{2}(a + a^*)$ is a symmetric matrix.

Exercise 6. True of false? The square of a symmetric matrix is symmetric.

Exercise 7. Prove that for any 3×3 matrix m, the product $m(m^*)$ is symmetric. (Recall that $(mn)^* = n^* m^*$.)

We shall need a general formula involving the transpose of a matrix. Let a be any matrix.

$$a = \begin{bmatrix} a_{11} & a_{12} & a_{13} \\ a_{21} & a_{22} & a_{23} \\ a_{31} & a_{32} & a_{33} \end{bmatrix}.$$

Then

$$a^* = \begin{bmatrix} a_{11} & a_{21} & a_{31} \\ a_{12} & a_{22} & a_{32} \\ a_{13} & a_{23} & a_{33} \end{bmatrix}.$$

Lemma 1. *Let A be the linear transformation with matrix a and A^* the linear transformation with matrix a^*. Then for every pair of vectors $\mathbf{X}$, $\mathbf{Y}$,*

$$A(\mathbf{X}) \cdot \mathbf{Y} = \mathbf{X} \cdot A^*(\mathbf{Y}). \tag{2}$$

PROOF OF LEMMA 1. Set $\mathbf{X} = \begin{bmatrix} x_1 \\ x_2 \\ x_3 \end{bmatrix}$, $\mathbf{Y} = \begin{bmatrix} y_1 \\ y_2 \\ y_3 \end{bmatrix}$. Then

$$A(\mathbf{X}) \cdot \mathbf{Y} = \begin{bmatrix} a_{11}x_1 + a_{12}x_2 + a_{13}x_3 \\ a_{21}x_1 + a_{22}x_2 + a_{23}x_3 \\ a_{31}x_1 + a_{32}x_2 + a_{33}x_3 \end{bmatrix} \cdot \begin{bmatrix} y_1 \\ y_2 \\ y_3 \end{bmatrix}$$

$$= \begin{matrix} a_{11}x_1y_1 + a_{12}x_2y_1 + a_{13}x_3y_1 \\ + a_{21}x_1y_2 + a_{22}x_2y_2 + a_{23}x_3y_2 \\ + a_{31}x_1y_3 + a_{32}x_2y_3 + a_{33}x_3y_3 \end{matrix} . \tag{3}$$

$$\mathbf{X} \cdot A^*(\mathbf{Y}) = \begin{bmatrix} x_1 \\ x_2 \\ x_3 \end{bmatrix} \cdot \begin{bmatrix} a_{11} & a_{21} & a_{31} \\ a_{12} & a_{22} & a_{32} \\ a_{13} & a_{23} & a_{33} \end{bmatrix} \begin{bmatrix} y_1 \\ y_2 \\ y_3 \end{bmatrix}$$

$$= \begin{bmatrix} x_1 \\ x_2 \\ x_3 \end{bmatrix} \cdot \begin{bmatrix} a_{11}y_1 + a_{21}y_2 + a_{31}y_3 \\ a_{12}y_1 + a_{22}y_2 + a_{32}y_3 \\ a_{13}y_1 + a_{23}y_2 + a_{33}y_3 \end{bmatrix}$$

$$= \begin{matrix} a_{11}x_1y_1 + a_{21}x_1y_2 + a_{31}x_1y_3 \\ + a_{12}x_2y_1 + a_{22}x_2y_2 + a_{32}x_2y_3 \\ + a_{13}x_3y_1 + a_{23}x_3y_2 + a_{33}x_3y_3 \end{matrix} . \tag{4}$$

The first line of the sum (3) is the same as the first column of (4), and similarly for the other two lines. So the sums (3) and (4) consist of the same terms in different arrangements, and thus $A(\mathbf{X}) \cdot \mathbf{Y} = \mathbf{X} \cdot A^*(\mathbf{Y})$.

An immediate consequence of Lemma 1 is:

Lemma 2. *If m is a symmetric matrix, and M is the corresponding linear transformation, then for all vectors* $\mathbf{X}$, $\mathbf{Y}$,

$$M(\mathbf{X}) \cdot \mathbf{Y} = \mathbf{X} \cdot M(\mathbf{Y}). \tag{5}$$

We saw in Theorem 2.10 in Chapter 2.6, that eigenvectors of a symmetric 2×2 matrix corresponding to distinct eigenvalues are orthogonal. We now prove the analogous result in 3 dimensions.

Theorem 3.10. *Let m be a symmetric 3×3 matrix and let M be the corresponding linear transformation. Let t_1, t_2 be distinct eigenvalues of M, and let* $\mathbf{X}_1$, $\mathbf{X}_2$ *be corresponding eigenvectors. Then* $\mathbf{X}_1 \cdot \mathbf{X}_2 = 0$.

PROOF.

$$M(\mathbf{X}_1) \cdot \mathbf{X}_2 = (t_1\mathbf{X}_1) \cdot \mathbf{X}_2 = t_1(\mathbf{X}_1 \cdot \mathbf{X}_2),$$

$$\mathbf{X}_1 \cdot M(\mathbf{X}_2) = \mathbf{X}_1 \cdot (t_2\mathbf{X}_2) = t_2(\mathbf{X}_1 \cdot \mathbf{X}_2).$$

Then by (5),

$$t_1(\mathbf{X}_1 \cdot \mathbf{X}_2) = t_2(\mathbf{X}_1 \cdot \mathbf{X}_2).$$

If $\mathbf{X}_1 \cdot \mathbf{X}_2 \neq 0$, we can divide by $\mathbf{X}_1 \cdot \mathbf{X}_2$ and get $t_1 = t_2$, which contradicts our assumption. Therefore $\mathbf{X}_1 \cdot \mathbf{X}_2 = 0$.

EXAMPLE 1. Let us calculate the eigenvalues and eigenvectors of the linear transformation M with matrix

$$m = \begin{pmatrix} 1 & 1 & -1 \\ 1 & 0 & 0 \\ -1 & 0 & 0 \end{pmatrix}.$$

The characteristic equation of m is

$$\begin{vmatrix} 1-t & 1 & -1 \\ 1 & -t & 0 \\ -1 & 0 & -t \end{vmatrix} = (1-t)t^2 - 1(-t) - 1(-t) = 0$$

or

$$-t^3 + t^2 + 2t = 0,$$

i.e.,

$$-t(t^2 - t - 2) = 0.$$

Since $t^2 - t - 2 = (t-2)(t+1)$, the eigenvalues of m are

$$t_1 = 0, \qquad t_2 = 2, \qquad t_3 = -1.$$

Let $\mathbf{X}_i$ denote an eigenvector corresponding to t_i, for $i = 1, 2, 3$.

$$M(\mathbf{X}_1) = 0\mathbf{X}_1 = \mathbf{0},$$

so setting

$$\mathbf{X}_1 = \begin{pmatrix} x \\ y \\ z \end{pmatrix}, \qquad \begin{pmatrix} 1 & 1 & -1 \\ 1 & 0 & 0 \\ -1 & 0 & 0 \end{pmatrix} \begin{pmatrix} x \\ y \\ z \end{pmatrix} = \begin{pmatrix} 0 \\ 0 \\ 0 \end{pmatrix}$$

or

$$\begin{aligned} x + y - z &= 0, \\ x &= 0, \\ -x &= 0. \end{aligned}$$

Therefore $x = 0$, $y = z$, so $\mathbf{X}_1 = \begin{pmatrix} 0 \\ 1 \\ 1 \end{pmatrix}$ is an eigenvector corresponding to t_1.

Similarly, if

$$\mathbf{X}_2 = \begin{pmatrix} x \\ y \\ z \end{pmatrix}, \qquad \begin{pmatrix} 1 & 1 & -1 \\ 1 & 0 & 0 \\ -1 & 0 & 0 \end{pmatrix} \begin{pmatrix} x \\ y \\ z \end{pmatrix} = 2 \begin{pmatrix} x \\ y \\ z \end{pmatrix},$$

so

$$\begin{aligned} x + y - z &= 2x, \\ x &= 2y, \\ -x &= 2z. \end{aligned}$$

Setting $x = 2$, we must take $y = 1$ and $z = -1$. Then $x + y - z = 2 + 1 + 1$

$= 4 = 2x$, so the first equation is also satisfied. Thus $\mathbf{X}_2 = \begin{pmatrix} 2 \\ 1 \\ -1 \end{pmatrix}$. Similarly

we find $\mathbf{X}_3 = \begin{pmatrix} 1 \\ -1 \\ 1 \end{pmatrix}$ is an eigenvector corresponding to $t_3 = -1$. Each pair

of two out of our three eigenvectors

$$\begin{pmatrix} 0 \\ 1 \\ 1 \end{pmatrix}, \quad \begin{pmatrix} 2 \\ 1 \\ -1 \end{pmatrix}, \quad \begin{pmatrix} 1 \\ -1 \\ 1 \end{pmatrix}$$

is indeed orthogonal, as stated in Theorem 3.10.

Note: The eigenvectors corresponding to a given eigenvalue fill a line in this case. For instance, the set of eigenvectors of m corresponding to $t_2 = 2$ is the set of all vectors

$$t\mathbf{X}_2 = t\begin{pmatrix} 2 \\ 1 \\ -1 \end{pmatrix} = \begin{pmatrix} 2t \\ t \\ -t \end{pmatrix}, \quad -\infty < t < \infty.$$

Exercise 8. Find the eigenvalues of the matrix $m = \begin{pmatrix} 3 & 4 & 0 \\ 4 & 3 & 0 \\ 0 & 0 & 1 \end{pmatrix}$ and find all corresponding eigenvectors.

Every 3×3 matrix has at least one eigenvalue, as we showed in Proposition 1 of Chapter 3.6. However, in general, we cannot say more.

Exercise 9. Give an example of a 3×3 matrix having as its only eigenvalue the number 1 and such that the corresponding eigenvectors make up a line.

For symmetric matrices, the situation is much better, as the following fundamental theorem shows:

Spectral Theorem in $\mathbb{R}^3$. *Let m be a symmetric 3×3 matrix and let M be the corresponding linear transformation. Then we can find three orthogonal unit vectors $\mathbf{X}_1$, $\mathbf{X}_2$, and $\mathbf{X}_3$ such that each $\mathbf{X}_i$ is an eigenvector of M.*

PROOF. M has at least one eigenvalue t_1, as we know. Let $\mathbf{X}_1$ be a corresponding eigenvector of length 1. Denote by π the plane through the origin which is perpendicular to $\mathbf{X}_1$. We wish to show that we can find two further mutually perpendicular eigenvectors of M lying in π. We claim that π is *invariant* under M, i.e., that if a vector $\mathbf{X}$ is in π, then $M(\mathbf{X})$ also lies in π (see Fig. 3.21). Suppose $\mathbf{X}$ belongs to π. Then by (5),

$$M(\mathbf{X}) \cdot \mathbf{X}_1 = \mathbf{X} \cdot M(\mathbf{X}_1) = \mathbf{X} \cdot t_1\mathbf{X}_1 = t_1(\mathbf{X} \cdot \mathbf{X}_1).$$

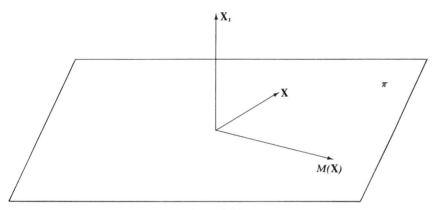

Figure 3.21

But $\mathbf{X} \cdot \mathbf{X}_1 = 0$, since $\mathbf{X}$ lies in π. Hence $M(\mathbf{X}) \cdot \mathbf{X}_1 = 0$, so $M(\mathbf{X})$ is in π, and our claim is proved.

We denote by A the transformation of π defined by

$$A(\mathbf{X}) = M(\mathbf{X}) \qquad \text{for } \mathbf{X} \text{ in } \pi.$$

We shall now go on to show that π may be identified with the plane $\mathbb{R}^2$. Also, when we make this identification, A turns into a linear transformation of $\mathbb{R}^2$ having a symmetric matrix. Using the results we found in Chapter 2.6, we shall then find two eigenvectors for this 2×2 symmetric matrix and these will turn out to give the "missing" eigenvectors in $\mathbb{R}^3$ for our original transformation M.

Let $\mathbf{F}_1$, $\mathbf{F}_2$ be vectors in π which are orthogonal and have length 1. If $\mathbf{X}$ and $\mathbf{Y}$ are vectors in π,

$$A(\mathbf{X}) \cdot \mathbf{Y} = M(\mathbf{X}) \cdot \mathbf{Y} = \mathbf{X} \cdot M(\mathbf{Y}) = \mathbf{X} \cdot A(\mathbf{Y}),$$

so A satisfies

$$A(\mathbf{X}) \cdot \mathbf{Y} = \mathbf{X} \cdot A(\mathbf{Y}) \tag{6}$$

whenever $\mathbf{X}$, $\mathbf{Y}$ lie in π. Each vector $\mathbf{X}$ in π can be expressed as

$$\mathbf{X} = x_1 \mathbf{F}_1 + x_2 \mathbf{F}_2,$$

where $x_1 = \mathbf{X} \cdot \mathbf{F}_1$, $x_2 = \mathbf{X} \cdot \mathbf{F}_2$. We identify $\mathbf{X}$ with the vector $\begin{pmatrix} x_1 \\ x_2 \end{pmatrix}$ in $\mathbb{R}^2$, and in this way π becomes identified with $\mathbb{R}^2$. Also, since A takes π into itself, A gives rise to a linear transformation A^0 of $\mathbb{R}^2$. For each $\mathbf{X} = x_1 \mathbf{F}_1 + x_2 \mathbf{F}_2$ in π, $A(\mathbf{X})$ is identified with $A^0 \begin{pmatrix} x_1 \\ x_2 \end{pmatrix}$ in $\mathbb{R}^2$ (see Fig. 3.22). What is the matrix of A^0? Since $A(\mathbf{F}_1)$ and $A(\mathbf{F}_2)$ lie in π, we have

$$A(\mathbf{F}_1) = a\mathbf{F}_1 + b\mathbf{F}_2,$$

$$A(\mathbf{F}_2) = c\mathbf{F}_1 + d\mathbf{F}_2,$$

where $a = A(\mathbf{F}_1) \cdot \mathbf{F}_1$, $b = A(\mathbf{F}_1) \cdot \mathbf{F}_2$, $c = A(\mathbf{F}_2) \cdot \mathbf{F}_1$, and $d = A(\mathbf{F}_2) \cdot \mathbf{F}_2$. By

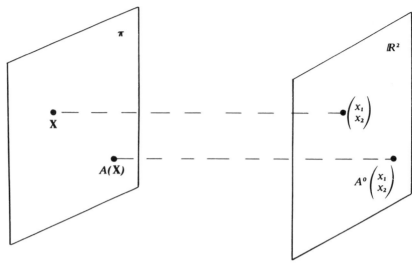

Figure 3.22

(6), we get

$$b = A(\mathbf{F}_1) \cdot \mathbf{F}_2 = \mathbf{F}_1 \cdot A(\mathbf{F}_2) = c. \tag{7}$$

Choose $\mathbf{X} = x_1\mathbf{F}_1 + x_2\mathbf{F}_2$ and set

$$A(\mathbf{X}) = x_1^1\mathbf{F}_1 + x_2^1\mathbf{F}_2.$$

Then $A^0\begin{pmatrix} x_1 \\ x_2 \end{pmatrix} = \begin{pmatrix} x_1^1 \\ x_2^1 \end{pmatrix}$. Also,

$$A(\mathbf{X}) = x_1 A(\mathbf{F}_1) + x_2 A(\mathbf{F}_2) = x_1(a\mathbf{F}_1 + b\mathbf{F}_2) + x_2(c\mathbf{F}_1 + d\mathbf{F}_2)$$
$$= (ax_1 + cx_2)\mathbf{F}_1 + (bx_1 + dx_2)\mathbf{F}_2.$$

Hence,

$$A^0\begin{pmatrix} x_1 \\ x_2 \end{pmatrix} = \begin{pmatrix} x_1^1 \\ x_2^1 \end{pmatrix} = \begin{pmatrix} ax_1 + cx_2 \\ bx_1 + dx_2 \end{pmatrix} = \begin{pmatrix} a & c \\ b & d \end{pmatrix}\begin{pmatrix} x_1 \\ x_2 \end{pmatrix}.$$

Thus the matrix of A^0 is $\begin{pmatrix} a & c \\ b & d \end{pmatrix} = \begin{pmatrix} a & b \\ b & d \end{pmatrix}$, because of (7). We note that A^0 has a symmetric matrix. By Theorem 2.10, Chapter 2.6, there exist two orthogonal nonzero eigenvectors $\begin{pmatrix} u_1 \\ u_2 \end{pmatrix}$ and $\begin{pmatrix} v_1 \\ v_2 \end{pmatrix}$ for A^0. Then, for a certain scalar t, $A^0\begin{pmatrix} u_1 \\ u_2 \end{pmatrix} = t\begin{pmatrix} u_1 \\ u_2 \end{pmatrix}$.

The vector $\mathbf{U} = u_1\mathbf{F}_1 + u_2\mathbf{F}_2$ in π is identified with $\begin{pmatrix} u_1 \\ u_2 \end{pmatrix}$ and $A(\mathbf{U})$ is identified with $A^0\begin{pmatrix} u_1 \\ u_2 \end{pmatrix} = t\begin{pmatrix} u_1 \\ u_2 \end{pmatrix}$ in $\mathbb{R}^2$. Also, $t\mathbf{U}$ is identified with $t\begin{pmatrix} u_1 \\ u_2 \end{pmatrix}$. So $M(\mathbf{U}) = A(\mathbf{U}) = t\mathbf{U}$, and so $\mathbf{U}$ is an eigenvector of M lying in π. Similarly,

$V = v_1F_1 + v_2F_2$ is an eigenvector of M lying in π. Finally,

$$U \cdot V = (u_1F_1 + u_2F_2) \cdot (v_1F_1 + v_2F_2) = u_1v_1 + u_2v_2 = \begin{pmatrix} u_1 \\ u_2 \end{pmatrix} \cdot \begin{pmatrix} v_1 \\ v_2 \end{pmatrix} = 0.$$

It follows that the three vectors

$$X_1, \quad \frac{U}{|U|}, \quad \frac{V}{|V|}$$

are an orthonormal set in $\mathbb{R}^3$ consisting of eigenvectors of M.

Note: Although we have just shown that the transformation M has three mutually orthogonal eigenvectors X_1, X_2, X_3, we have not shown that the corresponding eigenvalues are distinct. Indeed, this need not be the case. If M is the linear transformation with matrix $\begin{bmatrix} 1 & 0 & 0 \\ 0 & 1 & 0 \\ 0 & 0 & 2 \end{bmatrix}$, then M has only two distinct eigenvalues, 1 and 2, although it has three orthogonal eigenvectors E_1, E_2, E_3. Here, E_1 and E_2 both correspond to the eigenvalue 1, while E_3 corresponds to the eigenvalue 2.

EXAMPLE 2. Let us find all eigenvalues and eigenvectors for the linear transformation M with matrix

$$m = \begin{bmatrix} 1 & 0 & 5 \\ 0 & 1 & 3 \\ 5 & 3 & 1 \end{bmatrix}.$$

The characteristic equation is

$$\begin{vmatrix} 1-t & 0 & 5 \\ 0 & 1-t & 3 \\ 5 & 3 & 1-t \end{vmatrix} = (1-t)\left[(1-t)^2 - 9\right] + 5(-5(1-t)) = 0$$

or $(1-t)[(1-t)^2 - 9 - 25] = 0$. So the eigenvalues are $t_1 = 1$ and the roots of $(1-t)^2 - 34 = 0$, which are $t_2 = 1 + \sqrt{34}$ and $t_3 = 1 - \sqrt{34}$. We seek an eigenvector $X_1 = \begin{pmatrix} x \\ y \\ z \end{pmatrix}$ corresponding to $t = 1$. So we must solve

$$M\begin{pmatrix} x \\ y \\ z \end{pmatrix} = \begin{pmatrix} x \\ y \\ z \end{pmatrix}, \text{ i.e., } \begin{bmatrix} 1 & 0 & 5 \\ 0 & 1 & 3 \\ 5 & 3 & 1 \end{bmatrix}\begin{pmatrix} x \\ y \\ z \end{pmatrix} = \begin{pmatrix} x \\ y \\ z \end{pmatrix} \text{ or}$$

$$\begin{aligned} x + 5z &= x, \\ y + 3z &= y, \\ 5x + 3y + z &= z. \end{aligned}$$

The first two equations give $z = 0$. Then the third gives $5x + 3y = 0$. Hence, $x = -3$, $y = 5$, $z = 0$ solves all three equations. So we take

$$X_1 = \begin{bmatrix} -3 \\ 5 \\ 0 \end{bmatrix}.$$

The plane π through the origin and orthogonal to $\mathbf{X}_1$, which occurred in the proof of the Spectral Theorem, here has the equation

$$\begin{pmatrix} x \\ y \\ z \end{pmatrix} \cdot \begin{bmatrix} -3 \\ 5 \\ 0 \end{bmatrix} = -3x + 5y = 0 \quad \text{or} \quad y = \tfrac{3}{5}x.$$

By the proof of the Spectral Theorem, we can find in π a second eigenvector $\mathbf{X}_2$ of M, corresponding to $t_2 = 1 + \sqrt{34}$. $\mathbf{X}_2 = \begin{pmatrix} x \\ y \\ z \end{pmatrix}$ satisfies

$$y = \tfrac{3}{5}x, \tag{8}$$

since $\mathbf{X}_2$ is on π. Also,

$$M(\mathbf{X}_2) = (1 + \sqrt{34})\mathbf{X}_2 \quad \text{or} \quad \begin{bmatrix} 1 & 0 & 5 \\ 0 & 1 & 3 \\ 5 & 3 & 1 \end{bmatrix} \begin{bmatrix} x \\ y \\ z \end{bmatrix} = (1 + \sqrt{34}) \begin{bmatrix} x \\ y \\ z \end{bmatrix},$$

so

$$x + 5z = (1 + \sqrt{34})x, \tag{9}$$

as well as two further equations. However, Eqs. (8) and (9) suffice to give

$$y = \frac{3}{5}x, \qquad z = \frac{\sqrt{34}}{5}x,$$

so

$$\mathbf{X}_2 = \begin{bmatrix} x \\ y \\ z \end{bmatrix} = \begin{bmatrix} x \\ (3/5)x \\ (\sqrt{34}/5)x \end{bmatrix} = x \begin{bmatrix} 1 \\ 3/5 \\ \sqrt{34}/5 \end{bmatrix}.$$

In particular, taking $x = 5$, we find

$$\mathbf{X}_2 = \begin{bmatrix} 5 \\ 3 \\ \sqrt{34} \end{bmatrix}.$$

Each scalar multiple $t\mathbf{X}_2$ is an eigenvector of M corresponding to $t_2 = 1 + \sqrt{34}$. An eigenvector $\mathbf{X}_3$ corresponding to $t_3 = 1 - \sqrt{34}$ will be orthogonal to both $\mathbf{X}_1$ and $\mathbf{X}_2$, as we know by the Spectral Theorem. Hence, $\mathbf{X}_3$ is a scalar multiple of $\mathbf{X}_1 \times \mathbf{X}_2$.

$$\mathbf{X}_1 \times \mathbf{X}_2 = \begin{bmatrix} -3 \\ 5 \\ 0 \end{bmatrix} \times \begin{bmatrix} 5 \\ 3 \\ \sqrt{34} \end{bmatrix} = \begin{bmatrix} 5\sqrt{34} \\ 3\sqrt{34} \\ -34 \end{bmatrix} = \sqrt{34} \begin{bmatrix} 5 \\ 3 \\ -\sqrt{34} \end{bmatrix}.$$

So we can take $\mathbf{X}_3 = \begin{bmatrix} 5 \\ 3 \\ -\sqrt{34} \end{bmatrix}$ as eigenvector for $t_3 = 1 - \sqrt{34}$.

Note: The eigenvectors X_1, X_2, X_3 do not have unit length. However, the vectors $X_1/|X_1|$, etc., are also eigenvectors and do have unit length.

Exercise 10. For each of the following matrices, find an orthonormal set X_1, X_2, X_3 in $\mathbb{R}^3$ consisting of eigenvectors of that linear transformation with matrix m.

(a) $m = \begin{pmatrix} -4 & 0 & 0 \\ 0 & 2 & 3 \\ 0 & 3 & 2 \end{pmatrix}$,

(b) $m = \begin{pmatrix} 1 & 1 & 1 \\ 1 & 1 & 1 \\ 1 & 1 & 1 \end{pmatrix}$,

(c) $m = \begin{bmatrix} \sqrt{2} & 0 & 0 \\ 0 & \sqrt{3} & 0 \\ 0 & 0 & \sqrt{5} \end{bmatrix}$.

In Chapter 2.7, Theorem 2.11, we showed that if a 2×2 matrix m has two linearly independent eigenvectors corresponding to the eigenvalues t_1, t_2, then

$$m = pdp^{-1},$$

where d is the diagonal matrix $\begin{pmatrix} t_1 & 0 \\ 0 & t_2 \end{pmatrix}$ and p is a certain invertible matrix. We shall now prove the corresponding fact in $\mathbb{R}^3$.

Theorem 3.11. *Let M be a linear transformation having linearly independent eigenvectors X_1, X_2, X_3 corresponding to eigenvalues t_1, t_2, t_3. Let m be the matrix of M. Denote by p the matrix $(X_1|X_2|X_3)$ whose columns are the vectors X_i. Then p is invertible and, setting $d = \begin{bmatrix} t_1 & 0 & 0 \\ 0 & t_2 & 0 \\ 0 & 0 & t_3 \end{bmatrix}$,*

$$m = pdp^{-1}. \tag{10}$$

PROOF. Since X_1, X_2, X_3 are linearly independent by hypothesis, the matrix p has an inverse, by (ix), p. 166. We set $E_1 = \begin{bmatrix} 1 \\ 0 \\ 0 \end{bmatrix}$, $E_2 = \begin{bmatrix} 0 \\ 1 \\ 0 \end{bmatrix}$, $E_3 = \begin{bmatrix} 0 \\ 0 \\ 1 \end{bmatrix}$. Let P be the linear transformation with matrix p. Then

$$P(E_1) = X_1, \quad \text{so} \quad MP(E_1) = M(X_1) = t_1 X_1.$$

Similarly, $MP(E_2) = t_2 X_2$ and $MP(E_3) = t_3 X_3$. Also, if D is the linear transformation with matrix d,

$$D(\mathbf{E}_1) = t_1\mathbf{E}_1, \qquad \text{so} \quad PD(\mathbf{E}_1) = t_1\mathbf{X}_1.$$

Similarly, $PD(\mathbf{E}_2) = t_2\mathbf{X}_2$ and $PD(\mathbf{E}_3) = t_3\mathbf{X}_3$. Thus the transformations MP and PD give the same results when acting on $\mathbf{E}_1$, $\mathbf{E}_2$, $\mathbf{E}_3$, and so $MP = PD$. Hence, $M = PDP^{-1}$, and $m = pdp^{-1}$.

Corollary 1. *Let m be a symmetric 3×3 matrix and let M be the corresponding linear transformation. Let t_1, t_2, t_3 denote the eigenvalues of M, and set*
$$d = \begin{bmatrix} t_1 & 0 & 0 \\ 0 & t_2 & 0 \\ 0 & 0 & t_3 \end{bmatrix}. \textit{ Then there exists an orthogonal matrix } r \textit{ of } \mathbb{R}^3 \textit{ such that}$$

$$m = rdr^{-1}. \tag{11}$$

PROOF. By the Spectral Theorem, M has eigenvectors $\mathbf{X}_1$, $\mathbf{X}_2$, $\mathbf{X}_3$ with eigenvalues t_1, t_2, t_3 such that $\mathbf{X}_1$, $\mathbf{X}_2$, $\mathbf{X}_3$ form an orthonormal set in $\mathbb{R}^3$. In particular, $\mathbf{X}_1$, $\mathbf{X}_2$, $\mathbf{X}_3$ are linearly independent, and so we can make use of Theorem 3.11. We set $r = (\mathbf{X}_1 | \mathbf{X}_2 | \mathbf{X}_3)$. By Theorem 3.11, $m = rdr^{-1}$. The only thing left to prove is that r is an orthogonal matrix. But the columns of r are orthonormal vectors, so by Exercise 26 of Chapter 3.6, r is an orthogonal matrix.

EXAMPLE 3. In Example 2, we studied the matrix $m = \begin{bmatrix} 1 & 0 & 5 \\ 0 & 1 & 3 \\ 5 & 3 & 1 \end{bmatrix}$ and found the eigenvalues $t_1 = 1$, $t_2 = 1 + \sqrt{34}$, $t_3 = 1 - \sqrt{34}$ and corresponding (normalized) eigenvectors

$$\mathbf{X}_1 = \frac{1}{\sqrt{34}}\begin{bmatrix} -3 \\ 5 \\ 0 \end{bmatrix}, \qquad \mathbf{X}_2 = \frac{1}{\sqrt{68}}\begin{bmatrix} 5 \\ 3 \\ \sqrt{34} \end{bmatrix}, \qquad \mathbf{X}_3 = \frac{1}{\sqrt{68}}\begin{bmatrix} 5 \\ 3 \\ -\sqrt{34} \end{bmatrix}.$$

Here $r = (\mathbf{X}_1 | \mathbf{X}_2 | \mathbf{X}_3) = \begin{bmatrix} -3/\sqrt{34} & 5/\sqrt{68} & 5/\sqrt{68} \\ 5/\sqrt{34} & 3/\sqrt{68} & 3/\sqrt{68} \\ 0 & 1/\sqrt{2} & -1/\sqrt{2} \end{bmatrix}$. Formula (11) gives

$$\begin{bmatrix} 1 & 0 & 5 \\ 0 & 1 & 3 \\ 5 & 3 & 1 \end{bmatrix} = r\begin{bmatrix} 1 & 0 & 0 \\ 0 & 1+\sqrt{34} & 0 \\ 0 & 0 & 1-\sqrt{34} \end{bmatrix}r^{-1}.$$

We note a useful consequence of formula (11).

Corollary 2. *Let m be a symmetric* 3×3 *matrix with eigenvalues* t_1, t_2, t_3. *Let r be the matrix occurring in* (11). *Then for each positive integer k,*

$$(m)^k = r \begin{bmatrix} t_1^k & 0 & 0 \\ 0 & t_2^k & 0 \\ 0 & 0 & t_3^k \end{bmatrix} r^{-1}. \tag{12}$$

Exercise 11. Use (12) to find m^5, where $\begin{pmatrix} 1 & 0 & 5 \\ 0 & 1 & 3 \\ 5 & 3 & 1 \end{pmatrix}$ is the matrix of Example 3.

Exercise 12. For each of the matrices m in Exercise 10, obtain the form (11).

Classification of Quadric Surfaces

A *quadric surface* is the 3-dimensional generalization of a conic section. Such a surface is determined by an equation in the variables x, y, z so that each term is of second degree; for example,

$$x^2 + 2xy + 3z^2 = 1.$$

The general form of the equation of a quadric surface is

$$ax^2 + 2bxy + 2cxz + dy^2 + 2eyz + fz^2 = 1, \tag{1}$$

where the coefficients $a, b, c, d, e,$ and f are constants. We would like to predict the shape of the quadric surface in terms of the coefficients, much in the same way that we described a conic section in terms of the coefficients of an equation

$$ax^2 + 2bxy + cy^2 = 1$$

in two variables.

As in the 2-dimensional case, we may use the inner product and a symmetric matrix in order to describe the quadric surface. We may then use our analysis of symmetric matrices in order to get a classification of the associated quadric surfaces.

We denote by A the linear transformation with matrix

$$m = \begin{bmatrix} a & b & c \\ b & d & e \\ c & e & f \end{bmatrix} \quad \text{and} \quad \mathbf{X} = \begin{bmatrix} x \\ y \\ z \end{bmatrix}$$

so that

$$A(\mathbf{X}) = \begin{bmatrix} ax + by + cz \\ bx + dy + ez \\ cx + ey + fz \end{bmatrix}$$

and
$$\mathbf{X} \cdot A(\mathbf{X}) = ax^2 + byx + czx + bxy + dy^2 + ezy + cxz + eyz + fz^2$$
$$= ax^2 + 2bxy + 2cxz + dy^2 + 2eyz + fz^2.$$

We can therefore express relation (1) as
$$\mathbf{X} \cdot A(\mathbf{X}) = 1.$$

Observe that m is a symmetric matrix. Consider some examples: If m is a diagonal matrix so that

$$m = \begin{bmatrix} a & 0 & 0 \\ 0 & d & 0 \\ 0 & 0 & f \end{bmatrix},$$

then the equation has the form

$$ax^2 + dy^2 + fz^2 = 1.$$

If $a = d = f = (1/r)^2$ for some $r > 0$, then the equation becomes

$$x^2 + y^2 + z^2 = r^2,$$

and this is a *sphere* of radius r.

If a, d, and f are all positive, then we may write

$$a = \left(\frac{1}{\alpha}\right)^2 = d = \left(\frac{1}{\beta}\right)^2, \quad \text{and} \quad f = \left(\frac{1}{\gamma}\right)^2,$$

and we have the equation

$$\frac{x^2}{\alpha^2} + \frac{y^2}{\beta^2} + \frac{z^2}{\gamma^2} = 1.$$

This gives an *ellipsoid* with axes along the coordinate axes of $\mathbb{R}^3$.

If a and d are positive and f is negative, then

$$a = \left(\frac{1}{\alpha}\right)^2, \quad d = \left(\frac{1}{\beta}\right)^2,$$

and $f = -1/\gamma^2$ for some a, β, γ; so the equation becomes

$$\frac{x^2}{\alpha^2} + \frac{y^2}{\beta^2} - \frac{z^2}{\gamma^2} = 1.$$

This is a *hyperboloid of one sheet*.

If $a > 0$, but $d < 0$, $f < 0$, then

$$a = \left(\frac{1}{\alpha}\right)^2, \quad d = -\left(\frac{1}{\beta^2}\right), \quad f = -\left(\frac{1}{\gamma^2}\right) \qquad \text{for some} \quad a, \beta, \gamma,$$

so the equation becomes

$$\frac{x^2}{\alpha^2} - \frac{y^2}{\beta^2} - \frac{z^2}{\gamma^2} = 1.$$

This is a *hyperboloid of two sheets*.

If $a < 0$, $d < 0$, $f < 0$, there are no solutions, since the sum of three negative numbers can never be 1.

What if one or more of the diagonal entries are zero? If $f = 0$, we have $ax^2 + dy^2 = 1$, and this is either an *elliptical cylinder* (if $a > 0$, $d > 0$), a *hyperbolic cylinder* (if $a > 0$, $d < 0$), or no locus at all if $a < 0$, $d < 0$.

If $f = 0$ and $d = 0$, and $a > 0$, then we have $ax^2 = 1$, and this is a *pair of planes* $x = \pm 1/\sqrt{a}$.

If $a = 0 = d = f$, we have no locus.

This completes the classification of quadric surfaces corresponding to diagonal matrices.

What if the matrix m is not diagonal, or, in other words, if one of the cross-terms in Eq. (1), $2bxy$, $2cxz$, $2eyz$, is nonzero?

In this case, we shall introduce new coordinates $\begin{pmatrix} u \\ v \\ w \end{pmatrix}$ for each vector $\begin{pmatrix} x \\ y \\ z \end{pmatrix}$ in such a way that, expressed in terms of u, v, w, Eq. (1) takes on a simpler form. Recall that by formula (11) of Section 3.7, there exists an orthogonal matrix r such that

$$m = rdr^{-1},$$

where $d = \begin{bmatrix} t_1 & 0 & 0 \\ 0 & t_2 & 0 \\ 0 & 0 & t_3 \end{bmatrix}$ is the diagonal matrix formed with the eigenvalues t_1, t_2, t_3 of A. In other words, we have

$$A = RDR^{-1}, \tag{2}$$

where R and D are the linear transformations whose matrices are r and d. Note that since r is an orthogonal matrix, R is an isometry.

If $\mathbf{X} = \begin{pmatrix} x \\ y \\ z \end{pmatrix}$ is any vector, let $\mathbf{U} = \begin{pmatrix} u \\ v \\ w \end{pmatrix}$ be the vector defined by

$$\mathbf{U} = R^{-1}(\mathbf{X}). \tag{3a}$$

We regard u, v, w as new coordinates of $\mathbf{X}$. Then

$$\mathbf{X} = R(\mathbf{U}). \tag{3b}$$

By (2), $A(\mathbf{X}) = (RDR^{-1})(\mathbf{X}) = RD(\mathbf{U})$, so $\mathbf{X} \cdot A(\mathbf{X}) = R(\mathbf{U}) \cdot R(D(\mathbf{U}))$. Since R is an isometry, the right-hand side equals $\mathbf{U} \cdot D(\mathbf{U}) = \begin{pmatrix} u \\ v \\ w \end{pmatrix} \cdot \begin{bmatrix} t_1 & 0 & 0 \\ 0 & t_2 & 0 \\ 0 & 0 & t_3 \end{bmatrix} \begin{pmatrix} u \\ v \\ w \end{pmatrix} = t_1 u^2 + t_2 v^2 + t_3 w^2$. Expressing $\mathbf{X} \cdot A(\mathbf{X})$ in terms of x, y, z, we get:

Theorem 3.12.

$$ax^2 + 2bxy + 2cxz + dy^2 + 2eyz + fz^2 = t_1u^2 + t_2v^2 + t_3w^2. \qquad (4)$$

The quadric surface defined by (1) thus has, as its equation in u, v, w,

$$t_1u^2 + t_2v^2 + t_3w^2 = 1. \qquad (5)$$

Note: The new coordinates $\begin{pmatrix} u \\ v \\ w \end{pmatrix}$ of a vector $\mathbf{X} = \begin{pmatrix} x \\ y \\ z \end{pmatrix}$ are actually the coordinates of $\mathbf{X}$ relative to a system of orthogonal coordinate axes. Since R is an isometry, the vectors $R(\mathbf{E_1})$, $R(\mathbf{E_2})$, $R(\mathbf{E_3})$ are three mutually orthogonal unit vectors in $\mathbb{R}^3$.

$$\mathbf{U} = \begin{pmatrix} u \\ v \\ w \end{pmatrix} = u\mathbf{E_1} + v\mathbf{E_2} + w\mathbf{E_3},$$

so

$$\mathbf{X} = R(\mathbf{U}) = uR(\mathbf{E_1}) + vR(\mathbf{E_2}) + wR(\mathbf{E_3}).$$

Thus $\begin{pmatrix} u \\ v \\ w \end{pmatrix}$ are the coordinates of $\mathbf{X}$ in the system whose coordinate axes lie along the vectors $R(\mathbf{E_1})$, $R(\mathbf{E_2})$, $R(\mathbf{E_3})$.

EXAMPLE 1. We wish to classify the quadric surface

$$\Sigma: x^2 + 2xy - 2xz = 1.$$

The corresponding symmetric matrix m here is $\begin{bmatrix} 1 & 1 & -1 \\ 1 & 0 & 0 \\ -1 & 0 & 0 \end{bmatrix}$. By Example 1 of Chapter 3.7, the eigenvalues of m are $t_1 = 0$, $t_2 = 2$, $t_3 = -1$. We introduce new coordinates u, v, w as described above. By (5), we find that an equation for Σ in the new coordinates is

$$2v^2 - w^2 = 1. \qquad (6)$$

Hence Σ is a hyperbolic cylinder.

Question: How do we express the new coordinates $\begin{pmatrix} u \\ v \\ w \end{pmatrix}$ of a point $\mathbf{X} = \begin{pmatrix} x \\ y \\ z \end{pmatrix}$ in terms of the original coordinates here? We found in Example 1, Chapter 3.7, that the normalized eigenvectors of the matrix $m = \begin{bmatrix} 1 & 1 & -1 \\ 1 & 0 & 0 \\ -1 & 0 & 0 \end{bmatrix}$ are

$$\mathbf{X_1} = \frac{1}{\sqrt{2}}\begin{bmatrix} 0 \\ 1 \\ 1 \end{bmatrix}, \quad \mathbf{X_2} = \frac{1}{\sqrt{6}}\begin{bmatrix} 2 \\ 1 \\ -1 \end{bmatrix}, \quad \mathbf{X_3} = \frac{1}{\sqrt{3}}\begin{bmatrix} 1 \\ -1 \\ 1 \end{bmatrix}.$$

By the way the matrix r occurring in (11) of Chapter 3.7 was obtained, we now have

$$r = \begin{pmatrix} 0 & \dfrac{2}{\sqrt{6}} & \dfrac{1}{\sqrt{3}} \\[2mm] \dfrac{1}{\sqrt{2}} & \dfrac{1}{\sqrt{6}} & -\dfrac{1}{\sqrt{3}} \\[2mm] \dfrac{1}{\sqrt{2}} & -\dfrac{1}{\sqrt{6}} & \dfrac{1}{\sqrt{3}} \end{pmatrix}.$$

Since r is an orthogonal matrix, we have

$$r^{-1} = r^* = \begin{pmatrix} 0 & \dfrac{1}{\sqrt{2}} & \dfrac{1}{\sqrt{2}} \\[2mm] \dfrac{2}{\sqrt{6}} & \dfrac{1}{\sqrt{6}} & -\dfrac{1}{\sqrt{6}} \\[2mm] \dfrac{1}{\sqrt{3}} & -\dfrac{1}{\sqrt{3}} & \dfrac{1}{\sqrt{3}} \end{pmatrix}.$$

If $\mathbf{X}$ is any vector, $\begin{pmatrix} x \\ y \\ z \end{pmatrix}$ are its old coordinates and $\begin{pmatrix} u \\ v \\ w \end{pmatrix}$ its new coordinates, then by (3a),

$$\begin{pmatrix} u \\ v \\ w \end{pmatrix} = R^{-1}\begin{pmatrix} x \\ y \\ z \end{pmatrix} = \begin{pmatrix} 0 & \dfrac{1}{\sqrt{2}} & \dfrac{1}{\sqrt{2}} \\[2mm] \dfrac{2}{\sqrt{6}} & \dfrac{1}{\sqrt{6}} & -\dfrac{1}{\sqrt{6}} \\[2mm] \dfrac{1}{\sqrt{3}} & -\dfrac{1}{\sqrt{3}} & \dfrac{1}{\sqrt{3}} \end{pmatrix}\begin{pmatrix} x \\ y \\ z \end{pmatrix}. \qquad (7)$$

Equations (7) allow us to calculate the new coordinates for any given vector $\begin{pmatrix} x \\ y \\ z \end{pmatrix}$ in terms of x, y, and z.

Exercise 1. Classify the quadric surface:

$$x^2 + 10xz + y^2 + 6yz + z^2 = 1$$

(see Example 2, p. 197).

Exercise 2. Find an equation in new coordinates of the form

$$\lambda_1 u^2 + \lambda_2 v^2 + \lambda_3 w^2 = 1$$

for the quadric surface $-4x^2 + 2y^2 + 3yz + 2z^2 = 1$.

So far, we have been looking at surfaces with equation (1):

$$ax^2 + 2bxy + 2cxz + dy^2 + 2eyz + fz^2 = 1.$$

If we replace the constant 1 by the constant 0 in this equation, we get the equation

$$ax^2 + 2bxy + 2cxz + dy^2 + 2eyz + fz^2 = 0. \qquad (8)$$

(We exclude the case when all the coefficients equal zero.) What kind of locus is defined by this equation?

EXAMPLE 1a. $x^2 + y^2 + z^2 = 0$. This equation defines a *single point*, the origin.

EXAMPLE 1b. $x^2 + y^2 - z^2 = 0$. This defines a *(double) cone* with axis along the z-axis and vertex at the origin.

EXAMPLE 1c. $y^2 + z^2 = 0$. This defines a *straight line*, the x-axis.

EXAMPLE 1d. $x^2 = 0$. This defines a *plane*: $x = 0$.

Have we exhausted all the geometric possibilities by these examples? Let Σ denote the locus in $\mathbb{R}^3$ defined by equation (8). Using Theorem 3.12, we see that in suitable new coordinates u, v, w, equation (8) can be written

$$t_1 u^2 + t_2 v^2 + t_3 w^2 = 0, \qquad (9)$$

where t_1, t_2, t_3 are fixed scalars.

CASE 1. $t_1 > 0, t_2 > 0, t_3 > 0$.

Clearly then, only $u = v = w = 0$ satisfies equation (9), so Σ consists of one point, the origin.

CASE 2. $t_1 > 0, t_2 > 0, t_3 < 0$.

We write $t_3 = -k$ with $k > 0$. Then, (9) becomes

$$t_1 u^2 + t_2 v^2 = kw^2 \quad \text{or} \quad s_1 u^2 + s_2 v^2 = w^2,$$

where s_1, s_2 are positive constants. It follows that, in this case, Σ is an *elliptic cone*, with the axis along the w-axis and the vertex at the origin. The slice of Σ by the plane: $w = w_0$ where w_0 is a constant $\neq 0$, is the ellipse: $s_1 u^2 + s_2 v^2 = w_0^2$ in that plane (see Fig. 3.23).

Exercise 3. Examining the remaining cases, where each t_i is either positive, negative, or zero, shows that the only geometric possibilities for Σ, other than those in Cases 1 and 2 are:

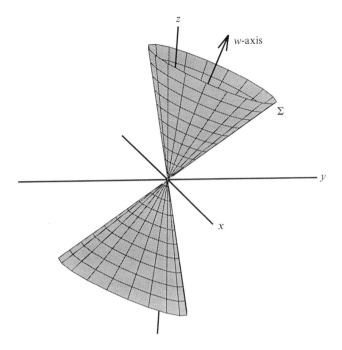

Figure 3.23

(a) a straight line through the origin;
(b) a plane through the origin; and
(c) a pair of planes, each of which passes through the origin.

Exercise 4. Describe the locus with the equation,

$$x^2 - 4y^2 - 4yz - z^2 = 0.$$

CHAPTER 4.0

Vector Geometry in n-Space, $n \geqslant 4$

§1. Introduction

In the preceding chapters, we have seen how the language and techniques of linear algebra can unify large parts of the geometry of vectors in 2 and 3 dimensions. What begins as an alternative way of treating problems in analytic geometry becomes a powerful tool for investigating increasingly complicated phenomena, such as eigenvectors or quadratic forms, which would be difficult to approach otherwise.

In the case of 4 dimensions and higher, linear algebra has to be used almost from the very beginning to define the concepts that correspond to geometric objects in 2 and 3 dimensions. We cannot visualize these higher-dimensional phenomena directly, but we can use the algebraic intuitions developed in 2 and 3 dimensions to guide us in the study of mathematical ideas that are not easily accessible. Many of the algebraic notions that we have used in lower dimensions can be transferred almost without change to dimensions of 4 and higher, and we will, therefore, continue to use familiar geometric terms, such as "vector," "dot product," "linear independence," and "eigenvector," when we study higher-dimensional geometry. For convenience, we will restrict most of our examples to 4 dimensions, but the same calculations work in n dimensions for $n \geqslant 4$.

§2. The Algebra of Vectors

A *vector in 4-space* is defined to be a 4-tuple of real numbers $[x_1, x_2, x_3, x_4]$, with x_i indicating the coordinate in the ith place. We denote this vector by a single capital letter $\mathbf{X}$, i.e., we write $\mathbf{X} = [x_1, x_2, x_3, x_4]$. The set of all vectors in 4-space is denoted by $\mathbb{R}^4$. In a similar way, we denote by $\mathbb{R}^n$ the set of all vectors in n-space, where each vector $\mathbf{X}$ is an n-tuple of real numbers $\mathbf{X} = [x_1, x_2, \ldots, x_n]$. We will sometimes write a vector as a row of real numbers rather than a column, when no confusion can arise. It is not easy to "picture" the vector $\mathbf{X}$ as an arrow beginning at the origin and ending at a point in 4-space or n-space. Nonetheless, the power of linear algebra is that it enables us to manipulate vectors in any dimension by using the same rules for addition and scalar multiplication that we used in dimensions 2 and 3. In $\mathbb{R}^n$, we add two vectors by adding their components, so if $\mathbf{X} = [x_1, x_2, \ldots, x_n]$ and $\mathbf{U} = [u_1, u_2, \ldots, u_n]$, then $\mathbf{X} + \mathbf{U} = [x_1 = u_1, x_2 + u_2, \ldots, x_n + u_n]$. We multiply a vector by a scalar r by multiplying each of the coordinates by r, so $r\mathbf{X} = [rx_1, rx_2, \ldots, rx_n]$.

In 4-space, we set $\mathbf{E}_1 = [1, 0, 0, 0]$, $\mathbf{E}_2 = [0, 1, 0, 0]$, $\mathbf{E}_3 = [0, 0, 1, 0]$, $\mathbf{E}_4 = [0, 0, 0, 1]$, and we call these the four *basis vectors* of 4-space. The *first coordinate axis* is then obtained by taking all multiples $x_1 \mathbf{E}_1 = [x_1, 0, 0, 0]$ of $\mathbf{E}_1$, and the ith *coordinate axis* is defined similarly for each $i = 2, 3, 4$. Any vector in 4-space may be uniquely expressed as a sum of vectors on the four coordinate axes $\mathbf{X} = x_1 \mathbf{E}_1 + x_2 \mathbf{E}_2 + x_3 \mathbf{E}_3 + x_4 \mathbf{E}_4$. Similarly, any vector in n-space can be expressed as a linear combination of the n basis vectors $\mathbf{E}_i$, where $\mathbf{E}_i$ has a 1 in the ith position and 0 elsewhere. Any vector $\mathbf{X}$ in $\mathbb{R}^n$ can then be uniquely written as $\mathbf{X} = x_1 \mathbf{E}_1 + x_2 \mathbf{E}_2 + \cdots + x_n \mathbf{E}_n$.

In dimension 3, we described $x_1 \mathbf{E}_1 + x_2 \mathbf{E}_2 + x_3 \mathbf{E}_3$ as a diagonal segment in a rectangular prism with edges parallel to the coordinates axes. We drew a picture that was completely determined as soon as we chose a position for each of the basis vectors. We can do the same thing in the case of a vector in 4 dimensions, although it is not so immediately clear what we mean by the analog of a 4-dimensional rectangular parallelepiped or its n-dimensional counterpart. The basic insight that enables us to represent a vector on a 2-dimensional page is that we can determine the picture of any vector once we have the pictures of the basis vectors. Once we know the picture of $\mathbf{E}_1$, we can find the picture of $x_1 \mathbf{E}_1$ simply by stretching it by a factor of x_1. Once we know the pictures of $x_1 \mathbf{E}_1$ and $x_2 \mathbf{E}_2$, we can obtain a picture of $x_1 \mathbf{E}_1 + x_2 \mathbf{E}_2$ just by finding the diagonal of the parallelogram they determine in the plane. We can then get a picture of $x_1 \mathbf{E}_1 + x_2 \mathbf{E}_2 + x_3 \mathbf{E}_3$ just by taking the diagonal of the parallelogram formed by the pictures of $x_1 \mathbf{E}_1 + x_2 \mathbf{E}_2$ and $x_3 \mathbf{E}_3$, and similarly for $x_1 \mathbf{E}_1 + x_2 \mathbf{E}_2 + x_3 \mathbf{E}_3 + x_4 \mathbf{E}_4$ (see Fig. 4.1). We may continue this process all the way to $x_1 \mathbf{E}_1 + x_2 \mathbf{E}_2 + \cdots + x_n \mathbf{E}_n$.

The *line* through $\mathbf{X}$ parallel to the non-zero vector $\mathbf{U}$ is defined to be the set of all vectors of the form $\mathbf{X} + t\mathbf{U}$ for all real numbers t.

Exercise 1. Let $\mathbf{X} = [1, 2, 0, -1]$ and $\mathbf{U} = [1, 1, 1, 2]$. Find the intersection of the

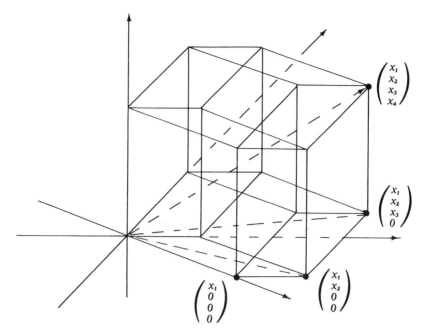

Figure 4.1

line through **X** parallel to **U** and the set of all vectors that have fourth coordinate 0.

Exercise 2. Show that the line of Exercise 1 is the same as the line through **Y** = [3, 4, 2, 3] parallel to **V** = [−2, −2, −2, −4]. (Show that every vector of the form **X** + *t***U** can be written in the form **Y** + *s***V** for some choice of *s*, and conversely, that every vector of the form **Y** + *s***V** can be written in the form **X** + *t***U** for some *t*.)

Exercise 3. Show that the line of Exercise 1 meets the line through **Z** = [0, 1, 2, 0] parallel to **W** = [−1, −1, −4, −5] at exactly one point. (Find *t* and *s* such that **X** + *t***U** = **Z** + *s***W**, and explain why there is only one such pair of scalars.)

Exercise 4. In $\mathbb{R}^5$, find the coordinates of the point on the line through **X** = [1, 2, 3, 4, 5] parallel to **U** = [5, 4, 3, 2, 1] that has its last coordinate equal to 0.

As in the case of $\mathbb{R}^2$ and $\mathbb{R}^3$, vectors in $\mathbb{R}^4$ and $\mathbb{R}^n$ satisfy the following algebraic properties: For all vectors **X**, **Y**, **U** and all scalars *r* and *s*, we have

(a) $(\mathbf{X} + \mathbf{Y}) + \mathbf{U} = \mathbf{X} + (\mathbf{Y} + \mathbf{U})$.
(b) $\mathbf{X} + \mathbf{U} = \mathbf{U} + \mathbf{X}$.
(c) There is a vector **0** with $\mathbf{X} + \mathbf{0} = \mathbf{X}$ for all **X**.
(d) For each **X**, there is a vector $-\mathbf{X}$ with $\mathbf{X} + (-\mathbf{X}) = \mathbf{0}$.

(e) $(r + s)\mathbf{X} = r\mathbf{X} + s\mathbf{X}$.
(f) $r(s(\mathbf{X})) = (rs)\mathbf{X}$.
(g) $r(\mathbf{X} + \mathbf{U}) = r\mathbf{X} + r\mathbf{U}$.
(h) $1\mathbf{X} = \mathbf{X}$ for all $\mathbf{X}$.

As in previous chapters, these properties may be verified componentwise.

We may use the properties of addition and scalar multiplication to define the notion of the *centroid* of a collection of vectors in $\mathbb{R}^4$ or $\mathbb{R}^n$. As before, we define the midpoint of a pair of vectors $\mathbf{X}$ and $\mathbf{U}$ to be $C(\mathbf{X}, \mathbf{U}) = (\mathbf{X} + \mathbf{U})/2$ and the centroid of a triplet of vectors $\mathbf{X}, \mathbf{Y}, \mathbf{U}$ to be $C(\mathbf{X}, \mathbf{Y}, \mathbf{U}) = (\mathbf{X} + \mathbf{Y} + \mathbf{U})/3$. Similarly, for any r-tuple of vectors $\mathbf{X}_1, \mathbf{X}_2, \dots, \mathbf{X}_r$ in $\mathbb{R}^n$, we define the centroid to be $C(\mathbf{X}_1, \mathbf{X}_2, \dots, \mathbf{X}_r) = (\mathbf{X}_1 + \mathbf{X}_2 + \cdots + \mathbf{X}_r)/r$.

EXAMPLE. $C(\mathbf{E}_1, \mathbf{E}_2, \mathbf{E}_3, \mathbf{E}_4) = [1, 1, 1, 1]/4 = [\frac{1}{4}, \frac{1}{4}, \frac{1}{4}, \frac{1}{4}]$.

Recall that, in elementary geometry, the centroid of a triangle can be found by going two-thirds of the way from a vertex to the midpoint of the opposite side. In vector form, this is equivalent to the statement that $C(\mathbf{X}, \mathbf{Y}, \mathbf{U}) = (\frac{1}{3})\mathbf{X} + (\frac{2}{3})C(\mathbf{Y} + \mathbf{U})$. A simple substitution shows that this is indeed correct.

Exercise 5. Show that the centroid of the tetrahedron determined by the four points $\mathbf{X}, \mathbf{Y}, \mathbf{U}, \mathbf{V}$ is three-fourths of the way from the vector $\mathbf{V}$ to the centroid of $\mathbf{X}, \mathbf{Y}$, and $\mathbf{U}$.

Exercise 6. Show that the centroid in Exercise 5 is the midpoint of the centroids $C(\mathbf{X}, \mathbf{Y})$ and $C(\mathbf{U}, \mathbf{V})$.

Exercise 7. Find a number t such that the centroid of $\mathbf{E}_1, \mathbf{E}_2, \mathbf{E}_3, \mathbf{E}_4$, and $t(\mathbf{E}_1 + \mathbf{E}_2 + \mathbf{E}_3 + \mathbf{E}_4)$ is $\mathbf{0}$.

Exercise 8. Show that the centroid of $\mathbf{X}, \mathbf{Y}, \mathbf{Z}, \mathbf{U}, \mathbf{V}$ is three-fifths of the way from the midpoint of $\mathbf{U}, \mathbf{V}$ to the centroid of $\mathbf{X}, \mathbf{Y}, \mathbf{Z}$.

§3. Dot Product, Length, and Angle in $\mathbb{R}^4$ and $\mathbb{R}^n$

In 4-space, we may define the length of the vector $\mathbf{X} = [x_1, x_2, x_3, x_4]$ to be $\sqrt{(x_1^2 + x_2^2 + x_3^2 + x_4^2)}$. In general, in $\mathbb{R}^n$, we may define the length of any vector to be the square root of the sum of the squares of its coordinates. We denote this length by $|\mathbf{X}|$, a real number, which is never negative, and which equals 0 if, and only if, $\mathbf{X} = \mathbf{0}$. Moreover $|r\mathbf{X}| = |r||\mathbf{X}|$ for any scalar r and any vector $\mathbf{X}$. If $\mathbf{X} \neq \mathbf{0}$, then we may write $\mathbf{X} = |\mathbf{X}|\mathbf{U}$, where $\mathbf{U} = \mathbf{X}/|\mathbf{X}|$ is a vector with unit length. The vectors of length 1 in $\mathbb{R}^4$ determine the *unit sphere* in $\mathbb{R}^4$, and more generally, the vectors of length 1 in $\mathbb{R}^n$ determine the *unit sphere* in $\mathbb{R}^n$.

Exercise 9. Show that for any choice of angles a and b, the vector $\mathbf{U} = (1/\sqrt{2})[\cos(a), \sin(a), \cos(b), \sin(b)]$ is a unit vector in $\mathbb{R}^4$.

Exercise 10. Show that for any choice of angles a, b, and c, the vector $\mathbf{U} = [\cos(a)\cos(c), \sin(a)\cos(c), \cos(b)\sin(c), \sin(b)\sin(c)]$ is a unit vector in $\mathbb{R}^4$.

Exercise 11. Show that for any choice of angles a, b, and c, the vector $\mathbf{U} = [\cos(a)\cos(b)\cos(c), \cos(a)\cos(b)\sin(c), \cos(a)\sin(b), \sin(a)]$ is a unit vector in $\mathbb{R}^4$.

Exercise 12. Find the length of the vector $\cos(a)\mathbf{E}_1 + \sin(a)\mathbf{E}_2 + 2\cos(b)\mathbf{E}_3 + 2\sin(b)\mathbf{E}_4 + 3\mathbf{E}_5$ in $\mathbb{R}^5$ and express this vector as a scalar multiple of a unit vector.

Definition. Five points in $\mathbb{R}^n$ are the vertices of a *regular 4-simplex* if the distance between any two points is the same.

Exercise 13. Find a number t such that the points $\mathbf{E}_1$, $\mathbf{E}_2$, $\mathbf{E}_3$, $\mathbf{E}_4$, and $t(\mathbf{E}_1 + \mathbf{E}_2 + \mathbf{E}_3 + \mathbf{E}_4)$ form the vertices of a regular 4-simplex.

Exercise 14. Show that the endpoints of the five basis vectors in $\mathbb{R}^5$ form the vertices of a regular 4-simplex.

Exercise 15. Show that for any s, the vector $s\mathbf{E}_4$ has the same distance from the four points $[1, 1, 1, 0]$, $[1, -1, -1, 0]$, $[-1, 1, -1, 0]$, and $[-1, -1, 1, 0]$. For which s will these five points form the vertices of a regular 4-simplex?

Exercise 16. Show that the distance between any two of the six vectors $\mathbf{E}_1 + \mathbf{E}_2$, $\mathbf{E}_1 + \mathbf{E}_3$, $\mathbf{E}_1 + \mathbf{E}_4$, $\mathbf{E}_2 + \mathbf{E}_3$, $\mathbf{E}_2 + \mathbf{E}_4$, $\mathbf{E}_3 + \mathbf{E}_4$ is either 2 or $\sqrt{2}$.

As in dimensions 2 and 3, we may define a notion of *dot product* in $\mathbb{R}^4$ or $\mathbb{R}^n$, which enables us to develop many important geometric ideas in linear algebra. In $\mathbb{R}^4$, we define $\mathbf{X} \cdot \mathbf{U} = [x_1, x_2, x_3, x_4] \cdot [u_1, u_2, u_3, u_4] = x_1 u_1 + x_2 u_2 + x_3 u_3 + x_4 u_4$ and, more generally, in $\mathbb{R}^n$, we define $\mathbf{X} \cdot \mathbf{U} = [x_1, x_2, \ldots, x_n] \cdot [u_1, u_2, \ldots, u_n] = x_1 u_1 + x_2 u_2 + \cdots + x_n u_n$. As before, $|\mathbf{X}| = \sqrt{\mathbf{X} \cdot \mathbf{X}}$, and $|\mathbf{X}| = 0$ if and only if $\mathbf{X} = \mathbf{0}$. Moreover, $|\mathbf{X}|^2 + |\mathbf{U}|^2 = |\mathbf{X} + \mathbf{U}|^2$ if, and only if, $\mathbf{X} \cdot \mathbf{U} = 0$. Thus, the vectors $\mathbf{X}$ and $\mathbf{U}$ form the legs of a right triangle if and only if $\mathbf{X} \cdot \mathbf{U} = 0$, so the condition that $\mathbf{X}$ and $\mathbf{U}$ be perpendicular or *orthogonal* is that their dot product is zero.

In $\mathbb{R}^3$, the set of vectors perpendicular to a fixed non-zero vector form a plane. In $\mathbb{R}^4$ or $\mathbb{R}^n$, the vectors orthogonal to a fixed non-zero vector form a *hyperplane*. For example, in $\mathbb{R}^4$, the set of vectors orthogonal to $\mathbf{E}_4$ is the hyperplane of all vectors with fourth coordinate zero.

Exercise 17. Describe the hyperplane in $\mathbb{R}^4$ consisting of all vectors that are orthogonal to the vector $\mathbf{E}_1 + \mathbf{E}_2 - \mathbf{E}_3 - \mathbf{E}_4$.

Using componentwise arguments, we may establish the following properties of the dot product for any vectors $\mathbf{X}, \mathbf{U}, \mathbf{V}$ and any scalar r:

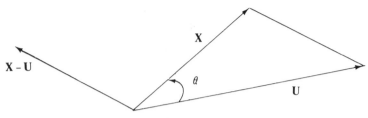

Figure 4.2

$$U \cdot X = X \cdot U,$$
$$(rX) \cdot U = r(X \cdot U),$$
$$X \cdot (U + V) = X \cdot U + X \cdot V.$$

As in dimensions 2 and 3, we wish to define the angle between two vectors in such a way that the law of cosines will hold, i.e., for any two nonzero vectors X and U (see Fig. 4.2), we wish to have

$$|X - U|^2 = |X|^2 + |U|^2 - 2|X||U| \cos\theta.$$

But by the properties of dot product, we have

$$|X - U|^2 = (X - U) \cdot (X - U) = X \cdot X - 2X \cdot U + U \cdot U,$$

so

$$|X - U|^2 = |X|^2 + |U|^2 - 2X \cdot U.$$

We would then like to define $\cos\theta$ by the condition

$$|X||U|\cos\theta = X \cdot U \quad \text{and} \quad 0 \le \theta \le \pi,$$

but to do this, we must have $|\cos\theta| \le 1$, i.e., $\cos^2\theta \le 1$. Thus we must show that for any nonzero X and U, we have

$$\left(\frac{X \cdot U}{|X| \cdot |U|}\right)^2 \le 1.$$

(This inequality is known as the *Cauchy–Schwarz Inequality*.)

One case is easy: If $U = tX$ for some t, then

$$\frac{X \cdot U}{|X||U|} = \frac{X \cdot tX}{|X||tX|} = \frac{tX \cdot X}{|t||X|^2} = \frac{t}{|t|},$$

and $-1 \le t/|t| \le 1$, since $t/|t| = 1$ if $t > 0$ and $t/|t| = -1$ if $t < 0$.

If $U - tX \ne 0$ for all t, then we can use the quadratic formula to provide the proof. We have

$$0 < |U - tX|^2 = (U - tX) \cdot (U - tX) = (U \cdot U) - 2(U \cdot X)t + (X \cdot X)t^2$$

for all t. But if $(U \cdot X)^2 - (U \cdot U)(X \cdot X)$ were positive or zero, we would have solutions t of the equation

$$0 = (U \cdot U) - 2(U \cdot X)t + (X \cdot X)t^2,$$

given by

$$t = \frac{-(-2\mathbf{U} \cdot \mathbf{X}) \pm \sqrt{4(\mathbf{U} \cdot \mathbf{X})^2 - 4(\mathbf{U} \cdot \mathbf{U})(\mathbf{X} \cdot \mathbf{X})}}{2(\mathbf{X} \cdot \mathbf{X})}.$$

Since we cannot have any such solutions, we must conclude that

$$(\mathbf{U} \cdot \mathbf{X})^2 < (\mathbf{U} \cdot \mathbf{U})(\mathbf{X} \cdot \mathbf{X})$$

i.e.,

$$\frac{(\mathbf{U} \cdot \mathbf{X})^2}{|\mathbf{U}|^2 |\mathbf{X}|^2} < 1.$$

We then define θ by the equation

$$\cos \theta = \frac{\mathbf{X} \cdot \mathbf{U}}{|\mathbf{X}| \, |\mathbf{U}|} \qquad \text{for} \quad 0 \leqslant \theta \leqslant \pi.$$

If $\mathbf{U} = t\mathbf{X}$ for $t > 0$, then $\cos \theta = 1$ and $\theta = 0$. If $\mathbf{U} = t\mathbf{X}$ for $t < 0$, then $\cos \theta = -1$ and $\theta = \pi$. If $\mathbf{X} \cdot \mathbf{U} = 0$, then $\theta = \pi/2$ and we say that the vectors $\mathbf{X}$ and $\mathbf{U}$ are *orthogonal*.

Exercise 18. Show that for any θ, the vectors $\cos \theta \, \mathbf{E}_1 - \sin \theta \, \mathbf{E}_3$ and $\sin \theta \, \mathbf{E}_1 + \cos \theta \, \mathbf{E}_3$ are orthogonal in $\mathbb{R}^4$.

Exercise 19. Find a real number t such that $\begin{pmatrix} 1 \\ 2 \\ 1 \\ -1 \end{pmatrix}$ is orthogonal to $\begin{pmatrix} t \\ -t \\ 1-t \\ 2t-1 \end{pmatrix}$.

We say that a collection of vectors is *orthonormal* if each vector has unit length and if any two distinct vectors in the set are orthogonal. For example, the basis vectors $\{\mathbf{E}_1, \mathbf{E}_2, \mathbf{E}_3, \mathbf{E}_4\}$ form an orthonormal set.

Exercise 20. Show that for any angle θ, the vectors $\{\cos \theta \, \mathbf{E}_1 + \sin \theta \, \mathbf{E}_4, \mathbf{E}_2, \mathbf{E}_3, -\sin \theta \, \mathbf{E}_1 + \cos \theta \, \mathbf{E}_4, \}$ form an orthonormal set.

Exercise 21. Show that the four vectors $\begin{pmatrix} x \\ y \\ u \\ v \end{pmatrix}, \begin{pmatrix} -y \\ x \\ -v \\ u \end{pmatrix}, \begin{pmatrix} u \\ -v \\ -x \\ y \end{pmatrix}, \begin{pmatrix} v \\ u \\ -y \\ -x \end{pmatrix}$ are mutually orthogonal and all have the same length.

Exercise 22. Show that for all angles θ and ϕ, the vectors

$$\{(\cos \theta \, \mathbf{E}_1 + \sin \theta \, \mathbf{E}_2)\cos \phi + \sin \phi \, \mathbf{E}_3,$$

$$-(\cos \theta \, \mathbf{E}_1 + \sin \theta \, \mathbf{E}_2)\sin \phi + \cos \phi \, \mathbf{E}_3,$$

$$-\sin \theta \, \mathbf{E}_1 + \cos \theta \, \mathbf{E}_2, \mathbf{E}_4\}$$

form an orthonormal set.

Exercise 23. Find the angle between the vectors $\mathbf{E}_1 + \mathbf{E}_2$ and $\mathbf{E}_1 + \mathbf{E}_2 + \mathbf{E}_3 + \mathbf{E}_4$.

Exercise 24. Find the angle between the vectors $E_3 - \frac{1}{2}(E_1 + E_2)$ and $E_4 - \frac{1}{2}(E_1 + E_2)$.

Exercise 25. Find the angle between the vectors $\begin{pmatrix} 1 \\ 1 \\ 1 \\ 1 \end{pmatrix}$ and $\begin{pmatrix} 1 \\ 1 \\ 1 \\ -1 \end{pmatrix}$. What are the

possible cosines of angles between the vector $\begin{pmatrix} 1 \\ 1 \\ 1 \\ 1 \end{pmatrix}$ and the other vectors which have

each coordinate 1 or -1?

Exercise 26. Show that if U, V, and W are distinct vectors with each coordinate 1 or -1 and if V and W each differ from U by exactly one coordinate, then $V - U$ and $W - U$ are orthogonal and they have the same length.

The collection of vectors $X = \begin{pmatrix} x_1 \\ x_2 \\ x_3 \\ x_4 \end{pmatrix}$ with $-1 \leq x_i \leq 1$ for $i = 1, 2, 3, 4$ is

called the 4-*cube centered at the origin.*

CHAPTER 4.1

Transformations of n-Space, $n \geqslant 4$

By a *transformation of 4-space*, we mean a rule T which assigns to each vector X of $\mathbb{R}^4$ some vector $T(X)$ of $\mathbb{R}^4$. The vector $T(X)$ is called the *image* of X under T, and the collection of all images of vectors in $\mathbb{R}^4$ under the transformation T is called the *range* of T. We continue to denote transformations by capital letters such as P, Q, R, S, T.

Examples of transformations are:

(1) Projection to the line along $\mathbf{U} \neq \mathbf{0}$ defined by

$$P(\mathbf{X}) = \left(\frac{\mathbf{X} \cdot \mathbf{U}}{\mathbf{U} \cdot \mathbf{U}} \right) \mathbf{U}.$$

(2) Reflection through the line along $\mathbf{U} \neq \mathbf{0}$ defined by

$$S(\mathbf{X}) = 2P(\mathbf{X}) - \mathbf{X}.$$

(3) Multiplication by a scalar t defined by

$$D_t(\mathbf{X}) = t\mathbf{X}.$$

(4) Projection to the hyperplane perpendicular to $\mathbf{U} \neq \mathbf{0}$ defined by

$$Q(\mathbf{X}) = \mathbf{X} - P(\mathbf{X}).$$

EXAMPLE 1. Let $\mathbf{U} = \begin{pmatrix} 1 \\ 1 \\ 1 \\ 1 \end{pmatrix}$ in $\mathbb{R}^4$; then

$$P(\mathbf{X}) = \left(\frac{x_1 + x_2 + x_3 + x_4}{4} \right) \begin{pmatrix} 1 \\ 1 \\ 1 \\ 1 \end{pmatrix} = \frac{1}{4} \begin{pmatrix} x_1 + x_2 + x_3 + x_4 \\ x_1 + x_2 + x_3 + x_4 \\ x_1 + x_2 + x_3 + x_4 \\ x_1 + x_2 + x_3 + x_4 \end{pmatrix},$$

$$S(\mathbf{X}) = 2P(\mathbf{X}) - \mathbf{X} = \frac{1}{2} \begin{pmatrix} -x_1 + x_2 + x_3 + x_4 \\ x_1 - x_2 + x_3 + x_4 \\ x_1 + x_2 - x_3 + x_4 \\ x_1 + x_2 + x_3 - x_4 \end{pmatrix},$$

$$Q(\mathbf{X}) = \frac{1}{4} \begin{pmatrix} 3x_1 - x_2 - x_3 - x_4 \\ -x_1 + 3x_2 - x_3 - x_4 \\ -x_1 - x_2 + 3x_3 - x_4 \\ -x_1 - x_2 - x_3 + 3x_4 \end{pmatrix}.$$

Exercise 1. In each of the following problems, let P denote projection to the line along $\mathbf{U}$. Find a formula for the coordinates of the image $P(\mathbf{X})$ in terms of the coordinates of $\mathbf{X}$.

(a) $\mathbf{U} = \begin{pmatrix} 1 \\ 0 \\ 0 \\ 0 \end{pmatrix}$,

(b) $\mathbf{U} = \begin{pmatrix} 1 \\ 1 \\ 0 \\ 0 \end{pmatrix}$,

(c) $\mathbf{U} = \begin{pmatrix} 1 \\ -1 \\ 1 \\ -1 \end{pmatrix}$,

(d) $\mathbf{U} = \begin{pmatrix} 1 \\ 0 \\ 2 \\ -1 \end{pmatrix}$.

Exercise 2. For each of the vectors **U** in Exercise 1, find a formula for the image of the reflection $S(\mathbf{X})$ through the line along **U**.

Exercise 3. For each of the vectors **U** in Exercise 1, find a formula for $Q(\mathbf{X})$, where Q is the projection to the hyperplane orthogonal to **U**.

Using the description of a transformation in terms of its coordinates, we can define further transformations, such as:

(5) Rotation in the $x_1 x_2$ plane by angle θ defined by

$$R_\theta^{12} \begin{bmatrix} x_1 \\ x_2 \\ x_3 \\ x_4 \end{bmatrix} = \begin{bmatrix} \cos\theta x_1 - \sin\theta x_2 \\ \sin\theta x_1 + \cos\theta x_2 \\ x_3 \\ x_4 \end{bmatrix}.$$

Exercise 4. In terms of the coordinate of $\mathbf{X} = \begin{bmatrix} x_1 \\ x_2 \\ x_3 \\ x_4 \end{bmatrix}$, calculate the images $R_\pi^{12}(\mathbf{X})$,

$R_{\pi/4}^{12}(\mathbf{X})$, $R_{-\pi/4}^{12}(\mathbf{X})$.

Similarly, we have the images of R_θ^{ij} in the $x_i x_j$ plane by setting $x_k' = x_k$ for all $k \neq i, j$ and by defining

$$x_i' = \cos\theta x_i - \sin\theta x_j,$$

$$x_j' = \sin\theta x_i + \cos\theta x_j.$$

Exercise 5. Calculate the images $R_\pi^{34}(\mathbf{X})$, $R_\theta^{34}(R_\phi^{12}(\mathbf{X}))$, $R_\phi^{12}(R_\theta^{34}(\mathbf{X}))$, $R_\theta^{23}(R_\phi^{12}(\mathbf{X}))$,

$R_\phi^{12}(R_\theta^{23}(\mathbf{X}))$, where $\mathbf{X} = \begin{bmatrix} x_1 \\ x_2 \\ x_3 \\ x_4 \end{bmatrix}$.

Just as in the case of objects in 3 dimensions, we may picture objects in 4-space by projecting them down to a 2-dimensional plane. The easiest such projection is simply the projection to the first two coordinates in $\mathbb{R}^4$, i.e.,

$$T \begin{bmatrix} x_1 \\ x_2 \\ x_3 \\ x_4 \end{bmatrix} = \begin{pmatrix} x_1 \\ x_2 \end{pmatrix}.$$

We call this the projection to the 1-2-coordinate plane. Even though this transformation takes a vector in $\mathbb{R}^4$ and sends it to a vector in $\mathbb{R}^2$, it possesses the properties of a linear transformation since $T(\mathbf{X} + t\mathbf{U}) =$

$T(\mathbf{X}) + tT(\mathbf{U})$ for any $\mathbf{X}$, $\mathbf{U}$ in $\mathbb{R}^4$ and any real number t. In particular, the images of a line is another line if $T(\mathbf{U}) \neq \mathbf{0}$ and the image is a point if $T(\mathbf{U}) = \mathbf{0}$. This fact makes it easy to draw 2-dimensional pictures of objects in 4-space which are composed of segments—we simply find the images of the vertices of the object and connect the image points by a segment in $\mathbb{R}^2$ if the original vertices are connected by a segment in $\mathbb{R}^4$.

EXAMPLE 2. In $\mathbb{R}^4$, the points $\mathbf{U} = \begin{bmatrix} 1 \\ 1 \\ 0 \\ 1 \end{bmatrix}$, $\mathbf{V} = \begin{bmatrix} 1 \\ 0 \\ 1 \\ 1 \end{bmatrix}$, and $\mathbf{W} = \begin{bmatrix} 0 \\ 1 \\ 1 \\ 1 \end{bmatrix}$ determine an

equilateral triangle. The image points are $T(\mathbf{U}) = \begin{pmatrix} 1 \\ 1 \end{pmatrix}$, $T(\mathbf{V}) = \begin{pmatrix} 1 \\ 0 \end{pmatrix}$,

$T(\mathbf{W}) = \begin{pmatrix} 0 \\ 1 \end{pmatrix}$ (see Fig. 4.3).

Note that the image itself is not equilateral.

EXAMPLE 3. The tetrahedron in $\mathbb{R}^4$ determined by the vertices $\mathbf{U}_1 = \begin{bmatrix} 1 \\ 1 \\ 0 \\ 1 \end{bmatrix}$,

$\mathbf{U}_2 = \begin{bmatrix} 1 \\ 0 \\ 1 \\ 1 \end{bmatrix}$, $\mathbf{U}_3 = \begin{bmatrix} 0 \\ 1 \\ 1 \\ 1 \end{bmatrix}$, $\mathbf{U}_4 = \begin{bmatrix} 1 \\ 1 \\ 1 \\ 0 \end{bmatrix}$ has the *same image* as in Example 2 since

$T(\mathbf{U}_1) = T(\mathbf{U}_4) = \begin{pmatrix} 1 \\ 1 \end{pmatrix}$.

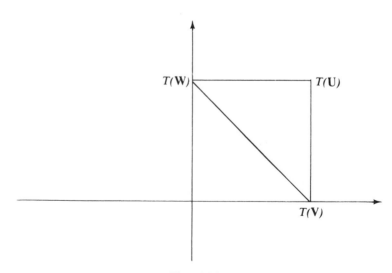

Figure 4.3

EXAMPLE 4. The tetrahedron in $\mathbb{R}^4$ determined by the vertices $V_1 = \begin{bmatrix} 2 \\ 2 \\ 1 \\ -1 \end{bmatrix}$,

$V_2 = \begin{bmatrix} 1 \\ -1 \\ 0 \\ 3 \end{bmatrix}$, $V_3 = \begin{bmatrix} -1 \\ 1 \\ 6 \\ 4 \end{bmatrix}$, $V_4 = \begin{bmatrix} -\frac{1}{2} \\ -\frac{1}{2} \\ 7 \\ 7 \end{bmatrix}$ has the image given by Figure 4.4.

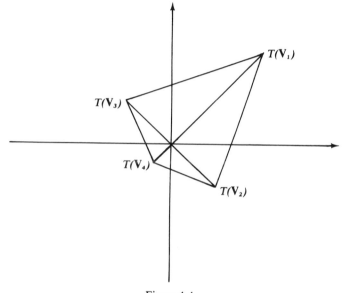

Figure 4.4

EXAMPLE 5. Consider the 4-cube centered at the origin with vertices given by the vectors with all coordinates either 1 or -1. The projection T of this 4-cube to the plane has only four distinct vertices $\begin{pmatrix} 1 \\ 1 \end{pmatrix}$, $\begin{pmatrix} 1 \\ -1 \end{pmatrix}$, $\begin{pmatrix} -1 \\ 1 \end{pmatrix}$, $\begin{pmatrix} -1 \\ -1 \end{pmatrix}$, even though the 4-cube has 16 vertices. For example, the four

vertices $\begin{bmatrix} 1 \\ 1 \\ 1 \\ 1 \end{bmatrix}$, $\begin{bmatrix} 1 \\ 1 \\ 1 \\ -1 \end{bmatrix}$, $\begin{bmatrix} 1 \\ 1 \\ -1 \\ -1 \end{bmatrix}$, $\begin{bmatrix} 1 \\ 1 \\ -1 \\ 1 \end{bmatrix}$ are all sent to $\begin{pmatrix} 1 \\ 1 \end{pmatrix}$ under T.

In order to get more useful pictures of an object like the 4-cube, we first rotate the object before projecting to the 1-2-coordinate plane. For example, if we rotate the 4-cube θ degrees in the 1-3-plane, we get

$$R_\theta^{13} \begin{bmatrix} x_1 \\ x_2 \\ x_3, \\ x_4 \end{bmatrix} = \begin{bmatrix} \cos\theta x_1 - \sin\theta x_3 \\ x_2 \\ \sin\theta x_1 + \cos\theta x_3 \\ x_4 \end{bmatrix},$$

so

$$TR_\theta^{13} \begin{bmatrix} x_1 \\ x_2 \\ x_3 \\ x_4 \end{bmatrix} = \begin{pmatrix} \cos\theta x_1 - \sin\theta x_3 \\ x_2 \end{pmatrix}.$$

If $\theta = 30°$, we then have $TR_{30}^{13} \begin{bmatrix} 1 \\ 1 \\ 1 \\ 1 \end{bmatrix} = \begin{pmatrix} \sqrt{3}/2 - 1/2 \\ 1 \end{pmatrix}$.

The picture opens up a certain amount, but we still see only eight distinct vertex images.

If we first rotate in the 1-3-plane by θ degrees and then in the 2-4-plane by ϕ degrees we, get

$$R_\phi^{24} R_\theta^{13} \begin{bmatrix} x_1 \\ x_2 \\ x_3 \\ x_4 \end{bmatrix} = \begin{bmatrix} \cos\theta x_1 - \sin\theta x_3 \\ \cos\phi x_2 - \sin\phi x_4 \\ \sin\theta x_1 + \cos\theta x_3 \\ \sin\phi x_2 + \cos\phi x_4 \end{bmatrix},$$

so

$$TR_\phi^{24} R_\theta^{13} \begin{bmatrix} x_1 \\ x_2 \\ x_3 \\ x_4 \end{bmatrix} = \begin{pmatrix} \cos\theta x_1 - \sin\theta x_3 \\ \cos\phi x_2 - \sin\phi x_4 \end{pmatrix}.$$

Thus

$$TR_{45}^{24} R_{30}^{13} \begin{bmatrix} 1 \\ -1 \\ -1 \\ -1 \end{bmatrix} = \begin{bmatrix} \sqrt{3}/2 + 1/2 \\ -\sqrt{2}/2 + \sqrt{2}/2 \end{bmatrix},$$

while

$$TR_{30}^{24} R_{30}^{13} \begin{bmatrix} 1 \\ -1 \\ -1 \\ -1 \end{bmatrix} = \begin{bmatrix} \sqrt{3}/2 + 1/2 \\ -\sqrt{3}/2 + 1/2 \end{bmatrix}$$

We get 16 different images for the 16 vertices of the 4-cube, but again it is difficult to interpret the image of the whole 4-cube.

If instead we first rotate in the 1-3-plane, then the 2-4-plane, then the 1-4-plane, we get a general position.

$$TR_\alpha^{14}R_\phi^{24}R_\theta^{13}\begin{bmatrix}x_1\\x_2\\x_3\\x_4\end{bmatrix} = \begin{pmatrix}\cos\alpha(\cos\theta x_1 - \sin\theta x_3) - \sin\alpha(\sin\phi x_2 + \cos\phi x_4)\\ \cos\phi x_2 - \sin\phi x_4\end{pmatrix}.$$

Then

$$TR_{30}^{14}R_{30}^{24}R_{30}^{13}\begin{bmatrix}x_1\\x_2\\x_3\\x_4\end{bmatrix} = \begin{bmatrix}\dfrac{\sqrt{3}}{2}\left(\dfrac{\sqrt{3}}{2}x_1 - \dfrac{1}{2}x_3\right) - \dfrac{1}{2}\left(\dfrac{1}{2}x_2 + \dfrac{\sqrt{3}}{2}x_4\right)\\ \dfrac{\sqrt{3}}{2}x_2 - \dfrac{1}{2}x_4\end{bmatrix}$$

$$= \begin{bmatrix}\dfrac{3}{4}x_1 - \dfrac{1}{4}x_2 - \dfrac{\sqrt{3}}{4}x_3 - \dfrac{\sqrt{3}}{4}x_4\\ \dfrac{\sqrt{3}}{2}x_2 - \dfrac{1}{2}x_4\end{bmatrix}.$$

Finally, if we rotate by β degrees in the 2-3-plane, we have

$$TR_\beta^{23}R_\alpha^{14}R_\phi^{24}R_\theta^{13}\begin{bmatrix}x_1\\x_2\\x_3\\x_4\end{bmatrix}$$

$$= \begin{pmatrix}\cos\alpha\cos\theta x_1 - \sin\alpha\sin\phi x_2 - \cos\alpha\sin\theta x_3 - \sin\alpha\cos\phi x_4\\ -\sin\beta\sin\phi x_1 + \cos\beta\cos\phi x_2 - \sin\beta\cos\theta x_3 - \cos\beta\sin\phi x_4\end{pmatrix}.$$

In particular,

$$TR_{30}^{23}R_{30}^{14}R_{30}^{24}R_{30}^{13}\begin{bmatrix}x_1\\x_2\\x_3\\x_4\end{bmatrix} = \begin{bmatrix}\dfrac{3}{4}x_1 - \dfrac{1}{4}x_2 - \dfrac{\sqrt{3}}{4}x_3 - \dfrac{\sqrt{3}}{4}x_4\\ -\dfrac{3}{4}x_1 + \dfrac{1}{4}x_2 - \dfrac{\sqrt{3}}{4}x_3 - \dfrac{\sqrt{3}}{4}x_4\end{bmatrix}.$$

Now we have a picture in "general position" where no two images of coordinate axes are linearly dependent.

These are precisely the sorts of instructions which are used in producing computer graphics images (see Fig. 4.5), for example, in the film *The*

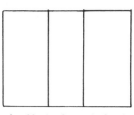

$\theta = 30, \phi = 0, \alpha = 0, \beta = 0$

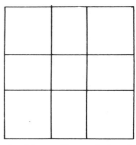

$\theta = 30, \phi = 30, \alpha = 0, \beta = 0$

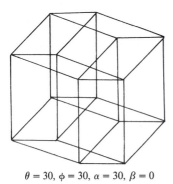

$\theta = 30, \phi = 30, \alpha = 30, \beta = 0$

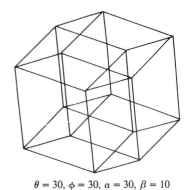

$\theta = 30, \phi = 30, \alpha = 30, \beta = 10$

Figure 4.5

Hypercube: *Projections and Slicing* by Thomas Banchoff and Charles Strauss. We include several different pictures of that object corresponding to other values of θ, ϕ, α, and β.

Linear Transformations and Matrices

In Chapter 4.1 we examined a number of transformations T of 4-space all of which have the property that the coordinates of $T(\mathbf{X})$ are given as *linear* functions of the coordinates of $\mathbf{X}$. In each case we have formulas of the sort

$$T\begin{bmatrix} x_1 \\ x_2 \\ x_3 \\ x_4 \end{bmatrix} = \begin{bmatrix} a_{11}x_1 + a_{12}x_2 + a_{13}x_3 + a_{14}x_4 \\ a_{21}x_1 + a_{22}x_2 + a_{23}x_3 + a_{24}x_4 \\ a_{31}x_1 + a_{32}x_2 + a_{33}x_3 + a_{34}x_4 \\ a_{41}x_1 + a_{42}x_2 + a_{43}x_3 + a_{44}x_4 \end{bmatrix}.$$

Any transformation which can be written in this form is called a *linear transformation of* 4-*space*.

The symbol $\begin{bmatrix} a_{11} & a_{12} & a_{13} & a_{14} \\ a_{21} & a_{22} & a_{23} & a_{24} \\ a_{31} & a_{32} & a_{33} & a_{34} \\ a_{41} & a_{42} & a_{43} & a_{44} \end{bmatrix}$ is called the *matrix* of the transforma-

tion T and is denoted $m(T)$. We abbreviate $m(T)$ by $((a_{ij}))$, where a_{ij} stands for the entry in the ith *row* and the jth *column*.

We can now list the matrices of the linear transformations in the

examples of Chapter 4.1 (with $\mathbf{U} = \begin{bmatrix} 1 \\ 1 \\ 1 \\ 1 \end{bmatrix}$):

$$m(P) = \begin{bmatrix} \frac{1}{4} & \frac{1}{4} & \frac{1}{4} & \frac{1}{4} \\ \frac{1}{4} & \frac{1}{4} & \frac{1}{4} & \frac{1}{4} \\ \frac{1}{4} & \frac{1}{4} & \frac{1}{4} & \frac{1}{4} \\ \frac{1}{4} & \frac{1}{4} & \frac{1}{4} & \frac{1}{4} \end{bmatrix}, \tag{1}$$

$$m(S) = \begin{pmatrix} -\frac{1}{2} & \frac{1}{2} & \frac{1}{2} & \frac{1}{2} \\ \frac{1}{2} & -\frac{1}{2} & \frac{1}{2} & \frac{1}{2} \\ \frac{1}{2} & \frac{1}{2} & -\frac{1}{2} & \frac{1}{2} \\ \frac{1}{2} & \frac{1}{2} & \frac{1}{2} & \frac{1}{2} \end{pmatrix}, \tag{2}$$

$$m(D_t) = \begin{pmatrix} t & 0 & 0 & 0 \\ 0 & t & 0 & 0 \\ 0 & 0 & t & 0 \\ 0 & 0 & 0 & t \end{pmatrix}, \tag{3}$$

$$m(Q) = \begin{pmatrix} \frac{3}{4} & -\frac{1}{4} & -\frac{1}{4} & -\frac{1}{4} \\ -\frac{1}{4} & \frac{3}{4} & -\frac{1}{4} & -\frac{1}{4} \\ -\frac{1}{4} & -\frac{1}{4} & \frac{3}{4} & -\frac{1}{4} \\ -\frac{1}{4} & -\frac{1}{4} & -\frac{1}{4} & \frac{3}{4} \end{pmatrix}, \tag{4}$$

$$m(R_\theta^{12}) = \begin{pmatrix} \cos\theta & -\sin\theta & 0 & 0 \\ \sin\theta & \cos\theta & 0 & 0 \\ 0 & 0 & 1 & 0 \\ 0 & 0 & 0 & 1 \end{pmatrix}. \tag{5}$$

As in dimensions 2 and 3, if T is the linear transformation with matrix $m(T) = ((a_{ij}))$, we then write

$$((a_{ij}))(X) = ((a_{ij})) \begin{bmatrix} x_1 \\ x_2 \\ x_3 \\ x_4 \end{bmatrix} = \begin{bmatrix} a_{11}x_1 + a_{12}x_2 + a_{13}x_3 + a_{14}x_4 \\ a_{21}x_1 + a_{22}x_2 + a_{23}x_3 + a_{24}x_4 \\ a_{31}x_1 + a_{32}x_2 + a_{33}x_3 + a_{34}x_4 \\ a_{41}x_1 + a_{42}x_2 + a_{43}x_3 + a_{44}x_4 \end{bmatrix} = \begin{bmatrix} y_1 \\ y_2 \\ y_3 \\ y_4 \end{bmatrix}$$

and we say that *the matrix $((a_{ij}))$ acts on the vector* X *to yield* $Y = \begin{bmatrix} y_1 \\ y_2 \\ y_3 \\ y_4 \end{bmatrix}$. We

may then write the equations for the coordinates of Y as

$$y_i = a_{i1}x_1 + a_{i2}x_2 + a_{i3}x_3 + a_{i4}x_4, \quad \text{for} \quad i = 1, 2, 3, 4.$$

As in dimensions 2 and 3, we now prove two crucial properties of linear transformations, which show how a matrix acts on sums and scalar prod-

ucts. If $X = \begin{bmatrix} x_1 \\ x_2 \\ x_3 \\ x_4 \end{bmatrix}$ and $U = \begin{bmatrix} u_1 \\ u_2 \\ u_3 \\ u_4 \end{bmatrix}$, and if $Y = ((a_{ij}))(X + U)$, then

$$y_i = a_{i1}(x_1 + u_1) + a_{i2}(x_2 + u_2) + a_{i3}(x_3 + u_3) + a_{i4}(x_4 + u_4)$$

$$= (a_{i1}x_1 + a_{i2}x_2 + a_{i3}x_3 + a_{i4}x_4) + (a_{i1}u_1 + a_{i2}u_2 + a_{i3}u_3 + a_{i4}u_4).$$

Therefore, $((a_{ij}))(X + U) = ((a_{ij}))X + ((a_{ij}))U$. It follows that $T(X + U) = T(X) + T(U)$.

Similarly, we may show that $T(r\mathbf{X}) = rT(\mathbf{X})$ for any scalar r.

Conversely, if T is a transformation such that $T(\mathbf{X} + \mathbf{U}) = T(\mathbf{X}) + T(\mathbf{U})$ and $T(r\mathbf{X}) = rT(\mathbf{X})$ for all vectors $\mathbf{X}$, $\mathbf{U}$ and scalars r, then

$$T\begin{bmatrix} x_1 \\ x_2 \\ x_3 \\ x_4 \end{bmatrix} = T\left(\begin{bmatrix} x_1 \\ 0 \\ 0 \\ 0 \end{bmatrix} + \begin{bmatrix} 0 \\ x_2 \\ 0 \\ 0 \end{bmatrix} + \begin{bmatrix} 0 \\ 0 \\ x_3 \\ 0 \end{bmatrix} + \begin{bmatrix} 0 \\ 0 \\ 0 \\ x_4 \end{bmatrix}\right)$$

$$= T(x_1\mathbf{E}_1 + x_2\mathbf{E}_2 + x_3\mathbf{E}_3 + x_4\mathbf{E}_4)$$

$$= x_1T(\mathbf{E}_1) + x_2T(\mathbf{E}_2) + x_3T(\mathbf{E}_3) + x_4T(\mathbf{E}_4).$$

We define a_{ij} by setting $T(\mathbf{E}_j) = \begin{bmatrix} a_{1j} \\ a_{2j} \\ a_{3j} \\ a_{4j} \end{bmatrix}$. Then $T\begin{bmatrix} x_1 \\ x_2 \\ x_3 \\ x_4 \end{bmatrix} = ((a_{ij}))\begin{bmatrix} x_1 \\ x_2 \\ x_3 \\ x_4 \end{bmatrix}$. Hence,

T is the linear transformation with matrix $((a_{ij}))$.

In summary, we have:

Theorem 4.1. *A transformation T of 4-space is a linear transformation if and only if $T(\mathbf{X} + \mathbf{U}) = T(\mathbf{X}) + T(\mathbf{U})$ and $T(r\mathbf{X}) = rT(\mathbf{X})$ for any vectors $\mathbf{X}$ and $\mathbf{U}$ and scalars r.*

In much the same way as in dimensions 2 and 3, we may define the notions of products of transformations and of their corresponding matrices, of inverses, determinants, and eigenvalues. These procedures lead to systems of equations in four and more variables which we will take up in the next chapter. We do mention two facts which are important differences between dimension 4 and dimension 3 to help the student in pursuing the subject of linear algebra beyond the material in this book.

First of all, although every linear transformation in $\mathbb{R}^3$ had at least one eigenvalue, this property no longer holds in $\mathbb{R}^4$ (as indeed it did not in $\mathbb{R}^2$). For example, if we consider the double rotation $R_\theta^{34}R_\phi^{12}$, we have

$$R_\theta^{34}R_\phi^{12}\begin{bmatrix} x_1 \\ x_2 \\ x_3 \\ x_4 \end{bmatrix} = \begin{bmatrix} \cos\theta x_1 - \sin\theta x_2 \\ \sin\theta x_1 + \cos\theta x_2 \\ \cos\phi x_3 - \sin\phi x_4 \\ \sin\phi x_3 + \cos\phi x_4 \end{bmatrix},$$

and if $R_\theta^{34}R_\phi^{12}(\mathbf{X}) = \lambda\mathbf{X}$ for some $\lambda \neq 0$, we have, first of all,

$$\cos\theta x_1 - \sin\theta x_2 = \lambda x_1,$$

$$\sin\theta x_1 + \cos\theta x_2 = \lambda x_2;$$

so unless x_1 and $x_2 = 0$,

$$(\cos\theta - \lambda)^2 + \sin^2\theta = 0,$$

and, therefore,

$$1 + \lambda^2 - 2\lambda \cos \theta = 0.$$

The only solutions then are $\lambda = (2 \cos \theta \pm \sqrt{4 \cos^2\theta - 4})/2$. But this has solutions only if $\cos^2\theta = 1$, $\theta = 0, \pi$. Similarly, the last two equations express the condition that

$$\cos \phi x_3 - \sin \phi x_4 = \lambda x_3,$$

$$\sin \phi x_3 + \cos \phi x_4 = \lambda x_4,$$

which can only occur if $\phi = 0$ or π or if both x_3 and $x_4 = 0$. Therefore, in the case where neither θ nor ϕ is 0 or π, the transformation $R_\theta^{34} R_\phi^{12}$ will have no (real) eigenvalues or eigenvectors.

The definition of the determinant of a matrix in $\mathbb{R}^4$ is analogous to the definition in $\mathbb{R}^2$ or $\mathbb{R}^3$. We recall that we can express the 3×3 determinant in terms of 2×2 determinants as follows.

$$\begin{vmatrix} a_1 & b_1 & c_1 \\ a_2 & b_2 & c_2 \\ a_3 & b_3 & c_3 \end{vmatrix} = a_1 \begin{vmatrix} b_2 & c_2 \\ b_3 & c_3 \end{vmatrix} - a_2 \begin{vmatrix} b_1 & c_1 \\ b_3 & c_3 \end{vmatrix} + a_3 \begin{vmatrix} b_1 & c_1 \\ b_2 & c_2 \end{vmatrix}.$$

In $\mathbb{R}^4$, we define a 4×4 determinant in terms of 3×3 determinants:

$$\begin{vmatrix} a_1 & b_1 & c_1 & d_1 \\ a_2 & b_2 & c_2 & d_2 \\ a_3 & b_3 & c_3 & d_3 \\ a_4 & b_4 & c_4 & d_4 \end{vmatrix} = a_1 \begin{vmatrix} b_2 c_2 d_2 \\ b_3 c_3 d_3 \\ b_4 c_4 d_4 \end{vmatrix} - a_2 \begin{vmatrix} b_1 c_1 d_1 \\ b_3 c_3 d_3 \\ b_4 c_4 d_4 \end{vmatrix} + a_3 \begin{vmatrix} b_1 c_1 d_1 \\ b_2 c_2 d_2 \\ b_4 c_4 d_4 \end{vmatrix} - a_4 \begin{vmatrix} b_1 c_1 d_1 \\ b_2 c_2 d_2 \\ b_3 c_3 d_3 \end{vmatrix}.$$

For example,

$$\begin{vmatrix} \cos \theta - \lambda & -\sin \theta & 0 & 0 \\ \sin \theta & \cos \theta - \lambda & 0 & 0 \\ 0 & 0 & \cos \phi - \lambda & -\sin \phi \\ 0 & 0 & \sin \phi & \cos \phi - \lambda \end{vmatrix}$$

$$= (\cos \theta - \lambda) \begin{vmatrix} \cos \theta - \lambda & 0 & 0 \\ 0 & \cos \phi - \lambda & -\sin \phi \\ 0 & \sin \phi & \cos \phi - \lambda \end{vmatrix}$$

$$- \sin \theta \begin{vmatrix} -\sin \theta & 0 & 0 \\ 0 & \cos \phi - \lambda & -\sin \phi \\ 0 & \sin \phi & \cos \phi - \lambda \end{vmatrix}$$

$$= \left((\cos \theta - \lambda)^2 + \sin^2\theta \right) \left((\cos \phi - \lambda)^2 + \sin^2\phi \right)$$

$$= (\lambda^2 - 2\lambda \cos \theta + 1)(\lambda^2 - 2\lambda \cos \phi + 1).$$

The only cases in which this determinant is zero occur when either θ or ϕ is 0 or π.

The determinant of an $n \times n$ matrix is defined inductively. If we assume that the determinant of an $(n - 1) \times (n - 1)$ matrix has already been defined, then we may obtain the determinant of an $n \times n$ matrix by using the following formula:

$$
\begin{vmatrix} a_{11} & a_{12} & a_{13} & \cdots & a_{1n} \\ a_{21} & a_{22} & a_{23} & \cdots & a_{2n} \\ a_{31} & a_{32} & a_{33} & \cdots & a_{1n} \\ \vdots & & & & \\ a_{n1} & a_{n2} & a_{n3} & \cdots & a_{nn} \end{vmatrix} = a_{11} \begin{vmatrix} a_{22} & a_{23} & \cdots & a_{2n} \\ a_{32} & a_{33} & \cdots & a_{3n} \\ \vdots & & & \\ a_{n2} & a_{n3} & \cdots & a_{nn} \end{vmatrix}
$$

$$
- a_{21} \begin{vmatrix} a_{12} & a_{13} & \cdots & a_{1n} \\ a_{22} & a_{23} & \cdots & a_{2n} \\ \vdots & & & \\ a_{n2} & a_{n3} & \cdots & a_{nn} \end{vmatrix} + a_{31} \begin{vmatrix} a_{12} & a_{13} & \cdots & a_{1n} \\ a_{22} & a_{23} & \cdots & a_{2n} \\ \vdots & & & \\ a_{n2} & a_{n3} & \cdots & a_{nn} \end{vmatrix}
$$

$$
+ \cdots + (-1)^{n-1} \begin{vmatrix} a_{12} & a_{13} & \cdots & a_{1n} \\ a_{22} & a_{23} & \cdots & a_{2n} \\ \vdots & & & \\ a_{n-12} & a_{n-13} & \cdots & a_{n-1n} \end{vmatrix} .
$$

Let A be a linear transformation of $\mathbb{R}^n$ with matrix $\mathbf{m}$ with entries a_{ij}. It can be shown that A has an inverse if and only if the determinant of $\mathbf{m} \neq 0$.

Homogeneous Systems of Equations in n-Space

In Chapters 2.4 and 3.4, we studied systems of linear equations in 2 and 3 unknowns. In this chapter, we will apply the techniques of linear algebra to systems of linear equations in n unknowns where $n \geqslant 4$.

EXAMPLE 1. Let us look at the system

$$\begin{cases} x_1 + 2x_2 + 3x_3 - x_4 = 0, \\ x_2 + x_3 + x_4 = 0, \end{cases} \tag{1}$$

in four unknowns. By a solution of (1) we mean a vector $\mathbf{X}$ in $\mathbb{R}^4$,

$$\mathbf{X} = \begin{bmatrix} x_1 \\ x_2 \\ x_3 \\ x_4 \end{bmatrix},$$

which satisfies the two equations in system (1). Thus

$$\begin{bmatrix} -1 \\ -1 \\ 1 \\ 0 \end{bmatrix} \quad \text{and} \quad \begin{bmatrix} 0 \\ 0 \\ 0 \\ 0 \end{bmatrix}$$

are two solutions of (1).

Let us find all solutions of (1). Assume

$$\mathbf{X} = \begin{bmatrix} x_1 \\ x_2 \\ x_3 \\ x_4 \end{bmatrix}$$

is a solution of (1). Subtracting twice the bottom equation from the top, we get

$$(x_1 + 2x_2 + 3x_3 - x_4) - 2(x_2 + x_3 + x_4) = 0 - 2(0)$$

or

$$x_1 + x_3 - 3x_4 = 0.$$

So we have

$$\begin{cases} x_1 + x_3 - 3x_4 = 0, \\ x_2 + x_3 + x_4 = 0, \end{cases} \tag{2}$$

which we rewrite in the form:

$$\begin{cases} x_1 = -x_3 + 3x_4, \\ x_2 = -x_3 - x_4. \end{cases} \tag{3}$$

We just saw that every solution of the system (1) satisfies the system (3). Conversely, retracing our steps, we see that if X is a solution of (3), then X is also a solution of (1). But now (3) can be solved directly. We give arbitrary values to x_3 and x_4 and then use (3) to calculate x_1 and x_2. For instance, set $x_3 = -10$, $x_4 = 3$. Then, by (3),

$$x_1 = 10 + 3 \cdot 3 = 19,$$

and

$$x_2 = 10 - 3 = 7.$$

Using these values for x_i, $i = 1, 2, 3, 4$, we get

$$X = \begin{pmatrix} 19 \\ 7 \\ -10 \\ 3 \end{pmatrix}.$$

One can directly check that X is a solution of (1).

More generally, fix two numbers u, v. Set $x_3 = u$, $x_4 = v$ and define x_1 and x_2 by (3). Then

$$x_1 = -u + 3v,$$

$$x_2 = -u - v.$$

Set

$$X = \begin{pmatrix} x_1 \\ x_2 \\ x_3 \\ x_4 \end{pmatrix} = \begin{pmatrix} -u + 3v \\ -u - v \\ u \\ v \end{pmatrix} = u \begin{pmatrix} -1 \\ -1 \\ 1 \\ 0 \end{pmatrix} + v \begin{pmatrix} 3 \\ -1 \\ 0 \\ 1 \end{pmatrix}. \tag{4}$$

Letting u, v take on all possible scalar values, formula (4) then delivers all solutions of (1).

Next, we shall study an arbitrary system of k equations in n unknowns having the following form:

$$\begin{cases} a_{11}x_1 + a_{12}x_2 + \cdots + a_{1n}x_n = 0, \\ a_{21}x_1 + a_{22}x_2 + \cdots + a_{2n}x_n = 0, \\ \quad\quad\quad\quad\vdots \\ a_{k1}x_1 + a_{k2}x_2 + \cdots + a_{kn}x_n = 0. \end{cases} \tag{H}$$

(H) is called a *homogeneous system of linear equations*.

Here a_{ij}, $1 \leq i \leq k$, $1 \leq j \leq n$, are certain given scalars called the *coefficients* of the system (H) and $x_1, \ldots, x_n$ are the unknowns. A *solution* of (H) is an n-tuple of numbers

$$\mathbf{X} = \begin{bmatrix} x_1 \\ x_2 \\ \vdots \\ x_n \end{bmatrix}$$

such that each of the k equations in (H) is satisfied by these n numbers.

We recall the discussion in Chapter 4.0, §2: We define an n-tuple $\mathbf{X}$ as a *vector in n-space*, and we denote the totality of all such vectors by $\mathbb{R}^n$. *Addition* of vectors in $\mathbb{R}^n$ and *multiplication of a vector by a scalar* is defined by analogy with the definitions given for the case $n = 4$, and the same basic algebraic properties hold which we noted in that case. Furthermore, the *dot product* of two vectors in $\mathbb{R}^n$ is defined as in Chapter 4.0, §3, by

$$\mathbf{X} \cdot \mathbf{U} = x_1u_1 + x_2u_2 + \cdots + x_nu_n,$$

where

$$\mathbf{X} = \begin{bmatrix} x_1 \\ x_2 \\ \vdots \\ x_n \end{bmatrix}, \quad \mathbf{U} = \begin{bmatrix} u_1 \\ u_2 \\ \vdots \\ u_n \end{bmatrix}.$$

Using dot-product notation, the system (H) can be written concisely as

$$\begin{cases} \mathbf{A}_1 \cdot \mathbf{X} = 0, \\ \mathbf{A}_2 \cdot \mathbf{X} = 0, \\ \quad\vdots \\ \mathbf{A}_k \cdot \mathbf{X} = 0, \end{cases} \tag{H}$$

where **X** is the unknown vector $\begin{pmatrix} x_1 \\ \vdots \\ x_n \end{pmatrix}$ and

$$\mathbf{A}_1 = \begin{pmatrix} a_{11} \\ a_{12} \\ \vdots \\ a_{1n} \end{pmatrix}, \quad \mathbf{A}_2 = \begin{pmatrix} a_{21} \\ a_{22} \\ \vdots \\ a_{2n} \end{pmatrix}, \quad \text{etc.}$$

Using the dot product, we may give a geometric interpretation of homogeneous equations that allows us to solve a number of geometric problems in n-space analogous to those we can solve in spaces of 2 or 3 dimensions. The statement that $\mathbf{A} \cdot \mathbf{X} = 0$ can be interpreted as saying that **X** is in the hyperplane orthogonal to **A**. The statement that $\mathbf{A}_1 \cdot \mathbf{X} = 0$ and $\mathbf{A}_2 \cdot \mathbf{X} = 0$ then means that **X** is in the intersection of two hyperplanes. This intersection might be a hyperplane itself, in case $\mathbf{A}_1$ and $\mathbf{A}_2$ are linearly dependent, but if $\mathbf{A}_1$ and $\mathbf{A}_2$ are linearly independent, any vector orthogonal to both of them will be orthogonal to any linear combination of them. Thus, the solutions of this system of two linear equations can be interpreted as the collection of vectors orthogonal to the plane determined by $\mathbf{A}_1$ and $\mathbf{A}_2$. In the case $n = 4$, the collection of vectors orthogonal to the plane containing $\mathbf{A}_1$ and $\mathbf{A}_2$ is itself a plane, and when we solve the system of linear equations, we are finding a basis for this plane. If we have a system of three homogeneous linear equations in 4-space, and if the vectors $\mathbf{A}_1$, $\mathbf{A}_2$, $\mathbf{A}_3$ are linearly independent, then they span a hyperplane, and the solution of the system consists of all vectors in the line orthogonal to this hyperplane. Lines, planes, and hyperplanes through the origin (as well as $\mathbb{R}^4$ itself and the set consisting of the zero vector) are the only subsets of 4-dimensional space that are closed under addition and scalar multiplication, and any of these sets can be the set of solutions of a system of homogeneous equations. In Chapter 5, we will generalize this notion to higher dimensions by introducing the concept of a *subspace of a vector space*.

Next, we will describe a procedure for solving any system of k homogeneous equations in n unknowns. This procedure is known as *Gaussian elimination* and it is the basis for most algorithms used by computers in solving such systems.

To find all the solutions **X** of a given system (H), consider a second system

$$\begin{cases} a'_{11}x_1 + a'_{12}x_2 + \cdots + a'_{1n}x_n = 0, \\ \qquad\qquad\qquad \vdots \\ a'_{k1}x_1 + a'_{k2}x_2 + \cdots + a'_{kn}x_n = 0. \end{cases} \tag{H$'$}$$

We say that the systems (H) and (H$'$) are *equivalent* if every solution of (H) is a solution of (H$'$), and conversely.

To solve (H) it will be enough if we can find a system (H') which is equivalent to (H) and which is easy to solve. Note that we did just that in Example 1 when we found the system (2) which was equivalent to (1).

In the next example, $k = 3$ and $n = 4$.

EXAMPLE 2. To solve

$$\begin{cases} x_1 - 2x_2 + 3x_3 = 0, \\ x_1 + x_2 + x_4 = 0, \\ 4x_3 - x_4 = 0, \end{cases} \tag{5}$$

subtract the top line from the middle line and leave the other lines alone. We get the new system

$$\begin{cases} x_1 - 2x_2 + 3x_3 = 0, \\ 3x_2 - 3x_3 + x_4 = 0, \\ 4x_3 - x_4 = 0. \end{cases} \tag{5a}$$

(5a) is equivalent to (5).

Next, add $\frac{2}{3}$ of the middle line in (5a) to the top line. We get

$$\begin{cases} x_1 + x_3 + \frac{2}{3}x_4 = 0, \\ 3x_2 - 3x_3 + x_4 = 0, \\ 4x_3 - x_4 = 0. \end{cases} \tag{5b}$$

(5b) is equivalent to (5a), and so it follows that (5b) is equivalent to (5).

Next we add $-\frac{1}{4}$ times the bottom line of (5b) to the top line, getting

$$\begin{cases} x_1 + (\frac{2}{3} + \frac{1}{4})x_4 = 0, \\ 3x_2 - 3x_3 + x_4 = 0, \\ 4x_3 - x_4 = 0. \end{cases} \tag{5c}$$

Finally, add $\frac{3}{4}$ of the bottom line of (5c) to the middle line, getting

$$\begin{cases} x_1 + \frac{11}{12}x_4 = 0, \\ 3x_2 + \frac{1}{4}x_4 = 0, \\ 4x_3 - x_4 = 0. \end{cases} \tag{5d}$$

As before, (5c) and (5d) are equivalent to (5). But (5d) can be solved at once. Give an arbitrary value to x_4 and then use (5d) to calculate x_1, x_2, x_3. We find

$$x_1 = -\tfrac{11}{12}x_4,$$
$$x_2 = (-\tfrac{1}{12})x_4,$$
$$x_3 = \tfrac{1}{4}x_4.$$

Hence, we get, as a solution of (5d):

$$\mathbf{X} = \begin{bmatrix} x_1 \\ x_2 \\ x_3 \\ x_4 \end{bmatrix} = \begin{bmatrix} -\frac{11}{12}x_4 \\ -\frac{1}{12}x_4 \\ \frac{1}{4}x_4 \\ x_4 \end{bmatrix} = x_4 \begin{bmatrix} -\frac{11}{12} \\ -\frac{1}{12} \\ \frac{1}{4} \\ 1 \end{bmatrix}. \tag{6}$$

For different choices of x_4, (6) gives all solutions of (5d) and, therefore, all solutions of (5).

The method just used in Example 2 can be applied to any system of the form (H). By a succession of steps in which a scalar times one line of the system is added to some other line, while the remaining lines are left unchanged, and, possibly, by relabeling the unknowns x_i, we finally obtain a system (H') of the following form:

$$\begin{cases} x_1 & + b_{11}x_{l+1} + b_{12}x_{l+2} + \cdots + b_{1,n-l}x_n = 0, \\ & x_2 & + b_{21}x_{l+1} + b_{22}x_{l+2} + \cdots + b_{2,n-l}x_n = 0, \\ & \ddots \\ & & x_l & + b_{l1}x_{l+1} + b_{l2}x_{l+2} + \cdots + b_{l,n-l}x_n = 0, \end{cases} \tag{H'}$$

where l is some integer, depending on the system (H), with $1 \leq l \leq n$, and b_{ij} are certain constants, such that the original system (H) and this system (H') are equivalent. To find all solutions of (H'), and, therefore, of (H), we need only fix numbers $u_1, u_2, \ldots, u_{n-l}$, set $x_{l+1} = u_1, \ldots, x_n = u_{n-l}$, and then find $x_1, x_2, \ldots, x_l$ from (H'). We get

$$x_1 = -b_{11}u_1 - b_{12}u_2 - \cdots - b_{1,n-l}u_{n-l},$$
$$x_2 = -b_{21}u_1 - b_{22}u_2 - \cdots - b_{2,n-l}u_{n-l},$$
$$\vdots$$
$$x_l = -b_{l1}u_1 - b_{l2}u_2 - \cdots - b_{l,n-l}u_{n-l}.$$

The solution $\mathbf{X}$ of (H) is then given by

$$\mathbf{X} = \begin{bmatrix} x_1 \\ x_2 \\ \vdots \\ x_l \\ x_{l+1} \\ x_{l+2} \\ \vdots \\ x_n \end{bmatrix} = \begin{bmatrix} -b_{11}u_1 - \cdots - b_{1,n-l}u_{n-l} \\ -b_{21}u_1 - \cdots - b_{2,n-l}u_{n-l} \\ \vdots \\ -b_{l1}u_1 - \cdots - b_{l,n-l}u_{n-l} \\ u_1 \\ u_2 \\ \vdots \\ u_{n-l} \end{bmatrix}.$$

In other words,

$$\mathbf{X} = u_1 \begin{bmatrix} -b_{11} \\ \vdots \\ -b_{l1} \\ 1 \\ 0 \\ \vdots \\ 0 \end{bmatrix} + u_2 \begin{bmatrix} -b_{12} \\ \vdots \\ -b_{l2} \\ 0 \\ 1 \\ \vdots \\ 0 \end{bmatrix} + \cdots + u_{n-l} \begin{bmatrix} -b_{1,n-1} \\ \vdots \\ -b_{l,n-l} \\ 0 \\ 0 \\ \vdots \\ 1 \end{bmatrix}. \tag{7}$$

Letting $u_1, \ldots, u_l$ take on all possible scalar values, (7) gives us all solutions of (H), and each choice of $u_1, \ldots, u_l$ provides a solution of (H).

EXAMPLE 3. Find a nonzero vector $\begin{bmatrix} x_1 \\ x_2 \\ x_3 \end{bmatrix}$ in $\mathbb{R}^3$ which is orthogonal to each

of the vectors $\begin{bmatrix} 2 \\ 3 \\ 0 \end{bmatrix}$ and $\begin{bmatrix} 1 \\ 1 \\ 1 \end{bmatrix}$.

The condition on x_1, x_2, x_3 is

$$\begin{cases} 2x_1 + 3x_2 = 0, \\ x_1 + x_2 + x_3 = 0. \end{cases} \tag{8}$$

So we must solve the system (8), of two equations in three unknowns. Subtracting $\frac{1}{2}$ the top line from the bottom line, we get the equivalent system

$$\begin{cases} 2x_1 + 3x_2 = 0, \\ -\frac{1}{2}x_2 + x_3 = 0. \end{cases} \tag{8a}$$

Adding 6 times the bottom line to the top one, we get the equivalent system

$$\begin{cases} 2x_1 + + 6x_3 = 0, \\ -\frac{1}{2}x_2 + x_3 = 0. \end{cases} \tag{8b}$$

(8b) can now be solved to give

$$x_1 = -3x_3,$$
$$x_2 = 2x_3.$$

So the solution $\mathbf{X} = \begin{bmatrix} x_1 \\ x_2 \\ x_3 \end{bmatrix}$ of (8) is given by

$$\mathbf{X} = \begin{bmatrix} -3x_3 \\ 2x_3 \\ x_3 \end{bmatrix} = x_3 \begin{bmatrix} -3 \\ 2 \\ 1 \end{bmatrix}.$$

Here x_3 is an arbitrary scalar. In particular, taking $x_3 = -1$, we get: *The vector* $\begin{bmatrix} 3 \\ -2 \\ -1 \end{bmatrix}$ *is orthogonal to* $\begin{bmatrix} 2 \\ 3 \\ 0 \end{bmatrix}$ *and to* $\begin{bmatrix} 1 \\ 1 \\ 1 \end{bmatrix}$.

Note: We could have solved this problem by using the cross product.

Exercise 1. Find all solutions of the system in four unknowns:

$$\begin{cases} x_1 & + x_4 = 0, \\ x_1 & - x_4 = 0, \\ x_1 + x_2 + x_3 + x_4 = 0. \end{cases} \tag{9}$$

Exercise 2. Find all solutions of the system:

$$\begin{cases} x_1 & + 2x_4 = 0, \\ x_1 + x_2 + x_3 + x_4 = 0. \end{cases} \tag{10}$$

Exercise 3. Find all solutions of the system in x_1, x_2, x_3, x_4:

$$\begin{cases} x_1 + 2x_2 & = 0, \\ x_2 + x_3 + x_4 = 0, \\ x_1 + x_2 - x_3 & = 0. \end{cases} \tag{11}$$

Exercise 4. Find all solutions of the system consisting of one equation in five variables:

$$2x_1 - x_2 + x_3 - 4x_4 + x_5 = 0.$$

Exercise 5. Find all vectors in $\mathbb{R}^4$ which are orthogonal in $\mathbb{R}^4$:

(a) to the vector $\begin{bmatrix} 1 \\ -1 \\ 1 \\ -1 \end{bmatrix}$;

(b) to the vectors $\begin{bmatrix} 1 \\ -1 \\ 1 \\ -1 \end{bmatrix}, \begin{bmatrix} 0 \\ 1 \\ -1 \\ 0 \end{bmatrix}$;

(c) to the vectors $\begin{bmatrix} 1 \\ -1 \\ 1 \\ -1 \end{bmatrix}, \begin{bmatrix} 0 \\ 1 \\ -1 \\ 0 \end{bmatrix}, \begin{bmatrix} 1 \\ 0 \\ 0 \\ 1 \end{bmatrix}$;

(d) to the vectors $\begin{bmatrix} 1 \\ -1 \\ 1 \\ 1 \end{bmatrix}, \begin{bmatrix} 0 \\ 1 \\ -1 \\ 0 \end{bmatrix}, \begin{bmatrix} 1 \\ 0 \\ 0 \\ 1 \end{bmatrix}$;

(e) to the vectors $\begin{bmatrix} 1 \\ -1 \\ 1 \\ 1 \end{bmatrix}, \begin{bmatrix} 0 \\ 1 \\ 1 \\ 0 \end{bmatrix}, \begin{bmatrix} 1 \\ 0 \\ 0 \\ 1 \end{bmatrix}, \begin{bmatrix} 1 \\ 0 \\ 0 \\ -1 \end{bmatrix}$.

Exercise 6. Find all vectors in $\mathbb{R}^4$ which are orthogonal to the vectors: $\begin{bmatrix} 1 \\ 2 \\ 3 \\ 4 \end{bmatrix}, \begin{bmatrix} 5 \\ 6 \\ 7 \\ 8 \end{bmatrix}$.

Inhomogeneous Systems of Equations in n-Space

§1. Solutions of Systems of Equations

The procedures that we used to solve homogeneous systems of linear equations can be modified to solve systems of equations that are not homogeneous. Once we have any one solution to an inhomogeneous system of linear equations, we will be able to obtain all other solutions just by adding the solutions to the associated homogeneous system of linear equations. In the case of 4-dimensional space, the geometric inter-pretation of the solution set of a homogeneous system in terms of lines, plane, and hyperplanes through the origin leads to a corresponding de-scription of the solution sets of inhomogeneous systems as lines, planes, and hyperplanes not passing through the origin.

Let a_{ij}, $1 \leqslant i \leqslant k$, $1 \leqslant j \leqslant n$ be a set of constants, and fix k constants $u_1, u_2, \ldots, u_k$. The system

$$\begin{cases} a_{11}x_1 + a_{12}x_2 + \cdots + a_{1n}x_n = u_1, \\ a_{21}x_1 + a_{22}x_2 + \cdots + a_{2n}x_n = u_2, \\ \qquad\qquad\qquad \vdots \\ a_{k1}x_1 + a_{k2}x_2 + \cdots + a_{kn}x_n = u_k \end{cases} \tag{I}$$

is called an *inhomogeneous system of linear equations*. If all the $u_i = 0$, (I) turns into the homogeneous system (H) which we studied in Chapter 4.3. We set

$$\mathbf{U} = \begin{bmatrix} u_1 \\ u_2 \\ \vdots \\ u_k \end{bmatrix}, \qquad \mathbf{X} = \begin{bmatrix} x_1 \\ x_2 \\ \vdots \\ x_n \end{bmatrix}.$$

X is a *solution* of (I) if the equations in (I) are satisfied. How can we solve such an inhomogeneous system? For $n = k = 2$ and for $n = k = 3$, we have studied such systems in earlier chapters. In the general case of arbitrary k and n, we can proceed as in our solution of homogeneous systems in Chapter 4.3 to find a succession of systems that are *equivalent* to (I), until we reach a system (I') of the following form, where we may have relabeled the x_i:

$$\begin{cases} x_1 & + b_{11}x_{l+1} + b_{12}x_{l+2} + \cdots + b_{1,n-l}x_n = v_1, \\ \quad x_2 & + b_{21}x_{l+1} + b_{22}x_{l+2} + \cdots + b_{2,n-l}x_n = v_2, \\ \qquad \ddots & \qquad\qquad\qquad\qquad\qquad \vdots \\ \qquad\quad x_l & + b_{l1}x_{l+1} + b_{l2}x_{l+2} + \cdots + b_{l,n-l}x_n = v_l, \end{cases} \tag{I'}$$

where $v_1, \ldots, v_l$ is a new sequence of constants, constructed out of the u_i. We then solve the system (I') directly by choosing $x_{l+1}, \ldots, x_n$ arbitrarily and solving for $x_1, x_2, \ldots, x_l$, using (I'). In this way, we find all solutions of (I'), and hence all solutions of (I).

EXAMPLE 1. Solve the system

$$\begin{cases} 2x + 3y + z = u, \\ x - y - z = v, \\ 3x + 2y \quad\;\; = w, \end{cases} \tag{1}$$

where u, v, w are given numbers. Subtracting twice the middle line from the top line, and then three times the middle line from the bottom line, we get the equivalent system:

$$\begin{cases} x - y - z = v, \\ 5y + 3z = u - 2v, \\ 5y + 3z = w - 3v. \end{cases} \tag{1'}$$

By a similar procedure, we get the following system (1''), equivalent to (1'), and, hence, also equivalent to (1):

$$\begin{cases} x - y - z = v, \\ 5y + 3z = u - 2v, \\ 0 = (w - 3v) - (u - 2v) = w - u - v. \end{cases} \tag{1''}$$

Observe that (1'') does not have solutions for every choice of u, v, w. The bottom line in (1'') implies that if (1'') has a solution, then $w = v + u$. One more step, adding $\frac{1}{5}$ times the middle of line (1'') to the top line, gives

$$\begin{cases} x \quad\; - \tfrac{2}{5}z = v + \tfrac{1}{5}(u - 2v) = \tfrac{1}{5}u + \tfrac{3}{5}v, \\ y + \tfrac{3}{5}z = \tfrac{1}{5}u - \tfrac{2}{5}v, \\ 0 = w - u - v. \end{cases} \tag{1'''}$$

(1''') has the form (I') discussed above. To solve (1'''), we must have $w = u + v$. Under this assumption, we give z an arbitrary value t and find

$$\begin{cases} x = \frac{1}{5}u + \frac{3}{5}v + \frac{2}{5}t, \\ y = \frac{1}{5}u - \frac{2}{5}v - \frac{3}{5}t, \\ z = \qquad\qquad t. \end{cases} \tag{2}$$

For different choices of t, (2) provides us with all solutions of $(1''')$ and, hence, all solutions of our original system (1). In particular, take $u = 5$, $v = 10$, $w = 15$. Then the system

$$\begin{cases} 2x + 3y + z = 5, \\ x - y - z = 10, \\ 3x + 2y = 15 \end{cases} \tag{3}$$

is solved by fixing a value t and setting

$$\begin{aligned} x &= 1 + 6 + \tfrac{2}{5}t = \quad 7 + \tfrac{2}{5}t, \\ y &= 1 - 4 - \tfrac{3}{5}t = -3 - \tfrac{3}{5}t, \\ z &= \qquad\qquad = \qquad t. \end{aligned}$$

One can check this by inserting these values in the system (3).

Exercise 1. Find all solutions of the system

$$\begin{aligned} 2x + 3y + z &= 5, \\ x - y - z &= 10. \end{aligned}$$

Exercise 2. Find conditions on u_1, u_2, u_3, u_4 under which there exists a solution x_1, x_2, x_3, x_4 of the system:

$$\begin{aligned} x_1 - x_2 &= u_1, \\ 2x_1 + x_3 &= u_2, \\ x_1 - x_4 &= u_3, \\ x_2 - x_4 &= u_4. \end{aligned}$$

Assuming these conditions are satisfied, find all solutions of the system.

EXAMPLE 2. Solve the system in four unknowns:

$$\begin{cases} x_1 + x_2 = u_1, \\ x_2 + x_3 = u_2, \\ x_3 + x_4 = u_3, \\ 2x_1 - 3x_2 = u_4. \end{cases} \tag{4}$$

We can easily see that this system is equivalent to

$$\begin{cases} -5x_2 = u_4 - 2u_1, \\ x_2 + x_3 = u_2, \\ x_3 + x_4 = u_3, \\ 2x_1 - 3x_2 = u_4, \end{cases} \tag{4'}$$

and (4'), in turn, is equivalent to

$$\begin{cases} -5x_2 = u_4 - 2u_1, \\ x_3 = \frac{1}{5}(u_4 - 2u_1) + u_2 = -\frac{2}{5}u_1 + u_2 + \frac{1}{5}u_4, \\ 2x_1 = u_4 - \frac{3}{5}(u_4 - 2u_1) = \frac{6}{5}u_1 + \frac{2}{5}u_4, \\ x_3 + x_4 = u_3, \end{cases} \qquad (4'')$$

and, at last, $(4')$ is equivalent to the system obtained from $(4'')$ by keeping the three top lines and replacing the bottom line by

$$x_4 = u_3 - \left(-\frac{2}{5}u_1 + u_2 + \frac{1}{5}u_4\right)$$
$$= \frac{2}{5}u_1 - u_2 + u_3 - \frac{1}{5}u_4.$$

Thus the solution of (4) is *unique*, with given u_i, and is as follows:

$$\begin{cases} x_1 = \frac{6}{10}u_1 + \frac{1}{5}u_4, \\ x_2 = \frac{2}{5}u_1 - \frac{1}{5}u_4, \\ x_3 = -\frac{2}{5}u_1 + u_2 + \frac{1}{5}u_4, \\ x_4 = \frac{2}{5}u_1 - u_2 + u_3 - \frac{1}{5}u_4. \end{cases} \qquad (5)$$

inserting these values for the x_i in (4), we can check that our solution is correct.

Note: In Example 1, the solution of the system (1) was not unique. Also, in Example 1, the solution exists only for certain choices of u, v, w. In Example 2, the solution was unique and exists for every choice of u_1, u_2, u_3, u_4. What can be said about the *existence* and *uniqueness* of the solutions for the system (I)? We state, without proof, the following basic result for the case $k = n$. Let us denote by (H) the homogeneous system corresponding to (I), obtained by setting $u_i = 0$, $i = 1, \ldots, k$, in (I).

Theorem 4.2. *Let* $k = n$. *We distinguish two cases:*

(i) (H) *has only the trivial solution* $\mathbf{0}$. *Then the inhomogeneous system* (I) *has a unique solution for every choice of the* u_i.

(ii) (H) *has a non-trivial solution. Then for certain* u_i, (I) *has no solution. Also, the solution of* (I) *is never unique.*

A proof of this theorem will be given in Chapter 5.2.

The following example will illustrate how Theorem 4.2 can be used in proofs.

EXAMPLE 3. Given three points in the plane: (x_1, y_1), (x_2, y_2), (x_3, y_3), which are not collinear, show that there exists a circle which passes through the three points.

The circle C with center (x_0, y_0) and radius R has the equation

$$(x - x_0)^2 + (y - y_0)^2 = R^2$$

or

$$x^2 - 2xx_0 + x_0^2 + y^2 - 2yy_0 + y_0^2 = R^2 .$$

We can rewrite this in the form

$$x^2 + y^2 + ax + by + c = 0,$$

where a, b, c are certain constants. This circle passes through our three given points if and only if

$$x_i^2 + y_i^2 + ax_i + by_i + c = 0, \qquad i = 1, 2, 3.$$

Note that x_i and y_i are given numbers and a, b, and c are numbers to be found. We can rewrite this system in the form:

$$\begin{cases} x_1 a + y_1 b + c = u_1, \\ x_2 a + y_2 b + c = u_2, \\ x_3 a + y_3 b + c = u_3, \end{cases} \tag{6}$$

where $u_i = -x_i^2 - y_i^2$. We regard (6) as an inhomogeneous system of equations in the unknowns a, b, c. The corresponding homogeneous system is the following:

$$\begin{cases} x_1 a + y_1 b + c = 0, \\ x_2 a + y_2 b + c = 0, \\ x_3 a + y_3 b + c = 0, \end{cases} \tag{7}$$

Suppose that this system has a nonzero solution a, b, c. Then the line defined by

$$ax + by + c = 0$$

passes through each of our three points. This contradicts the assumption that the points are not collinear. Hence, (7) has only the trivial solution. Thus we have case (i) in Theorem 4.2 for the system (6), and so (6) has a solution a, b, c. It follows that

$$x_i^2 + y_i^2 + x_i a + y_i b + c = 0, \qquad i = 1, 2, 3.$$

The equation

$$x^2 + y^2 + xa + yb + c = 0$$

is thus satisfied by (x_i, y_i), $i = 1, 2, 3$. This equation can be written

$$\left(x + \frac{a}{2} \right)^2 + \left(y + \frac{b}{2} \right)^2 = -c + \frac{a^2}{4} + \frac{b^2}{4} ,$$

which represents a circle which passes through each of the three points.

Note: The existence of this circle could also be shown by elementary geometry. However, the method we used, based on Theorem 4.2, is applicable to a wide variety of situations, some of which are given as exercises at the end of this chapter.

§2. Geometric Interpretation

What is the geometric interpretation of the set of solutions of an in-homogeneous system (I)? In the case of a homogeneous system of linear equations in n-dimensional space for $n \leqslant 4$, we have seen that the solution set is either empty, the whole space, or a line, a plane, or a hyperplane through the origin. Consider the following example in the plane.

EXAMPLE 4. Let L consist of all vectors $\mathbf{X} = (x, y)$ in $\mathbb{R}^2$, such that

$$2x + 3y = 5. \tag{8}$$

Clearly, L is a straight line that does not contain the origin. Let L_0 denote the solution set of the homogeneous equation obtained by replacing the right-hand side by 0:

$$2x + 3y = 0. \tag{9}$$

The set L_0 is a line through the origin. How are L are L_0 related? We can easily find one solution for the inhomogeneous equation (8), for example, the point $(1, 1)$. If (x, y) satisfies $2x + 3y = 5$, then $2(x - 1) + 3(y - 1) = 0$ so $(x, y) - (1, 1)$ is a solution of the homogeneous equation. We can obtain any solution (x, y) of the inhomogeneous equation by adding the particular solution $(1, 1)$ to a solution of the homogeneous equation. Geometrically speaking, we can obtain the solution set L of (I) by moving the solution set L_0 of (H) parallel to itself, *translating* the line L_0 by the fixed vector $(1, 1)$ (see Fig. 4.3).

This procedure generalizes to arbitrary dimensions.

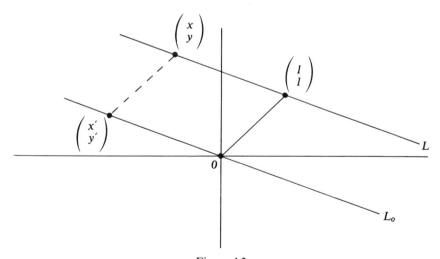

Figure 4.3

Proposition 1. *Let W be the set of solutions in $\mathbb{R}^n$ of the inhomogeneous system of equations (I). Then there is a vector $\mathbf{X}_0$, so that any element of W can be written as $\mathbf{X}_0 + \mathbf{Y}$, where $\mathbf{Y}$ is a solution of the associated homogeneous system (H). Thus, W is obtained by translating the solution set of (H) by the vector $\mathbf{X}_0$.*

To show this, we write the system (I) in the form:

$$\mathbf{A}_1 \cdot \mathbf{X} = u_1, \ldots, \mathbf{A}_k \cdot \mathbf{X} = u_k, \tag{I}$$

where $\mathbf{A}_1 = \begin{pmatrix} a_{11} \\ a_{12} \\ \vdots \\ a_{1n} \end{pmatrix}, \ldots, \mathbf{A}_k = \begin{pmatrix} a_{k1} \\ a_{k2} \\ \vdots \\ a_{kn} \end{pmatrix}$. Choose $\mathbf{X}^0 = \begin{pmatrix} x_1^0 \\ \vdots \\ x_n^0 \end{pmatrix}$ satisfying (I). Then if $\mathbf{X}$ satisfies (I),

$$\mathbf{A}_1 \cdot (\mathbf{X} - \mathbf{X}^0) = \mathbf{A}_1 \cdot \mathbf{X} - \mathbf{A}_1 \cdot \mathbf{X}^0 = \mathbf{A}_1 \cdot \mathbf{X} - u_1 = u_1 - u_1 = 0.$$

Hence,

$$\mathbf{A}_1 \cdot (\mathbf{X} - \mathbf{X}^0) = 0.$$

Similarly, $\mathbf{A}_j \cdot (\mathbf{X} - \mathbf{X}^0) = 0, j = 2, \ldots, k$. Thus $\mathbf{X} - \mathbf{X}^0$ satisfies the homogeneous system (H) which corresponds to (I).

We denote by S the set of solutions of the homogeneous system (H).

If $\mathbf{X}$ is a solution of (I), then $\mathbf{Y} = \mathbf{X} - \mathbf{X}^0$ is in S and $\mathbf{X} = \mathbf{Y} + \mathbf{X}^0$. Conversely, if $\mathbf{X}$ has this form, then

$$\mathbf{A}_1 \cdot \mathbf{X} = \mathbf{A}_1 \cdot (\mathbf{Y} + \mathbf{X}^0) = \mathbf{A}_1 \cdot \mathbf{Y} + \mathbf{A}_1 \cdot \mathbf{X}^0 = 0 + u_1 = u_1.$$

Similarly, $\mathbf{A}_j \cdot \mathbf{X} = u_j$ for all j. So $\mathbf{X}$ is in W. Thus we have shown: $\mathbf{X}$ is in W, i.e., $\mathbf{X}$ is a solution of (I), if and only if $\mathbf{X}$ lies in the translation of S by $\mathbf{X}^0$, and this is what Proposition 1 asserts.

Exercise 3. Let T be the subset of $\mathbb{R}^4$ defined by the equation

$$2x_1 - 3x_2 + x_3 + 5x_4 = 10.$$

Find a hyperplane S in $\mathbb{R}^4$ and a vector $\mathbf{X}^0$ in $\mathbb{R}^4$ such that T is the translate of S by $\mathbf{X}_0$.

§3. Exercises

Exercise 4. Let (x_1, y_1), (x_2, y_2), (x_3, y_3) be three non-collinear points in the plane with x_1, x_2, x_3 all different. Show that there exists a parabola P with equation

$$y = ax^2 + bx + c,$$

where a, b, c are constants and $a \neq 0$, such that each of the three given points lies on P.

Exercise 5. Find the coefficients a, b, c of a parabola $y = ax^2 + bx + c$ which passes through the points

$$(1,6), (2,4), (3,0).$$

Exercise 6. (a) Show that a sphere in $\mathbb{R}^3$ has an equation:

$$x^2 + y^2 + z^2 + ax + by + cz + d = 0,$$

where a, b, c, d are constants.

(b) Given four points (x_j, y_j, z_j), $j = 1, 2, 3, 4$ in $\mathbb{R}^3$ such that they do not all lie in a plane, show that there is a sphere passing through all four points.

Exercise 7. Find an equation for the sphere which passes through the points $(1, 0, 1)$, $(0, 2, 3)$, $(3, 0, 4)$, $(1, 1, 1)$.

Exercise 8. Find a cubic curve with the equation $y = ax^3 + bx^2 + cx + d$ passing through the four points: $(1, 0)$, $(2, 2)$, $(3, 12)$, $(4, 36)$.

§4. Partial Fractions Decomposition

EXAMPLE 5. We wish to express the function

$$f(x) = \frac{1}{(x - 1)(x - 2)(x - 3)}$$

in the form

$$f(x) = \frac{a}{x - 1} + \frac{b}{x - 2} + \frac{c}{x - 3}, \tag{10}$$

where a, b, c are constants to be found. Multiplying both sides by $(x - 1)$ $(x - 2)$ $(x - 3)$, we see that (10) is equivalent to

$$1 = a(x - 2)(x - 3) + b(x - 1)(x - 3) + c(x - 1)(x - 2)$$

which can be written as

$$1 = (a + b + c)x^2 + (-5a - 4b - 3c)x + 6a + 3b + 2c.$$

This is equivalent to the system

$$\begin{cases} a + b + c = 0, \\ -5a - 4b - 3c = 0, \\ 6a + 3b + 2c = 1. \end{cases} \tag{11}$$

We easily solve this and find:

$$a = \tfrac{1}{2}, \qquad b = -1, \qquad c = \tfrac{1}{2}.$$

Hence,

$$\frac{1}{(x - 1)(x - 2)(x - 3)} = \frac{1/2}{x - 1} + \frac{-1}{x - 2} + \frac{1/2}{x - 3}.$$

Exercise 9. Express the function

$$f(x) = \frac{1}{(x-1)(x+1)x}$$

in the form

$$f(x) = \frac{a}{x-1} + \frac{b}{x+1} + \frac{c}{x}.$$

Exercise 10. Find a, b, c, d such that

$$\frac{1}{(x-1)(x-2)(x-3)(x-4)} = \frac{a}{x-1} + \frac{b}{x-2} + \frac{c}{x-3} + \frac{d}{x-4}.$$

Exercise 11. Let $a_1, \ldots, a_n$ be n distinct numbers and set

$$f(x) = \frac{1}{(x-a_1)(x-a_2)\cdots(x-a_n)}.$$

An identity

$$f(x) = \frac{c_1}{x-a_1} + \frac{c_2}{x-a_2} + \cdots + \frac{c_n}{x-a_n} \tag{12}$$

is called a *partial fractions decomposition* of $f(x)$.

(a) Show that (12) is equivalent to an inhomogeneous system of n linear equations in the unknowns $c_1, \ldots, c_n$.
(b) Show that the corresponding homogeneous system has only the trivial solution.
(c) Use Theorem 5.3 to show that there exists constants $c_1, \ldots, c_n$ which satisfy (12).

Exercise 12. Does the system

$$\begin{cases} x_1 + 2x_2 + 3x_3 + 4x_4 = 1, \\ 2x_1 + 3x_2 + 4x_3 + 1x_4 = 0, \\ 3x_1 + 4x_2 + 1x_3 + 2x_4 = 0, \\ 4x_1 + 1x_2 + 2x_3 + 3x_4 = 0 \end{cases} \tag{13}$$

have a solution? If it does, find all solutions.

CHAPTER 5.0

Vector Spaces

We shall use the symbols

$$\in: \text{belongs to} \quad \text{and} \quad \notin: \text{does not belong to}.$$

For instance, the point $(3, 3) \in L$, where L is the line in the xy plane with equation $x = y$.

The basic notions of vector algebra that you have been studying in the spaces $\mathbb{R}^2$, $\mathbb{R}^3$, etc., make sense in a more general context, the context of *vector spaces*. A vector space V is a collection of objects called "vectors," which we denote by capital letters $\mathbf{X}$, $\mathbf{Y}$, $\mathbf{U}$, etc., together with the notions of "+" and "·" satisfying rules (4)–(11) given in Chapter 2.0. The objects that make up a vector space might be polynomials, trigonometric functions, 2×2 matrices, or many other kinds of things. Once we have defined addition, +, and scalar multiplication, ·, on these objects and verified rules (4)–(11), we can think of them as geometrical objects analogous to the familiar vectors in $\mathbb{R}^2$ and $\mathbb{R}^3$.

EXAMPLE 1. Fix an integer $n \geqslant 1$. The space $\mathbb{R}^n$, defined in Chapter 4.0, consists of n-tuples of real numbers and is a vector space.

EXAMPLE 2. Let S be a subset of $\mathbb{R}^n$ such that whenever $\mathbf{X}$, $\mathbf{Y} \in S$, then also $\mathbf{X} + \mathbf{Y} \in S$, and if $t \in \mathbb{R}$, also $t\mathbf{X} \in S$. Equivalently, we could say: $s\mathbf{X} + t\mathbf{Y} \in S$ for all s, $t \in \mathbb{R}$.

Then, S is called a *subspace of* $\mathbb{R}^n$. The space S inherits the notion of vector addition and scalar multiplication from $\mathbb{R}^n$, evidently obeys our rules (4)–(11), and so, is a vector space.

For instance, the plane $2x + y - 5z = 0$ in $\mathbb{R}^3$ is a subspace of $\mathbb{R}^3$.

EXAMPLE 3. Consider a homogeneous system of linear equations in $x_1, \ldots, x_n$:

$$\mathbf{A}_1 \cdot X = 0, \ldots, \mathbf{A}_k \cdot X = 0, \tag{H}$$

where $\mathbf{A}_1, \ldots, \mathbf{A}_k$ are given vectors in $\mathbb{R}^n$ and $X = (x_1, \ldots, x_n)$. Denote by S the subset of $\mathbb{R}^n$ consisting of all solutions X of this system (H).

Exercise 1. Show that S is a subspace of $\mathbb{R}^n$, and so, that S is a vector space under the usual addition, $+$, and scalar multiplication, $\cdot$.

EXAMPLE 4. Fix an integer $n \geq 1$. $\mathbb{P}_n$ denotes the set of all polynomials f of degree $\leq n$, with real coefficients:

$$f = a_0 + a_1 x + a_2 x^2 + \cdots + a_k x^k,$$

where $a_k \neq 0$ and $k \leq n$. Then k is the *degree* of f; (we write $k = \deg f$). If f and g both are in $\mathbb{P}_n$, then $\deg f \leq n$, $\deg g \leq n$, and so, $\deg(f + g) \leq n$, so $f + g \in \mathbb{P}_n$. If $f \in \mathbb{P}_n$, $t \in R$, then $tf \in \mathbb{P}_n$. It is easy to check that with these definitions, $\mathbb{P}_n$ is a vector space.

Put $\mathbb{Q}_n =$ set of all polynomials of degree exactly n, where n is some fixed integer, and add the elements of $\mathbb{Q}_n$ as polynomials. Notice that $\mathbb{Q}_n$ is *not* a vector space. For instance, $f = 2x + x^2$ and $g = 1 - 3x + 2x^2$ are both $\in \mathbb{Q}_2$, but $2f - g = 7x - 1$ has a degree of 1 and so $\notin \mathbb{Q}_2$.

EXAMPLE 5. Fix an integer $n \geq 1$. C_n denotes the set of all functions

$$\phi(x) = a_1 \cos x + a_2 \cos 2x + \cdots + a_n \cos nx,$$

where $a_1, a_2, \ldots, a_n \in R$. If

$$\psi(x) = b_1 \cos x + b_2 \cos 2x + \cdots + b_n \cos nx,$$

then

$$(\phi + \psi)(x) = (a_1 + b_1) \cos x + \cdots + (a_n + b_n) \cos nx,$$

so $\phi + \psi$ again $\in C_n$. Also, $t\phi \in C_n$ for $xt \in R$. So, C_n is a vector space.

EXAMPLE 6. Fix an integer $n \geq 1$. Let $\mathbb{T}_n$ denote the set of all functions

$$\gamma(x) = a_0 + \sum_{j=1}^{n} (a_j \cos jx + b_j \sin jx),$$

$a_j, b_j \in R$. Defining addition and scalar multiplication in the usual way, we see that $\mathbb{T}_n$ is a vector space. We shall refer to the "vectors" in $\mathbb{T}_n$, that is, to the functions γ as *trigonometric sums of order* $\leq n$.

In Example 2 we defined subspaces of $\mathbb{R}^n$. In a similar way, any subset S of a vector space V that is such that, whenever $X, Y \in S$ and $s, t \in \mathbb{R}$, then $sX + tY \in S$, is called a *subspace of V*.

Exercise 2. Show that, for fixed n, C_n is a subspace of $\mathbb{T}_n$.

EXAMPLE 7. Denote by $\mathbb{M}^2$ the collection of all 2×2 matrices $\begin{pmatrix} a & b \\ c & d \end{pmatrix}$.
Define

$$\begin{pmatrix} a & b \\ c & d \end{pmatrix} + \begin{pmatrix} a' & b' \\ c' & d' \end{pmatrix} = \begin{pmatrix} a + a' & b + b' \\ c + c' & d + d' \end{pmatrix}$$

and $t \begin{pmatrix} a & b \\ c & d \end{pmatrix} = \begin{pmatrix} ta & tb \\ tc & td \end{pmatrix}$. Then $\mathbb{M}^2$ is a vector space.

EXAMPLE 8. Let $\mathbb{C}$ denote the collection of all complex numbers $w = u + iv$, where $u, v \in \mathbb{R}$ and $i = \sqrt{-1}$. Add complex numbers in the usual way: $(u + vi) + (u' + iv') = (u + u') + i(v + v')$, and define $t(u + iv) = tu + itv$, for $t \in \mathbb{R}$. Then $\mathbb{C}$ is a vector space.

EXAMPLE 9. Let $\mathbb{P}$ denote the collection of all polynomials in the variable x, without restrictions on the degree. Add them in the usual way of adding polynomials, and similarly for scalar multiplication. Then $\mathbb{P}$ is a vector space. Each vector space $\mathbb{P}_n$ is a subspace of $\mathbb{P}$, for $n = 1, 2, \ldots$.

In Chapters 2.0 and 3.0, we discussed *linear dependence* and *linear independence* of sets of vectors in $\mathbb{R}^2$ and in $\mathbb{R}^3$.

Now, let V be a vector space and $\mathbf{Y}_1, \ldots, \mathbf{Y}_k$ be a set of k vectors in V. We say that the set is *linearly dependent* if one of the $\mathbf{Y}_i$ is a linear combination of the rest. Equivalently, we can say that $\mathbf{Y}_1, \ldots, \mathbf{Y}_k$ is linearly dependent if there exist scalars $s_1, \ldots, s_k$ not all 0, such that

$$s_1 \mathbf{Y}_1 + \cdots + s_k \mathbf{Y}_k = 0. \tag{1}$$

To see this, note that if (1) holds and one of the s_i, say, s_2, is not 0, we can solve for $\mathbf{Y}_2$, getting

$$\mathbf{Y}_2 = -\frac{s_1}{s_2} \mathbf{Y}_1 - \frac{s_3}{s_2} \mathbf{Y}_3 - \cdots - \frac{s_k}{s_2} \mathbf{Y}_k,$$

and so, $\mathbf{Y}_2$ is a linear combination of the remaining $\mathbf{Y}_i$ and, thus, the set $\mathbf{Y}_1, \mathbf{Y}_2, \ldots, \mathbf{Y}_k$ is linearly dependent.

Conversely, suppose the set $\mathbf{Y}_1, \ldots, \mathbf{Y}_k$ is linearly dependent. Then, for some i,

$$\mathbf{Y}_i = c_1 \mathbf{Y}_1 + \cdots + c_{i-1} \mathbf{Y}_{i-1} + c_{i+1} \mathbf{Y}_{i+1} + \cdots + c_k \mathbf{Y}_k,$$

and so, we have

$$c_1 \mathbf{Y}_1 + \cdots + c_{i-1} \mathbf{Y}_{i-1} + (-1)\mathbf{Y}_i + c_{i+1} \mathbf{Y}_{i+1} + \cdots + c_k \mathbf{Y}_k = 0.$$

Thus, (1) can be solved by the set of scalars $c_1, \ldots, c_{i-1}, -1, c_{i+1}, \ldots, c_k$, which are not all 0. Thus, deciding about linear dependence amounts to seeing whether (1) can be satisfied by scalars $s_1, \ldots, s_k$, which are not all 0.

A set of vectors $\mathbf{A}_1, \ldots, \mathbf{A}_k$ in V, which is *not* linearly dependent, is called *linearly independent*.

CHAPTER 5.1

Bases and Dimension

Let V be a vector space. Let $\mathbf{Y}_1, \ldots, \mathbf{Y}_l$ be a set of vectors in V. The *span* of this set of vectors is defined as the collection of all vectors $\mathbf{Y}$ in V of the form

$$\mathbf{Y} = s_1\mathbf{Y}_1 + s_2\mathbf{Y}_2 + \cdots + s_l\mathbf{Y}_l, \qquad s_1, \ldots, s_l \in \mathbb{R}.$$

The span of $\mathbf{Y}_1, \ldots, \mathbf{Y}_l$ is denoted

$$[\mathbf{Y}_1, \ldots, \mathbf{Y}_l].$$

A set of vectors $\mathbf{X}_1, \ldots, \mathbf{X}_k$ in V is called a *basis of* V if it has the following two properties:

(i) The span of $\mathbf{X}_1, \ldots, \mathbf{X}_k$ is V, and
(ii) the set $\mathbf{X}_1, \ldots, \mathbf{X}_k$ is linearly independent.

Suppose that $\mathbf{X}_1, \ldots, \mathbf{X}_k$ is a basis of V and consider a vector $\mathbf{X}$ in V. By (i), there exist scalars $c_1, \ldots, c_k$, such that

$$\mathbf{X} = c_1\mathbf{X}_1 + \cdots + c_k\mathbf{X}_k. \tag{1}$$

If $c'_1, \ldots, c'_k$ is a second set of scalars, such that

$$\mathbf{X} = c'_1\mathbf{X}_1 + \cdots + c'_k\mathbf{X}_k,$$

then

$$0 = (c_1 - c'_1)\mathbf{X}_1 + \cdots + (c_k - c'_k)\mathbf{X}_k.$$

It now follows from (ii) that $c_1 - c'_1 = 0, \ldots, c_k - c'_k = 0$, and so, $c_1 = c'_1$, $c_2 = c'_2, \ldots, c_n = c'_n$. Thus, each vector $\mathbf{X}$ in V has one and only one representation in the form (1). Conversely, if $\mathbf{X}_1, \ldots, \mathbf{X}_k$ is a set of vectors in V, such that every vector $\mathbf{X}$ in V has one and only one representation in the form (1), then properties (i) and (ii) hold, and so, $\mathbf{X}_1, \ldots, \mathbf{X}_k$ is a basis of V.

Exercise 1. Prove that if $\mathbf{Y}_1, \ldots, \mathbf{Y}_l$ is any set of vectors in V, then $[\mathbf{Y}_1, \ldots, \mathbf{Y}_l]$ is a subspace of V.

Examples of Bases.

EXAMPLE 1. The vectors $\mathbf{E}_1, \ldots, \mathbf{E}_n$ form a basis of $\mathbb{R}^n$, called *the standard basis of $\mathbb{R}^n$*.

EXAMPLE 2. Let V be the vector space of all vectors $\mathbf{X}$ in $\mathbb{R}^n$ such that $x_1 + x_2 + \cdots + x_n = 0$. The $(n-1)$-tuple of vectors

$$\begin{bmatrix} 1 \\ -1 \\ 0 \\ \vdots \\ 0 \end{bmatrix}, \begin{bmatrix} 1 \\ 0 \\ -1 \\ \vdots \\ 0 \end{bmatrix}, \ldots, \begin{bmatrix} 1 \\ 0 \\ 0 \\ \vdots \\ -1 \end{bmatrix}$$

is a basis of V.

Exercise 2. Verify that they do form such a basis.

EXAMPLE 3. The polynomials

$$1, x, x^2, \ldots, x^n$$

form a basis of the vector space $\mathbb{P}_n$.

EXAMPLE 4. The polynomials

$$1, 1 + x, 1 + x + x^2$$

form a basis for the vector space $\mathbb{P}_2$.

Exercise 3. Verify this.

Exercise 4. Exhibit a basis for the vector space T_n.

Note: In our definition of a basis, we require that a basis of V consist of a *finite* set of vectors in V. Most of the vector spaces we are interested in have a basis (in fact, they have many different bases). However, not every vector space has a basis in our sense.

EXAMPLE 5. The vector space $\mathbb{P}$ of all polynomials has no basis.

PROOF. Suppose the polynomials $P_1, \ldots, P_k$ form a basis of $\mathbb{P}$. Let d

denote the largest degree of these k polynomials. Since $x^{d+1} \in \mathbb{P}$, we have a representation

$$x^{d+1} = c_1 P_1 + \cdots + c_k P_k.$$

The right-hand side is a polynomial of degree $\leqslant d$, while the left-hand side has degree $d + 1$. This is impossible. So $\mathbb{P}$ has no basis.

Bases of $\mathbb{R}^3$. Let A_1, A_2, A_3 be any linearly independent triple of vectors in R^3. We claim that A_1, A_2, A_3 is a basis of $\mathbb{R}^3$. We can see this in the following way: the collection of all vectors $s_1 A_1 + s_2 A_2$ with $s_1, s_2 \in R$ is a plane Π. By hypothesis, A_3 sticks out of Π. If X is any vector in $\mathbb{R}^3$, we may draw a line L through the tip of X which is parallel to A_3. Then L intersects Π at a point P. We now have: The vector $OP = u_1 A_1 + u_2 A_2$ for certain scalars u_1, u_2. Also, the vector from P to the tip of $X = wA_3$ for some scalar w. Since $OP + PX = OX$, we have

$$u_1 A_1 + u_2 A_2 + wA_3 = X.$$

So our triple spans $\mathbb{R}^3$, and, being linearly independent by hypothesis, it forms a basis for $\mathbb{R}^3$, as asserted. The situation in $\mathbb{R}^n$ is similar for $n \geqslant 4$. In studying this situation in $\mathbb{R}^n$, we shall use a method of reasoning called the Principle of Mathematical Induction: A statement concerning $\mathbb{R}^n$ that is true for a certain value of n (say $n = 3$) and that, whenever it is true for $\mathbb{R}^n$ is also true for $\mathbb{R}^{n+1}$, is true for every value of $n \geqslant 3$. This is because, being true for 3, it must hold for 4, hence, for 5, and so on.

Theorem 5.1. *Let A_1, A_2, ..., A_n be a linearly independent n-tuple of vectors in $\mathbb{R}^n$. Then $A_1, ..., A_n$ is a basis of $\mathbb{R}^n$.*

PROOF. We shall show that if the statement of the theorem is true for a certain value of n, then it is also true for $n + 1$.

We then suppose it is true for n. We consider a linearly independent $(n + 1)$-tuple of vectors A_1, ..., A_{n+1} in $\mathbb{R}^{n+1}$. We have to show that $[A_1, ..., A_{n+1}] = \mathbb{R}^{n+1}$.

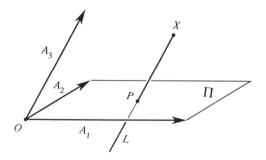

Figure 5.1

We can write

$$A_1 = \begin{pmatrix} U_1 \\ z_1 \end{pmatrix},$$

where $U_1 \in \mathbb{R}^n$, $z_1 \in \mathbb{R}$. Similarly,

$$A_j = \begin{pmatrix} U_j \\ z_j \end{pmatrix},$$

where $U_j \in \mathbb{R}^n$, $z_j \in \mathbb{R}$, $2 \leqslant j \leqslant n + 1$. Suppose that all $z_j = 0$. Then,

$$A_1 = \begin{pmatrix} U_1 \\ 0 \end{pmatrix}, \quad A_2 = \begin{pmatrix} U_2 \\ 0 \end{pmatrix}, \dots, A_{n+1} = \begin{pmatrix} U_{n+1} \\ 0 \end{pmatrix}.$$

The n-tuple $U_1, \dots, U_n$ in $\mathbb{R}^n$ is linearly independent, for if $\sum_{j=1}^n c_j U_j = 0$, then $\sum_{j=1}^n c_j A_j = 0$, and so, the $c_j = 0$ for all j. Since Theorem 5.1 is true in R^n (by assumption), $U_1, \dots, U_n$ is a basis for $\mathbb{R}^n$, and so, $U_{n+1} = \sum_{j=1}^n t_j U_j$ for certains scalars t_j. It follows that

$$A_{n+1} = \sum_{j=1}^{n} t_j A_j.$$

This is impossible. So the assumption that all $z_j = 0$ must be false. Some $z_j \neq 0$, and by renumbering the A_j, we get $z_{n+1} \neq 0$. Then,

$$A_1 = \begin{bmatrix} a_{11} \\ a_{12} \\ \vdots \\ a_{1n} \\ z_1 \end{bmatrix}, \dots, \quad A_{n+1} = \begin{bmatrix} a_{n+1,1} \\ a_{n+1,2} \\ a_{n+1,n} \\ z_{n+1} \end{bmatrix},$$

for certain scalars a_{ij}. The last entry in the vector

$$A_1 - \frac{z_1}{z_{n+1}} A_{n+1} \quad \text{is} \quad z_1 - \left(\frac{z_1}{z_{n+1}}\right) z_{n+1} = 0.$$

So,

$$A_1 - \left(\frac{z_1}{z_{n+1}}\right) A_{n+1} = \begin{pmatrix} B_1 \\ 0 \end{pmatrix}$$

for some B_1 in $\mathbb{R}^n$. Similarly,

$$A_j - \left(\frac{z_j}{z_{n+1}}\right) A_{n+1} = \begin{pmatrix} B_j \\ 0 \end{pmatrix}, \; B_j \quad \text{in } \mathbb{R}^n, \tag{2}$$

for $j = 1, \dots, n$.

The n-tuple $B_1, \dots, B_n$ in R^n is linearly independent, for otherwise some linear combination of $A_1, \dots, A_n$ will equal A_{n+1}. (Do the calculation.)

Since Theorem 5.1 is true in R^n, these exist scalars $c_1, \ldots, c_n$, with

$$\sum_{j=1}^{n} c_j \mathbf{B}_j = \begin{pmatrix} 1 \\ 0 \\ \vdots \\ 0 \end{pmatrix} \quad \text{in } R^n.$$

Hence,

$$\sum_{j=1}^{n} c_j \left(\mathbf{A}_j - \left(\frac{z_j}{z_{n+1}} \right) \mathbf{A}_{n+1} \right) = \begin{pmatrix} 1 \\ 0 \\ \vdots \\ 0 \\ 0 \end{pmatrix} = \mathbf{E}_1 \text{ in } \mathbb{R}^{n+1}. \tag{3}$$

Let us write S for the span $[\mathbf{A}_1, \ldots, \mathbf{A}_{n+1}]$. The left-hand side of (3) is in S, so $\mathbf{E}_1 \in S$. Similarly $\mathbf{E}_2, \ldots, \mathbf{E}_n \in S$. How can we capture $\mathbf{E}_{n+1}$? We have

$$\mathbf{A}_{n+1} = a_{n+1,\,1} \mathbf{E}_1 + \cdots + a_{n+1,\,n} \mathbf{E}_n + z_{n+1} \mathbf{E}_{n+1}.$$

So,

$$z_{n+1} \mathbf{E}_{n+1} = \mathbf{A}_{n+1} - a_{n+1,\,1} \mathbf{E}_1 - \cdots - a_{n+1,\,n} \mathbf{E}_n \in S.$$

Hence, $z_{n+1} \mathbf{E}_{n+1} \in S$, and so, $\mathbf{E}_{n+1} \in S$.

Thus, S contains every $\mathbf{E}_j$ in $\mathbb{R}^{n+1}$ and, hence, $S = \mathbb{R}^{n+1}$. We are done. We have shown that $\mathbf{A}_1, \ldots, \mathbf{A}_{n+1}$ is a basis in $\mathbb{R}^{n+1}$. So, Theorem 5.1 is true in $\mathbb{R}^{n+1}$. Since Theorem 5.1 is true in $\mathbb{R}^3$, it follows that the theorem is true in $\mathbb{R}^n$ for every $a \geqslant 3$. It is also true for $\mathbb{R}^2$ and $\mathbb{R}^1$. (Can you determine why this is true?) So it is true for all n.

Corollary. *Every $(n + 1)$-tuple of vectors in $\mathbb{R}^n$ is linearly dependent.*

Exercise 5. Deduce this corollary from Theorem 5.1.

We can use Theorem 5.1 and the corollary to get information about bases in arbitrary vector spaces.

Let V be a vector space, and let $\mathbf{C}_1, \ldots, \mathbf{C}_k$ be a basis of V. For each vector $\mathbf{X}$ in V, we can write

$$\mathbf{X} = \sum_{j=1}^{k} x_i \mathbf{C}_i \tag{4}$$

and the scalars x_i are uniquely determined by $\mathbf{X}$. We define a map Φ from V to R^k as follows: for each $\mathbf{X}$ in V,

$$\Phi(\mathbf{X}) = \begin{pmatrix} x_1 \\ x_2 \\ \vdots \\ x_k \end{pmatrix},$$

where the x_i are defined by (4).

Exercise 6. Prove that the map Φ has the following properties:

$$\Phi(s\mathbf{X} + t\mathbf{Y}) = s\Phi(\mathbf{X}) + t\Phi(\mathbf{Y}), \quad \forall \mathbf{X}, \mathbf{Y} \in V, s, t \in R \tag{5}$$

(We say: Φ is a *linear map*).

$$\text{If} \quad \mathbf{X} \neq \mathbf{Y}, \quad \text{then } \Phi(\mathbf{X}) \neq \Phi(\mathbf{Y}). \tag{6}$$

(We say: Φ is *one-to-one*.)

$$\text{If} \quad T \in \mathbb{R}^k, \quad \text{then there is an } \mathbf{X} \text{ in } V \text{ with } \Phi(\mathbf{X}) = T. \tag{7}$$

(We say, Φ *maps onto* $\mathbb{R}^k$).

Now, let V be a vector space with a basis $\mathbf{B}_1, \ldots, \mathbf{B}_n$ consisting of n vectors and another basis $\mathbf{C}_1, \ldots, \mathbf{C}_k$ consisting of k vectors.

Claim. $k \geqslant n$. Suppose the claim is false. Then $k < n$. We shall see that this leads to a contradiction.

For each $\mathbf{X}$ in V, we can write uniquely,

$$\mathbf{X} = \sum_{j=1}^{k} x_i \mathbf{C}_i, \, x_i \quad \text{in } \mathbb{R},$$

and we define a map Φ from V to $\mathbb{R}^k$ as above by

$$\Phi(\mathbf{X}) = \begin{bmatrix} x_1 \\ x_2 \\ \vdots \\ x_k \end{bmatrix} \quad \text{for all } \mathbf{X} \text{ in } V.$$

Consider the $(k + 1)$-tuple of vectors $\Phi(\mathbf{B}_1), \ldots, \Phi(\mathbf{B}_{k+1})$ in $\mathbb{R}^k$. By the corollary to Theorem 5.1, this $(k + 1)$-tuple is linearly dependent, so there exist scalars $t_1, \ldots, t_{k+1}$, not all 0, such that

$$\sum_{j=1}^{k+1} t_j \Phi(\mathbf{B}_j) = 0. \tag{8}$$

Using (5) in Exercise 6, we see that the left-hand side in (8) equals

$$\Phi\left(\sum_{j=1}^{k+1} t_j \mathbf{B}_j\right).$$

So

$$\Phi\left(\sum_{j=1}^{k+1} t_j \mathbf{B}_j\right) = 0$$

and, hence, by (6),

$$\sum_{j=1}^{k+1} t_j \mathbf{B}_j = 0.$$

This is impossible, since $\mathbf{B}_1, \ldots, \mathbf{B}_n$ is a basis of V and, hence, is linearly independent. We have arrived at a contradiction.

So the claim is true and $k \geqslant n$. A similar argument gives $n \geqslant k$. Hence, $n = k$. We have proven the following important fact about vector spaces.

Theorem 5.2. *Let V be a vector space. Every two bases of V consist of the same number of vectors.*

If V is a vector space that has a basis of n elements, then every other basis of V also has n elements. We define the *dimension* of V, denoted dim V, to be the integer n. We say: V is *n-dimensional*.

EXAMPLE 6. dim $\mathbb{R}^n = n$, because the standard basis of $\mathbb{R}^n$ has n elements.

EXAMPLE 7. dim $\mathbb{P}_n = n + 1$, because of the basis of Example 3.

Exercise 7. Calculate the dimension of the subspace of $\mathbb{P}_4$ consisting of all polynomials in $\mathbb{P}_4$ with $P(0) = P(1)$.

Do the same for the subspace of $\mathbb{P}_5$ consisting of all polynomials P in $\mathbb{P}_5$ with $P(0) = P'(0) = 0$, where P' denotes the derivative of P.

Exercise 8. Let V be a vector space of dimension n. Let $\mathbf{C}_1, \ldots, \mathbf{C}_k$ be a linearly independent k-tuple of vectors in V. Show that $k \leqslant n$.

Hint. Choose a basis $\mathbf{B}_1, \ldots, \mathbf{B}_n$ of V. As above, construct a one-to-one linear transformation Φ of V on $\mathbb{R}^n$. Consider the k-tuple of vectors $\Phi(\mathbf{C}_1), \ldots, \Phi(\mathbf{C}_k)$ in $\mathbb{R}^n$, and use the corollary of Theorem 5.1 to conclude that $k \leqslant n$.

Exercise 9. Let V be a vector space of dimension n. Let W be a subspace of V.

(a) Show that W has dimension $\leqslant n$.
(b) Show that if W has dimension n, then $W = V$.

Existence and Uniqueness of Solutions

Recall the inhomogeneous system (I) of k linear equations in n unknowns, which we studied in Chapter 4.4. We consider the case $k = n$, where the number of equations to be solved equals the number of unknowns, and we shall use our results about bases in $\mathbb{R}^n$ in the preceding chapter to study the question of existence and uniqueness of solutions of the system (I).

We consider the vectors

$$\mathbf{C}_1 = \begin{bmatrix} a_{11} \\ a_{21} \\ \vdots \\ a_{n1} \end{bmatrix}, \; \mathbf{C}_2 = \begin{bmatrix} a_{12} \\ a_{22} \\ \vdots \\ a_{n2} \end{bmatrix}, \; \dots, \; \mathbf{C}_n = \begin{bmatrix} a_{1n} \\ a_{2n} \\ \vdots \\ a_{nn} \end{bmatrix}, \quad \text{in } \mathbb{R}^n,$$

and call them the *column vectors* of the system (I). Putting $\mathbf{U} = (u_1, \dots, u_n)$ and $\mathbf{X} = (x_1, \dots, x_n)$, we can write (I) in the form

$$\sum_{j=1}^{n} x_i \mathbf{C}_i = \mathbf{U}. \tag{I'}$$

Claim 1. Fix $\mathbf{U}$ in $\mathbb{R}^n$. There exists a solution $\mathbf{X}$ of the system (I') with $\mathbf{U}$ as right-hand side if, and only if, $\mathbf{U} \in [\mathbf{C}_1, \dots, \mathbf{C}_n]$.

PROOF. If there is a solution of (I'), then $\mathbf{U} \in [\mathbf{C}_1, \dots, \mathbf{C}_n]$. Conversely, if $\mathbf{U} \in [\mathbf{C}_1, \dots, \mathbf{C}_n]$, then there exist scalars t_i so that $\mathbf{U} = \sum_{j=1}^{k} t_i \mathbf{C}_i$, and so, $t_1, \dots, t_n$ solves (I').

Claim 2. $[\mathbf{C}_1, \dots, \mathbf{C}_n] = \mathbb{R}^n$ if, and only if, the n-tuple $\mathbf{C}_1, \dots, \mathbf{C}_n$ is linearly independent.

PROOF. An n-tuple of vectors in $\mathbb{R}^n$ is a basis of $\mathbb{R}^n$ if, and only if, the n-tuple is linearly independent.

Claim 3. The n-tuple $\mathbf{C}_1, \ldots, \mathbf{C}_n$ is linearly independent if, and only if, the homogeneous system (H),

$$a_{11}x_1 + a_{12}x_2 + \cdots + a_{1n}x_n = 0,$$

$$\vdots$$

$$a_{n1}x_1 + a_{x2}x_2 + \cdots + a_{nn}x_n = 0,$$

has 0 as its only solution.

PROOF. (H) can be written in the form

$$\sum_{j=1}^{n} x_i \mathbf{C}_i = 0.$$

(H) has 0 as its only solution precisely when the n-tuple $\mathbf{C}_1, \ldots, \mathbf{C}_n$ is linearly independent.

Theorem 5.3. *If the homogeneous system* (H) *has 0 as its only solution, then the inhomogeneous system* (I) *has a unique solution for each choice of right-hand side* U.

PROOF. Existence of a solution follows from the claims 1, 2, 3. Uniqueness holds because if $\mathbf{X}$ and $\mathbf{X}'$ are solutions of (I) corresponding to the same right-hand side U, then $\mathbf{X} - \mathbf{X}'$ is a solution of (H) and so, by assumption, $= 0$.

Theorem 5.4. *If the homogeneous system* (H) *has a non-zero solution, then there exists* U *such that* (I) *has no solution for this* U.

PROOF. Claim 3 yields that the n-tuple $\mathbf{C}_1, \ldots, \mathbf{C}_n$ is not linearly independent, hence, does not span $\mathbb{R}^n$, so there exists U in $\mathbb{R}^n$, which does not belong to $[\mathbf{C}_1, \ldots, \mathbf{C}_n]$, and so, (I) has no solution for this U.

Exercise 1. Show that if the column vectors $\mathbf{C}_1, \ldots, \mathbf{C}_n$ are linearly independent, then (I) has a solution for each right-hand side U.

Exercise 2. Define *row vectors* $\mathbf{R}_1, \ldots, \mathbf{R}_n$ in $\mathbb{R}^n$ by $\mathbf{R}_1 = (a_{11}, \ldots, a_{1n})$, etc. Show that if the n-tuple of row vectors is linearly independent then (I) has a solution for every U.

The Matrix Relative to a Given Basis

In Chapter 3.2 we assigned to each linear transformation T of $\mathbb{R}^3$ a matrix $m(T)$ as follows:

$$\text{If} \quad T\begin{bmatrix} x_1 \\ x_2 \\ x_3 \end{bmatrix} = \begin{bmatrix} a_1x_1 + a_2x_2 + a_3x_3 \\ b_1x_1 + b_2x_2 + b_3x_3 \\ c_1x_1 + c_2x_2 + c_3x_3 \end{bmatrix}, \quad \text{then}$$

$$m(T) = \begin{bmatrix} a_1 & a_2 & a_3 \\ b_1 & b_2 & b_3 \\ c_1 & c_2 & c_3 \end{bmatrix}.$$

We shall now extend this definition to assign to each linear transformation T on an n-dimensional vector space V with a given basis $\mathbf{B}$ an $n \times n$ matrix called *the matrix of T relative to the basis* $\mathbf{B}$, *and denoted* $m_{\mathbf{B}}(T)$.

If $\mathbf{X}$ is a vector in V, then $\mathbf{X}$ has an n-tuple of coordinates $x_1, \ldots, x_n$ relative to $\mathbf{B}$ given by the equation

$$\mathbf{X} = \sum_{j=1}^{n} x_i \mathbf{B}_i, \tag{1}$$

where $\mathbf{B}_1, \ldots, \mathbf{B}_n$ are the vectors making up the basis $\mathbf{B}$. The vector $T\mathbf{X}$ has an n-tuple of coordinates $y_1, \ldots, y_n$, such that

$$T\mathbf{X} = \sum_{j=1}^{n} y_j \mathbf{B}_j. \tag{2}$$

We shall exhibit an $n \times n$ matrix

$$\begin{bmatrix} a_{11} & a_{12} & \cdots & a_{1n} \\ a_{21} & a_{22} & \cdots & a_{2n} \\ \vdots & & & \\ a_{n1} & a_{n2} & \cdots & a_{nn} \end{bmatrix},$$

such that

$$\begin{bmatrix} a_{11} & a_{12} & \cdots & a_{1n} \\ a_{21} & a_{22} & \cdots & a_{2n} \\ \vdots & & & \\ a_{n1} & a_{n2} & \cdots & a_{nn} \end{bmatrix} \begin{bmatrix} x_1 \\ x_2 \\ \vdots \\ x_n \end{bmatrix} = \begin{bmatrix} y_1 \\ y_2 \\ \vdots \\ y_n \end{bmatrix},$$

that is, such that we have

$$y_1 = \sum_{j=1}^{n} a_{1j} x_j$$

$$y_2 = \sum_{j=1}^{n} a_{2j} x_j$$

$$\vdots \tag{3}$$

$$y_n = \sum_{j=1}^{n} a_{nj} x_j$$

for every vector $\mathbf{X}$ given by (1).

We choose the numbers a_{ij} as follows: since $\mathbf{B}_1, \ldots, \mathbf{B}_n$ forms a basis, we can find scalars a_{ij}, $1 \leqslant i, j \leqslant n$, such that

$$T\mathbf{B}_1 = \sum_{k=1}^{n} a_{k1} \mathbf{B}_k,$$

$$T\mathbf{B}_2 = \sum_{k=1}^{n} a_{k2} \mathbf{B}_k,$$

$$\vdots \qquad \vdots$$

$$T\mathbf{B}_n = \sum_{k=1}^{n} a_{kn} \mathbf{B}_k.$$

Applying T to equation (1), we get

$$T\mathbf{X} = \sum_{j=1}^{n} x_j T\mathbf{B}_j = \sum_{j=1}^{n} x_j \sum_{k=1}^{n} a_{kj} \mathbf{B}_k = \sum_{k=1}^{n} \left(\sum_{j=1}^{n} a_{kj} x_j \right) \mathbf{B}_k.$$

It now follows from (2) and the fact that the $\mathbf{B}_k$ are linearly independent that

$$y_k = \sum_{j=1}^{n} a_{kj} x_j, \qquad k = 1, 2, \ldots, n.$$

Thus, (3) holds, as claimed.

Definition 1. The matrix

$$\begin{bmatrix} a_{11} & a_{12} & \cdots & a_{1n} \\ a_{21} & a_{22} & \cdots & a_{2n} \\ \vdots & & & \\ a_{n1} & a_{n2} & \cdots & a_{nn} \end{bmatrix}$$

is called the matrix of T relative to the basis $\mathbf{B}$, and is denoted $m_{\mathbf{B}}(T)$.

Exercise 1. Show that the columns of the matrix $m_B(T)$ are the coordinate n-tuples of the vectors $T\mathbf{B}_1, T\mathbf{B}_2, \ldots, T\mathbf{B}_n$, relative to the basis **B**.

EXAMPLE 1. Let V be $\mathbb{R}^2$ and let T be the linear transformation of $\mathbb{R}^2$ given by the equations

$$x' = 2x + 3y,$$
$$y' = x - y.$$

Take **B** to be the basis: $\mathbf{B}_1 = \begin{pmatrix} 1 \\ 1 \end{pmatrix}$, $\mathbf{B}_2 = \begin{pmatrix} -3 \\ 2 \end{pmatrix}$. Then

$$T\mathbf{B}_1 = \begin{pmatrix} 5 \\ 0 \end{pmatrix} = 2\begin{pmatrix} 1 \\ 1 \end{pmatrix} + (-1)\begin{pmatrix} -3 \\ 2 \end{pmatrix} = 2\mathbf{B}_1 + (-1)\mathbf{B}_2,$$

$$T\mathbf{B}_2 = \begin{pmatrix} 0 \\ -5 \end{pmatrix} = (-3)\begin{pmatrix} 1 \\ 1 \end{pmatrix} + (-1)\begin{pmatrix} -3 \\ 2 \end{pmatrix} = -3\mathbf{B}_1 + (-1)\mathbf{B}_2.$$

Thus, the coordinate pair of $T\mathbf{B}_1$ relative to the basis **B** is $(2, -1)$ and that of $T\mathbf{B}_2$ is $(-3, -1)$. By Exercise 1, then,

$$m_B(T) = \begin{pmatrix} 2 & -3 \\ -1 & -1 \end{pmatrix}.$$

EXAMPLE 2. Let S be the linear transformation of $\mathbb{R}^2$ given by equations:

$$x' = 3x + 4y,$$
$$y' = 4x - 3y.$$

Let **B'** be the basis $\mathbf{B}'_1, \mathbf{B}'_2$ where $\mathbf{B}'_1 = \begin{pmatrix} 2 \\ 1 \end{pmatrix}$ $\mathbf{B}'_2 = \begin{pmatrix} -1 \\ 2 \end{pmatrix}$. Then,

$$S\begin{pmatrix} 2 \\ 1 \end{pmatrix} = \begin{pmatrix} 10 \\ 5 \end{pmatrix} = 5\begin{pmatrix} 2 \\ 1 \end{pmatrix} + 0\begin{pmatrix} -1 \\ 2 \end{pmatrix},$$

$$S\begin{pmatrix} -1 \\ 2 \end{pmatrix} = \begin{pmatrix} 5 \\ -10 \end{pmatrix} = 0\begin{pmatrix} 2 \\ 1 \end{pmatrix} - 5\begin{pmatrix} -1 \\ 2 \end{pmatrix}.$$

Hence,

$$m_{B'}(S) = \begin{pmatrix} 5 & 0 \\ 0 & -5 \end{pmatrix}.$$

Note: $m_{B'}(S)$ is simpler than the matrix

$$\begin{pmatrix} 3 & 4 \\ 4 & -3 \end{pmatrix}$$

of S relative to the standard basis $\mathbf{E}_1, \mathbf{E}_2$. This is no accident. **B'** is a basis that is tailor-made for the transformation S. (Compare Example 4, Section 2.6). More generally, given a linear transformation T of a vector space V, a good choice of basis $\mathbf{B}$ may yield a matrix $m_{\mathbf{B}}(T)$, which has a simple form. In Section 7.0 below, we shall pursue such a good choice of basis for an important class of linear transformations, the *self-adjoint* transformations.

Theorem 5.5. *Let S and T be two linear transformations of V and $\mathbf{B}$ a basis of V. Then,*

$$m_{\mathbf{B}}(ST) = m_{\mathbf{B}}(S) \cdot m_{\mathbf{B}}(T). \tag{4}$$

PROOF. We put $t = m_{\mathbf{B}}(T)$ and $s = m_{\mathbf{B}}(S)$, and we write t_{ij} for the (i,j)-entry of the matrix t, and s_{ij} for the (i,j)-entry of s.

Fix a vector $\mathbf{X}$ in V and let $x_1, \ldots, x_n$ be the coordinates of $\mathbf{X}$. Also, let $y_1, \ldots, y_n$ be the coordinates of $T\mathbf{X}$ and $z_1, \ldots, z_n$ the coordinates of $(ST)(\mathbf{X}) = S(T\mathbf{X})$. Then,

$$y_i = \sum_{j=1}^{n} t_{ij}x_j, \quad i = 1, \ldots, n$$

and

$$z_k = \sum_{i=1}^{n} s_{ki}y_i, \quad k = 1, \ldots, n.$$

Hence,

$$z_k = \sum_{i=1}^{n} s_{ki} \sum_{j=1}^{n} t_{ij}x_j = \sum_{j=1}^{n} \left(\sum_{i=1}^{n} s_{ki}t_{ij} \right) x_j,$$

$$z_k = \sum_{j=1}^{n} (s \cdot t)_{kj}x_j.$$

It follows that the matrix $s \cdot t = m_{\mathbf{B}}(ST)$, and so,

$$m_{\mathbf{B}}(S) \cdot m_{\mathbf{B}}(T) = m_{\mathbf{B}}(ST),$$

that is, (4).

Change of Basis. Given two bases $\mathbf{B} = (\mathbf{B}_1, \ldots, \mathbf{B}_n)$ and $\mathbf{B}' = (\mathbf{B}'_1, \ldots, \mathbf{B}'_n)$ of the vector space V. Each vector $\mathbf{X}$ receives two n-tuples of coordinates $(x_1, \ldots, x_n)$ and $(x'_1, \ldots, x'_n)$, where

$$\mathbf{X} = \sum_{j=1}^{n} x_j\mathbf{B}_j \quad \text{and} \quad \mathbf{X} = \sum_{j=1}^{n} x'_j\mathbf{B}'_j.$$

Exercise 2. Show the following:

(a) There exist matrices $c = ((c_{ij}))$ and $d = ((d_{ij}))$, such that whenever $(x_1, \ldots, x_n)$ and $(x'_1, \ldots, x'_n)$ are n-tuples of coordinates of a vector $\mathbf{X}$, as above, then

$$x'_j = \sum_{k=1}^{n} c_{jk}x_k, \quad j = 1, \ldots, n, \tag{5a}$$

and

$$x_j = \sum_{l=1}^{n} d_{jl} x'_l, \quad j = 1, \ldots, n. \tag{5b}$$

(b) c and d are uniquely determined by the relations (5a) and (5b).
(c) $d = c^{-1}$.

We use this Exercise in the following proof.

Theorem 5.6. *Let* **B**, **B'** *be two bases of the vector space V and T be a linear transformation of V. Then*

$$m_{\mathbf{B'}}(T) = c m_{\mathbf{B}}(T) c^{-1}, \tag{6}$$

where c is the matrix in (5a).

PROOF. Let **X** be a vector in V and put $\mathbf{Y} = T\mathbf{X}$. Let $(x_1, \ldots, x_n)$ and $(y_1, \ldots, y_n)$ be the coordinate n-tuples of **X** and **Y** relative to **B**, and let $(x'_1, \ldots, x'_n)$ and $(y'_1, \ldots, y'_n)$ be the coordinate n-tuples of **X** and **Y** relative to **B'**. Put

$$a = m_{\mathbf{B}}(T) \quad \text{and} \quad a' = m_{\mathbf{B'}}(T),$$

and let a_{ij}, respectively, a'_{ij}, be the (i, j)-entry in a, respectively, a'. Fix i. Then

$$y'_i = \sum_{k=1}^{n} c_{ik} y_k = \sum_{k=1}^{n} c_{ik} \left(\sum_j a_{kj} x_j \right) = \sum_j \left(\sum_k c_{ik} a_{kj} \right) x_j$$

$$= \sum_j (ca)_{ij} x_j = \sum_j (ca)_{ij} \left(\sum_l d_{jl} x'_l \right) = \sum_l \left(\sum_j (ca)_{ij} d_{jl} \right) x'_l$$

$$= \sum_l (cad)_{il} x'_l.$$

Hence,

$$cad = a'.$$

So

$$c m_{\mathbf{B}}(T) c^{-1} = m_{\mathbf{B'}}(T),$$

that is, (6) holds. We are done.

EXAMPLE 3. $V = \mathbb{P}_2 = $ the space of polynomials x of degree $\leqslant 2$. Let D be the transformation of differentiation, that is, $Df = \dfrac{df}{dx}$ for each polynomial f in $\mathbb{P}_2$. Since the derivative of a polynomial of degree $\leqslant 2$ is another such polynomial, and since differentiation is a linear operation, D is a linear transformation of $\mathbb{P}_2$. We let **B** be the basis

$$\mathbf{B}_1 = 1, \quad \mathbf{B}_2 = x, \quad \mathbf{B}_3 = x^2.$$

Then,

$$DB_1 = 0 = 0B_1 + 0B_2 + 0B_3,$$
$$DB_2 = 1 = 1B_1 + 0B_2 + 0B_3,$$
$$DB_3 = 2x = 0B_1 + 2B_2 + 0B_3.$$

So,

$$m_B(D) = \begin{bmatrix} 0 & 1 & 0 \\ 0 & 0 & 2 \\ 0 & 0 & 0 \end{bmatrix}.$$

Exercise 3. T is the linear transformation of $\mathbb{R}^3$ given by equation

$$x_1' = x_1 - x_3,$$
$$x_2' = x_1 - x_3,$$
$$x_3' = x_2 - x_3.$$

Find a basis **B** of $\mathbb{R}^3$, such that

$$m_B(T) = \begin{bmatrix} 0 & 0 & 0 \\ 1 & 0 & 0 \\ 0 & 1 & 0 \end{bmatrix}.$$

Exercise 4. $V = \mathbb{P}_3$. Take as basis **B**: 1, x, x^2, x^3. Let T be the transformation: $f(x) \to f(x + 1)$ of $\mathbb{P}_3$ into $\mathbb{P}_3$. Find the matrix $m_B(T)$.

Exercise 5. V is the space of all functions

$$f(x) = a \cos(x) + b \sin(x), \quad a, b \quad \text{in } \mathbb{R}.$$

Let T be the transformation of V which sends $f(x)$ into $f(x + \pi)$ for each f, and let **B** be the basis of V with $B_1 = \cos x$, $B_2 = \sin x$.

(a) Show that T is a linear transformation of V.
(b) Calculate $m_B(T)$.

Vector Spaces with an Inner Product

We found it useful to introduce the dot product of two vectors in order to study the geometry of $\mathbb{R}^2$ and $\mathbb{R}^3$. There is a natural generalization of the dot product to an arbitrary vector space, which is called an *inner product*.

Let V be a vector space. An *inner product* $(\ ,\ )$ on V is a rule that assigns to each pair of vectors $\mathbf{X}$, $\mathbf{Y}$ in V a real number $(\mathbf{X}, \mathbf{Y})$ in such a way that

if $\mathbf{X}_1, \mathbf{X}_2, \mathbf{Y} \in V$ and $a, b \in \mathbb{R}$, then $\qquad\qquad\qquad$ (1)

$$(a\mathbf{X}_1 + b\mathbf{X}_2, \mathbf{Y}) = a(\mathbf{X}_1, \mathbf{Y}) + b(\mathbf{X}_2, \mathbf{Y});$$

if $\mathbf{X}, \mathbf{Y} \in V$, then $\qquad\qquad\qquad\qquad\qquad\qquad\qquad\qquad$ (2)

$$(\mathbf{X}, \mathbf{Y}) = (\mathbf{Y}, \mathbf{X});$$

$(\mathbf{X}, \mathbf{X}) > 0$ if $\mathbf{X} \neq \mathbf{0}$ and $(\mathbf{0}, \mathbf{0}) = \mathbf{0}$. $\qquad\qquad\qquad$ (3)

Properties (1) and (2) imply

If $\mathbf{X}, \mathbf{Y}_1, \mathbf{Y}_2 \in V$ and $a, b \in R$, then $\qquad\qquad\qquad\qquad$ (1′)

$$(\mathbf{X}, a\mathbf{Y}_1 + b\mathbf{Y}_2) = a(\mathbf{X}, \mathbf{Y}_1) + b(\mathbf{X}, \mathbf{Y}_2).$$

We express properties (1) and (1′) together by saying that the inner product is *bilinear*, property (2) by saying that the inner product is *symmetric*, and property (3) by saying that the inner product is *positive definite*.

We recognize these three properties as familiar rules for the dot product on $\mathbb{R}^2$ and $\mathbb{R}^3$.

Two vectors $\mathbf{X}$, $\mathbf{Y}$ in V are called *orthogonal* (we write: $\mathbf{X} \perp \mathbf{Y}$) if $(\mathbf{X}, \mathbf{Y}) = 0$. The *length* of the vector $\mathbf{X}$ in V, denoted $|\mathbf{X}|$, is defined by $|\mathbf{X}| = \sqrt{(\mathbf{X}, \mathbf{X})}$.

Examples of Inner Products

EXAMPLE 1. Let $V = \mathbb{R}^n$. For

$$\mathbf{X} = \begin{bmatrix} x_1 \\ \vdots \\ x_n \end{bmatrix}, \; \mathbf{Y} = \begin{bmatrix} y_1 \\ \vdots \\ y_n \end{bmatrix} \in \mathbb{R}^n,$$

we put

$$(\mathbf{X}, \mathbf{Y}) = x_1 y_1 + x_2 y_2 + \cdots + x_n y_n = \sum_{i=1}^{n} x_i y_i.$$

EXAMPLE 2. V is a subspace of $\mathbb{R}^n$. For $\mathbf{X}$, $\mathbf{Y}$ in V, define $(\mathbf{X}, \mathbf{Y})$ as in Example 1. Then (,) is an inner product on V.

EXAMPLE 3. $V = \mathbb{P}_n$. For

$$f = \sum_{j=0}^{n} a_j x^j, \quad g = \sum_{j=0}^{n} b_j x^j$$

in $\mathbb{P}_n$, define

$$(f, g) = \int_0^1 f(x)g(x)\, dx.$$

Verify that this definition makes (,) an inner product on $\mathbb{P}_n$.

Exercise 1. Express (f, g) in terms of the coefficients $a_0, \ldots, a_n, b_0, \ldots, b_n$.

EXAMPLE 4. $V = \mathbb{T}_n$. For

$$f = a_0 + \sum_{j=1}^{n} (a_j \cos jx + b_j \sin jx),$$

$$g = a_0' + \sum_{j=1}^{n} (a_j' \cos jx + b_j' \sin jx),$$

define $(f, g) = \int_0^{2\pi} f(x)g(x)\, dx$.

Exercise 2. (a) Show that (,) is an inner product on $\mathbb{T}_n$, (b) Express (f, g) in terms of the coefficients $a_j, b_j a_j', b_j'$.

Exercise 3. With the inner product on $\mathbb{P}_2$ given in Example 3, (a) find a non-zero vector h in $\mathbb{P}_2$ such that $h \perp 1$ and $h \perp x$. (b) Show that if $k \in \mathbb{P}_2$ and if $k \perp 1$, $k \perp x$ and $k \perp x^2$, then $k = 0$.

Orthonormal Bases

Let V be a vector space. Let $\mathbf{B}_1, \ldots, \mathbf{B}_n$ be a basis of V. To each vector $\mathbf{X}$, we let correspond a set of scalars $x_1, \ldots, x_n$ called the *coordinates of* $\mathbf{X}$ *relative to the basis* $\mathbf{B}_1, \ldots, \mathbf{B}_n$, by putting:

$$\mathbf{X} = x_1\mathbf{B}_1 + x_2\mathbf{B}_2 + \cdots + x_n\mathbf{B}_n.$$

Usually, it is laborious to compute the coordinates of a given vector.

EXAMPLE 5.

$$\begin{pmatrix} 2 \\ -1 \\ -2 \end{pmatrix}, \quad \begin{pmatrix} 1 \\ 1 \\ 0 \end{pmatrix}, \quad \begin{pmatrix} 3 \\ 5 \\ 1 \end{pmatrix}$$

is a basis of $\mathbb{R}^3$. What are the coordinates of the vector $\mathbf{X} = \begin{pmatrix} 1 \\ 0 \\ 0 \end{pmatrix}$ relative to this basis?

We write

$$\mathbf{X} = \begin{pmatrix} 1 \\ 0 \\ 0 \end{pmatrix} = x_1 \begin{pmatrix} 2 \\ -1 \\ -2 \end{pmatrix} + x_2 \begin{pmatrix} 1 \\ 1 \\ 0 \end{pmatrix} + x_3 \begin{pmatrix} 3 \\ 5 \\ 1 \end{pmatrix} \quad \text{or}$$

$$\begin{cases} 1 = & 2x_1 + x_2 + 3x_3 \\ 0 = & -x_1 + x_2 + 5x_3 \\ 0 = & -2x_1 + x_3 \end{cases} \tag{4}$$

Solving the system (4), we find

$$x_3 = 2x_1,$$

and so

$$1 = 2x_1 + x_2 + 6x_1 = 8x_1 + x_2$$

$$0 = -x_1 + x_2 + 10x_1 = 9x_1 + x_2.$$

So,

$$x_2 = -9x_1 \quad \text{and} \quad 1 = 8x_1 - 9x_1 = -x_1,$$

so

$$x_1 = -1, \, x_2 = 9, \, x_3 = -2.$$

So $-1, 9, -2$ are the coordinates of $\mathbf{X}$ relative to the given basis.

As you see, we can do it but it takes a bit of work. For a certain class of bases, the problem of calculating the coordinates of a given vector relative to the basis is very easy. These are the so-called *orthonormal bases*.

Let V be a vector space with an inner product (,). An *orthogonal set* of vectors in V is a set of vectors $\mathbf{X}_1, \ldots, \mathbf{X}_k$, such that $\mathbf{X}_i \perp \mathbf{X}_j$ whenever $i \neq j$.

Exercise 4. Let $\mathbf{X}_1, \ldots, \mathbf{X}_k$ be an orthogonal set of vectors in V, with $\mathbf{X}_i \neq 0$ for each i. Show that $\mathbf{X}_1, \ldots, \mathbf{X}_k$ is a linearly independent set in V.

An orthogonal set of vectors, each of which has length 1, is called an *ortho-normal set of vectors*. If, in addition, the set of vectors is a basis, we call it an *orthonormal basis of V*. Thus, $\mathbf{B}_1, \ldots, \mathbf{B}_n$ is an orthonormal basis of V provided that (a) $(\mathbf{B}_i, \mathbf{B}_j) = 0$ if $i \neq j$, (b) $(\mathbf{B}_i, \mathbf{B}_i) = 1$ for each i, and (c) $\mathbf{B}_1, \ldots, \mathbf{B}_n$ is a basis of V.

Theorem 6.1. *Let $\mathbf{B}_1, \ldots, \mathbf{B}_n$ be an orthonormal basis of V. Let $\mathbf{X}$ be a vector in V, and let $x_1, \ldots, x_n$ be the coordinates of $\mathbf{X}$ relative to the basis $\mathbf{B}_1, \ldots, \mathbf{B}_n$. Then,*

$$x_i = (\mathbf{X}, \mathbf{B}_i), \quad i = 1, \ldots, n. \tag{5}$$

PROOF. $\mathbf{X} = x_1 \mathbf{B}_1 + \cdots + x_n \mathbf{B}_n$. Take the inner product of this equation with $\mathbf{B}_i$ for some value of i and use the bilinearity of the inner product. Then

$$(\mathbf{X}, \mathbf{B}_i) = x_1(\mathbf{B}_1, \mathbf{B}_i) + x_2(\mathbf{B}_2, \mathbf{B}_i) + \cdots + x_n(\mathbf{B}_n, \mathbf{B}_i).$$

Since $(\mathbf{B}_j, \mathbf{B}_i) = 0$ whenever $j \neq i$ and $(B_i, B_i) = 1$, this gives $(\mathbf{X}, \mathbf{B}_i) = x_i$, as was to be shown.

EXAMPLE 6. V_0 is the vector space which is the plane $2x + 3y + 4z = 0$ in $\mathbb{R}^3$, and the inner product on V_0 is that given in Example 2. We seek an orthonormal basis of V_0: We have $(-3, 2, 0) \in V$ and $(2, 3, 4)$ is a normal vector to V_0. Hence, the cross-product of these two vectors: $(8, 12, -13)$ is in V_0 and is orthogonal to $(-3, 2, 0)$. Thus,

$$\mathbf{X}_1 = \frac{1}{\sqrt{9+4}}(-3, 2, 0) \quad \text{and} \quad \mathbf{X}_2 = \frac{1}{\sqrt{8^2 + 12^2 + (-13)^2}}(8, 12, -13)$$

is an orthonormal basis for V_0.

Theorem 6.2. *We use the notation of Theorem 6.1. Fix* $\mathbf{X} \in V$. *Then,*

$$|\mathbf{X}|^2 = \sum_{i=1}^n (\mathbf{X}, \mathbf{B}_i)^2. \tag{6}$$

PROOF. $\mathbf{X} = \sum_{i=1}^n x_i \mathbf{B}_i$, where $x_i = (\mathbf{X}, \mathbf{B}_i)$.

$$(\mathbf{X}, \mathbf{X}) = \left(\sum_i x_i \mathbf{B}_i, \sum_j x_j \mathbf{B}_j\right).$$

Using the bilinearity of the inner product, we get

$$|\mathbf{X}|^2 = (\mathbf{X}, \mathbf{X}) = \sum_{i,j} x_i x_j (\mathbf{B}_i, \mathbf{B}_j) = \sum_{i=1}^n x_i^2 = \sum_{i=1}^n (\mathbf{X}, \mathbf{B}_i)^2$$

as was to be proved.

Using the same method, we can prove a generalization of Theorem 6.2.

Theorem 6.3. *Let* $\mathbf{X}$, $\mathbf{Y}$ *be two vectors in* V. *Then,*

$$(\mathbf{X}, \mathbf{Y}) = \sum_{i=1}^n (\mathbf{X}, \mathbf{B}_i)(\mathbf{Y}, \mathbf{B}_i). \tag{7}$$

Give the proof of this Theorem.

Existence of an orthonormal basis. Let V be an n-dimensional vector space with an inner product $(,)$. Then V has a basis $\mathbf{B}_1, \ldots, \mathbf{B}_n$. We shall construct an orthonormal basis for V, and our method of construction will be *Mathematical Induction*, as in the proof of Theorem 5.1.

Claim. For each k, $1 \leqslant k \leqslant n$, the subspace $[\mathbf{B}_1, \ldots, \mathbf{B}_k]$ has an orthonormal basis.

PROOF OF CLAIM. For $k = 1$, the one-element basis $\dfrac{\mathbf{B}_1}{|\mathbf{B}_1|}$ is an orthonormal basis of $[\mathbf{B}_1]$.

Suppose that the claim is true for k. Let $\mathbf{G}_1, \ldots, \mathbf{G}_k$ be an orthonormal basis for $[\mathbf{B}_1, \ldots, \mathbf{B}_k]$. Put

$$\mathbf{C} = \mathbf{B}_{k+1} - \sum_{i=1}^k (\mathbf{B}_{k+1}, \mathbf{G}_i)\, \mathbf{G}_i.$$

If $C = 0$, $B_{k+1} = \sum_{i=1}^{k} (B_{k+1}, G_i)G_j$. So $B_{k+1} \in [G_1, \ldots, G_k] = [B_1, \ldots, B_k]$, contrary to the fact that B_{k+1} is not a linear combination of $B_1, \ldots, B_k$. So, $C \neq 0$. Put

$$G_{k+1} = \frac{C}{|C|}.$$

Then,

$$B_{k+1} = \sum_{i=1}^{k} (B_{k+1}, G_i)G_i + |C|G_{k+1}. \tag{8}$$

Hence,

$$B_{k+1} = D + |C|G_{k+1}, \tag{9}$$

where $D \in [G_1, \ldots, G_k] = [B_1, \ldots, B_k]$. Equation (8) yields $B_{k+1} \in [G_1, \ldots, G_{k+1}]$. Also, $B_1, \ldots, B_k \in [G_1, \ldots, G_{k+1}]$. Hence, $[B_1, \ldots, B_{k+1}] \subseteq [G_1, \ldots, G_{k+1}]$. Equation (9) gives that $G_{k+1} \in [B_1, \ldots, B_{k+1}]$. Also, $G_1, \ldots, G_k \in [B_1, \ldots, B_{k+1}]$. Hence, $[G_1, \ldots, G_{k+1}] \subseteq [B_1, \ldots, B_{k+1}]$.

If we are given two sets of vectors, each of which is contained in the other, then the two sets are equal. So

$$[G_1, \ldots, G_{k+1}] = [B_1, \ldots, B_{k+1}]. \tag{10}$$

For each $j \leqslant k$,

$$(C, G_j) = (B_{k+1}, G_j) - (B_{k+1}, G_j) = 0.$$

Since $G_{k+1} = \frac{1}{|C|}C$, hence, also $(G_{k+1}, G_j) = 0$. It follows that $G_1, \ldots, G_{k+1}$ is an orthonormal set of vectors. Because of (10), then, $G_1, G_2, \ldots, G_{k+1}$ is an orthonormal basis for $[B_1, \ldots, B_{k+1}]$.

So the claim is true for $k + 1$. Since the claim is true for $k = 1$, it follows that it is true for $k = 2, 3, \ldots, n$. Then $G_1, \ldots, G_n$ is an orthonormal basis for $[B_1, \ldots, B_n] = V$. We have proved the following:

Theorem 6.4. *Let V be an n-dimensional vector space with an inner product. Then V has an orthonormal basis $G_1, \ldots, G_n$.*

This process is called *Gram–Schmidt orthonormalization*.

Exercise 5. Find an orthonormal basis for each of the following subspaces V of $\mathbb{R}^4$, where the inner product on V is as in Example 2.

(a) V has the equation $x_4 = 0$;
(b) V has the equation $x_1 + x_2 + x_3 + x_4 = 0$;
(c) V is given by the equations $x_1 = x_2$ and $x_3 = x_4$.

Exercise 6. Find an orthonormal basis for the space $\mathbb{T}_2$ of all functions:

$$f(x) = a_0 + a_1 \cos x + b_1 \sin x + a_2 \cos(2x) + b_2 \sin(2x),$$

where $(f, g) = \int_0^{2\pi} f(x)g(x)\, dx$ for f, g in $\mathbb{T}_2$ gives the inner product.

Exercise 7. (a) Find an orthonormal basis for the space $\mathbb{P}_2$ consisting of all polynomials $P(x) = a + bx + cx^2$, where a, b, c are scalars, and $(P, Q) = \int_{-1}^{1} P(X)Q(x)\,dx$ for P, Q in $\mathbb{P}_2$.

(b) Expand the polynomial x^2 in the form:

$$x^2 = t_1 \mathbf{B}_1 + t_2 \mathbf{B}_2 + t_3 \mathbf{B}_3,$$

where $\mathbf{B}_1$, $\mathbf{B}_2$, $\mathbf{B}_3$ is the orthonormal basis you have chosen.

(c) Verify the formula (6) in this case, i.e., show that

$$(x^2, x^2) = t_1^2 + t_2^2 + t_3^2$$

by direct calculation.

CHAPTER 6.2

Orthogonal Decomposition of a Vector Space

Let Π be a plane through the origin in $\mathbb{R}^3$, and let L be the line through the origin that is orthogonal to Π. Then every vector $\mathbf{X}$ in $\mathbb{R}^3$ can be expressed in the form

$$\mathbf{X} = \mathbf{U} + \mathbf{V}, \tag{11}$$

where $\mathbf{U} \in \Pi$ and $\mathbf{V} \in L$.

Now, let V be a vector space with an inner product $(\ ,\ )$ and let Π be a subspace of V. We should like to have a generalization of formula (11) to this situation.

We say that a vector $\mathbf{Z}$ in V is *perpendicular to* Π if $\mathbf{Z} \perp \mathbf{B}$ for each $\mathbf{B}$ in Π, and we denote by $\Pi^\perp$ the set of all vectors in V that are perpendicular to Π. So $\Pi^\perp$ is a replacement for the line L above. The generalization of (11) that we seek is the formula

$$\mathbf{X} = \bar{\mathbf{X}} + \mathbf{Z}, \tag{12}$$

where $\mathbf{X}$ is a given vector in V, $\bar{\mathbf{X}}$ is a vector in Π, and $\mathbf{Z}$ is a vector in $\Pi^\perp$.

Exercise 1. Show that $\Pi^\perp$ is a subspace of V.

Let $\mathbf{X}_1, \ldots, \mathbf{X}_l$ be an orthonormal basis of Π.

Exercise 2. Let $\mathbf{X}$ be a vector in V and put

$$\mathbf{Z} = \mathbf{X} - \sum_{i=1}^{l} (\mathbf{X}, \mathbf{X}_i)\mathbf{X}_i.$$

Show that $\mathbf{Z}$ belongs to $\Pi^\perp$.

Fix $\mathbf{X} \in V$. Put $\bar{\mathbf{X}} = \sum_{i=1}^{l}(\mathbf{X}, \mathbf{X}_i)\mathbf{X}_i$. Then $\bar{\mathbf{X}} \in \Pi$. Also, by Exercise 2,

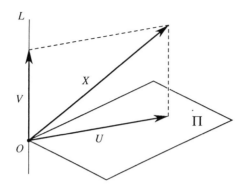

Figure 6.1

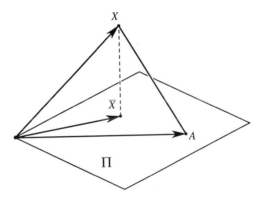

Figure 6.2

$\mathbf{Z} \in \Pi^{\perp}$. Finally,

$$\mathbf{X} = \sum_{i=1}^{l} (\mathbf{X}, \mathbf{X}_i)\mathbf{X}_i + \mathbf{Z} = \bar{\mathbf{X}} + \mathbf{Z}.$$

So we have the desired formula (12). Since (12) holds for each vector $\mathbf{X}$ in V, we regard (12) as giving an *orthogonal decomposition* of the whole space V into a sum of the two mutually orthogonal subspaces Π and $\Pi^{\perp}$.

Now consider an arbitrary vector $\mathbf{A}$ in Π. Then

$$\mathbf{A} = \sum_{i=1}^{l} (\mathbf{A}, \mathbf{X}_i)\mathbf{X}_i,$$

and so

$$\mathbf{X} - \mathbf{A} = \sum_{i=1}^{l} [(\mathbf{X}, \mathbf{X}_i) - (\mathbf{A}, \mathbf{X}_i)]\mathbf{X}_i + \mathbf{Z}.$$

Exercise 3. (a) Using the fact that $Z = X - \bar{X}$, show that

$$|X - A|^2 = \sum_{i=1}^{l} |(X, X_i) - (A, X_i)|^2 + |X - \bar{X}|^2.$$

(b) Deduce that $|X - A| > |X - \bar{X}|$ unless $A = \bar{X}$.
(c) Conclude that $\bar{X}$ is the vector in Π closest to X.

Exercise 4. Let Π be a subspace of V and let X be a vector in V. Let X^* be the vector in Π that is closest to X. Show that $X - X^*$ is perpendicular to Π.

Exercise 5. Let X be a vector in V. Show that the representation of X given in (12) is unique, in the sense that if

$$X = X' + Z' \quad \text{and} \quad X = X'' + Z'',$$

where X' and X'' belong to Π, and Z' and Z'' belong to $\Pi^\perp$, then $X' = X''$ and $Z' = Z''$.

Exercise 6. We use the preceding notation. Define a transformation P of the vector space V to itself by putting

$$P(X) = \sum_{i=1}^{l} (X, X_i) X_i$$

for each X in V. Show that the following are true:

(a) P is a linear transformation of V.
(b) $P^2 = P$.
(c) For all X, Y in V, we have $(PX, Y) = (X, PY)$.
(d) The range of P (i.e., the set of all values taken on by P) equals the subspace Π.

P is called the *orthogonal projection* of V on Π.

Symmetric Matrices in n Dimensions

In the cases of dimensions 2 and 3, we have seen that a special role is played by *symmetric matrices*, those matrices that are equal to their own transposes. The same definition works in $\mathbb{R}^n$ for $n \geqslant 4$, and as in the case of lower dimensions, these matrices have special properties that make them particularly valuable in the analysis of quadratic forms.

Recall that the *transpose* of an $n \times n$ matrix m with entry a_{ij} in the ith row and jth column is the matrix m^* with $a_{ij}^* = a_{ji}$. Thus, the matrix m^* is obtained by reflecting the entries of m across its diagonal. A matrix is *symmetric* if it is equal to its own transpose, so $m^* = m$, or, equivalently, $a_{ij} = a_{ji}$ for all i, j. We shall write $\langle \mathbf{X}, \mathbf{Y} \rangle$ for $\mathbf{X} \cdot \mathbf{Y}$ if $\mathbf{X}, \mathbf{Y} \in \mathbb{R}^n$.

A linear transformation T of $\mathbb{R}^n$ is said to be *self-adjoint* if for all vectors $\mathbf{X}$ and $\mathbf{Y}$ the inner product of $\mathbf{X}$ with $T\mathbf{Y}$ is the same as the inner product of $T\mathbf{X}$ with $\mathbf{Y}$. The matrix of a self-adjoint transformation (with respect to the standard basis) is symmetric. This is so because the jth column of the matrix $m(T)$ of T is given by $T(\mathbf{E}_j)$, so $\langle T\mathbf{E}_j, \mathbf{E}_i \rangle = a_{ij}$, and $\langle \mathbf{E}_j, T\mathbf{E}_i \rangle = a_{ji}$. Since T is self-adjoint, these are equal.

Exercise 1. Show, conversely, that if T is a linear transformation of $\mathbb{R}^n$ with a symmetric matrix, then T is self-adjoint.

In three dimensions, we proved an important result about the eigenvectors of a self-adjoint linear transformation, the fact that eigenvectors corresponding to distinct eigenvalues are perpendicular. That proof did not use coordinates, and it works just as well in n-dimensional space: If $T\mathbf{X} = t\mathbf{X}$ and $T\mathbf{Y} = s\mathbf{Y}$, where $s \neq t$, then

$$t\langle \mathbf{X}, \mathbf{Y} \rangle = \langle t\mathbf{X}, \mathbf{Y} \rangle = \langle T\mathbf{X}, \mathbf{Y} \rangle = \langle \mathbf{X}, T\mathbf{Y} \rangle = \langle \mathbf{X}, s\mathbf{Y} \rangle = s\langle \mathbf{X}, \mathbf{Y} \rangle.$$

But, since $s \neq t$, this can only happen when $\langle \mathbf{X}, \mathbf{Y} \rangle = 0$.

In three dimensions, we used the fact that every linear transformation T has a real eigenvalue to obtain an orthonormal basis for $\mathbb{R}^3$ consisting of eigenvectors of T, (Spectral Theorem in $\mathbb{R}^3$ in Chapter 3.7). As we have seen, however, not every linear transformation of $\mathbb{R}^4$ has a real eigenvalue.

Nonetheless, we shall show that every self-adjoint linear transformation T of $\mathbb{R}^4$ does have a real eigenvalue, and, moreover, that there is an orthonormal basis of $\mathbb{R}^4$ that consists of eigenvectors of T.

To prove all this, we shall use a device frequently used in mathematics when one wishes to construct a certain mathematical object: assume that we have the object already, find out what properties it will have and use those properties in our construction. Let us then consider a linear transformation T of $\mathbb{R}^4$ that possesses eigenvectors X_i with $TX_i = \lambda_i X_i$, $i = 1, 2, 3, 4$, where the λ_i are real scalars and the set $X_1, \ldots, X_4$ is an orthonormal basis of $\mathbb{R}^4$. We define the function $Q(X)$ on $\mathbb{R}^4$ by

$$Q(X) = \langle TX, X \rangle$$

for each X in $\mathbb{R}^4$. Since the X_i form an orthonormal basis of $\mathbb{R}^4$, we have

$$X = \sum_{i=1}^{4} \langle X, X_i \rangle X_i \quad \text{and} \quad |X|^2 = \sum_{i=1}^{4} \langle X, X_i \rangle^2$$

for each X in $\mathbb{R}^4$, and so $TX = \sum_{i=1}^{4} \lambda_i \langle X, X_i \rangle X_i$ and

$$Q(X) = \left\langle \sum_{i=1}^{4} \langle X, X_i \rangle \lambda_i X_i, \sum_{i=1}^{4} \langle X, X_i \rangle X_i \right\rangle = \sum_{i=1}^{4} \lambda_i \langle X, X_i \rangle^2. \qquad (1)$$

By suitable labeling of the eigenvectors, we can assure that λ_1 is the largest of the 4 eigenvalues, then, by (1) we have

$$Q(X) \leqslant \sum_{i=1}^{4} \langle X, X_i \rangle^2 \lambda_1 = \lambda_1 |X|^2. \qquad (2)$$

We denote by Γ the "unit sphere" in $\mathbb{R}^4$ consisting of all vectors X with $|X| = 1$. By (2), $Q(X) \leqslant \lambda_1$ for all X in Γ. Also, $Q(X_1) = \langle TX_1, X_1 \rangle = \langle \lambda_1 X_1, X_1 \rangle = \lambda_1 \langle X_1, X_1 \rangle = \lambda_1$. Thus, we have $Q(X) \leqslant Q(X_1)$ for all X in Γ. So we have found: the function $Q(X)$ attains its maximum value on the unit sphere Γ at the vector X_1.

Let us now turn this argument around: we consider a self-adjoint linear transformation T on $\mathbb{R}^4$. We form the function $Q(X) = \langle TX, X \rangle$ defined for all X in $\mathbb{R}^4$.

It can be shown that every continuous function defined on a sphere in $\mathbb{R}^4$ takes its maximum value at some point of the sphere. Let Z be a vector in Γ where $Q(X)$ takes on its maximum value on Γ, that is, such that $Q(X) \leqslant Q(Z)$ for all X in Γ.

Choose a unit vector Y perpendicular to Z. Then, for each ϕ in $\mathbb{R}$, $\cos(\phi)Z + \sin(\phi)Y$ is also a unit vector.

We define a function $f(\phi)$ by

$$f(\phi) = Q(\cos(\phi)Z + \sin(\phi)Y).$$

Since $Q(\mathbf{X})$ takes on its maximum value on Γ at $\mathbf{Z}$, $f(\phi)$ takes on its maximum value when $\phi = 0$. Now,

$$f(\phi) = \langle T(\cos(\phi)\mathbf{Z} + \sin(\phi)\mathbf{Y}), \cos(\phi)\mathbf{Z} + \sin(\phi)\mathbf{Y}\rangle$$

$$= \cos^2(\phi)\langle T\mathbf{Z}, \mathbf{Z}\rangle + \cos(\phi)\sin(\phi)[\langle T\mathbf{Z}, \mathbf{Y}\rangle + \langle \mathbf{Z}, T(\mathbf{Y})\rangle]$$

$$+ \sin^2(\phi)\langle T\mathbf{Y}, \mathbf{Y}\rangle.$$

Hence,

$$f'(\phi) = -2\cos(\phi)\sin(\phi)\langle T\mathbf{Z}, \mathbf{Z}\rangle + (\cos^2(\phi) - \sin^2(\phi))[\langle T\mathbf{Z}, \mathbf{Y}\rangle$$

$$+ \langle \mathbf{Z}, T\mathbf{Y}\rangle] + 2\cos(\phi)\sin(\phi)\langle T\mathbf{Y}, \mathbf{Y}\rangle.$$

Since $f(\phi)$ takes on its maximum value at $\phi = 0$, we have $f'(0) = 0$. It follows that $\langle T\mathbf{Z}, \mathbf{Y}\rangle + \langle \mathbf{Z}, T\mathbf{Y}\rangle = 0$.

Since T is self-adjoint, $\langle \mathbf{Z}, T\mathbf{Y}\rangle = \langle T\mathbf{Z}, \mathbf{Y}\rangle$, so $2\langle T\mathbf{Z}, \mathbf{Y}\rangle = 0$, and so $\langle T\mathbf{Z}, \mathbf{Y}\rangle = 0$. We have shown that every vector $\mathbf{Y}$ perpendicular to $\mathbf{Z}$ is also perpendicular to $T\mathbf{Z}$. This can only happen if $T\mathbf{Z}$ is a scalar multiple of $\mathbf{Z}$ (prove this!), and so $\mathbf{Z}$ is an eigenvector of T. Thus, we have found: if $Q(\mathbf{X}) = \langle T\mathbf{X}, \mathbf{X}\rangle$ takes its maximum value on Γ at $\mathbf{Z}$, then $\mathbf{Z}$ is an eigenvector of T.

Moreover, we have shown that the hyperplane consisting of all vectors $\mathbf{Y}$ perpendicular to $\mathbf{Z}$ is mapped into itself by T.

Next, let V be an n-dimensional vector space with an inner product $\langle \ , \ \rangle$ and let T be a self-adjoint linear transformation on V. We form, as before, the function $Q(\mathbf{X}) = \langle T\mathbf{X}, \mathbf{X}\rangle$ defined for all $\mathbf{X}$ in V, and we choose a point $\mathbf{Z}$ on the unit sphere Γ of V, such that $Q(\mathbf{X})$ assumes at $\mathbf{X} = \mathbf{Z}$ its maximum value on Γ. By exactly the same argument we used above in $\mathbb{R}^4$, $\mathbf{Z}$ is an eigenvector of T and, furthermore, if V' denotes the subspace of V consisting of all vectors in V that are perpendicular to $\mathbf{Z}$, then T maps V' into itself. The restriction of T to V' is then a self-adjoint linear transformation on V'. If we apply the preceding argument to this transformation on V', we obtain an eigenvector $\mathbf{X}'$ of T in V', with $|X'| = 1$, and further, T maps the subspace of V' consisting of all vectors perpendicular to $\mathbf{X}'$ into itself. Also, since $\mathbf{X}' \in V'$, $\mathbf{X}'$ is perpendicular to X. Continuing in this way, we obtain a succession of subspaces $V, V', V'', \ldots$, each contained in the preceding one and of dimension 1 less than it, and a succession of mutually orthogonal unit eigenvectors of T, $\mathbf{Z}, X', X'', \ldots$, such that $\mathbf{Z} \in V$, $X' \in V'$, $X'' \in V''$, and so. This succession must stop after n steps, when we have reached a subspace of dimension 1. In this way, we obtain n mutually perpendicular unit eigenvectors of T in V. We have proved the following:

Theorem 7.1 (*Spectral Theorem in Dimension n*). *Let V be an n-dimensional vector space with an inner product $\langle \ , \ \rangle$, and let T be a self-adjoint linear transformation on V. Then, there exists an orthonormal basis of V that consists of n eigenvectors of T.*

The characteristic equation. Recall that we found the eigenvalues of a symmetric matrix in 2 and 3 dimensions by solving a certain polynomial equation, called the characteristic equation. The situation is similar in $\mathbb{R}^n$ for $n \geq 4$.

Let T be a self-adjoint linear transformation of $\mathbb{R}^n$ with symmetric matrix $m = ((a_{ij}))$. If λ is an eigenvalue of T, then there is a corresponding vector $\mathbf{X} \neq \mathbf{0}$ in $\mathbb{R}^n$ with $T\mathbf{X} = \lambda\mathbf{X}$ or $(T - \lambda I)\mathbf{X} = \mathbf{0}$. Expressed in terms of coordinates, this becomes, if $\mathbf{X} = (x_1, \ldots, x_n)$,

$$\sum_{j=1}^{n} a_{ij}x_j - \lambda x_i = 0, \quad j = 1, 2, \ldots, n, \quad \text{or}$$

$$(a_{11} - \lambda)x_1 + a_{12}x_2 + \cdots + a_{1n}x_n = 0,$$

$$a_{21}x_1 + (a_{22} - \lambda)x_2 + \cdots + a_{2n}x_n = 0, \tag{3}$$

$$\vdots$$

$$a_{n1}x_1 + a_{n2}x_2 + \cdots + (a_{nn} - \lambda)x_n = 0.$$

Thus, the homogeneous system (3) has a non-zero solution, and so, the determinant of the corresponding matrix is 0, that is, we have

$$\begin{vmatrix} a_{11} - \lambda & a_{12} & \cdots & a_{1n} \\ a_{21} & a_{22} - \lambda & \cdots & a_{2n} \\ & & \vdots & \\ a_{n1} & a_{n2} & \cdots & a_{nn} - \lambda \end{vmatrix} = 0. \tag{4}$$

Each of the steps of this argument is reversible, and so, if (4) holds for a given number λ, then there exists a non-zero vector $\mathbf{X}$ such that $T\mathbf{X} = \lambda\mathbf{X}$.

The equation (4), thus, is a necessary and sufficient condition on a number λ in order that λ is an eigenvalue of T.

If we expand out the determinant, we see that (4) can be written in the form

$$(-1)^n\lambda^n + B_1\lambda^{n-1} + B_2\lambda^{n-2} + \cdots + B_{n-1}\lambda + B_n = 0, \tag{5}$$

where $B_1, B_2, \ldots, B_n$ are certain constants. Finding the eigenvalues of T then amounts to finding the roots (solutions) of the polynomial equation (5) of degree n. Equation (5) or, equivalently, equation (4), is called the *characteristic equation* of the transformation T.

Eigenvalue problems occur frequently in applications of linear algebra in science, engineering, statistics, economics, etc. For this reason, computer programs have been written to solve the characteristic equation (5) or to find alternative ways of calculating the eigenvalues of a given $n \times n$ matrix.

Let us compute the eigenvalues in a simple example to illustrate the preceding theory.

EXAMPLE 1. Let T be the transformation of $\mathbb{R}^4$ with matrix

$$m = \begin{pmatrix} 6 & 2 & 0 & 0 \\ 2 & 3 & 0 & 0 \\ 0 & 0 & 5 & 5 \\ 0 & 0 & 5 & 5 \end{pmatrix}.$$

The characteristic equation here is

$$\begin{vmatrix} 6-\lambda & 2 & 0 & 0 \\ 2 & 3-\lambda & 0 & 0 \\ 0 & 0 & 5-\lambda & 5 \\ 0 & 0 & 5 & 5-\lambda \end{vmatrix} = 0, \quad \text{or}$$

$$(6-\lambda)\begin{vmatrix} 3-\lambda & 0 & 0 \\ 0 & 5-\lambda & 5 \\ 0 & 5 & 5-\lambda \end{vmatrix} - 2\begin{vmatrix} 2 & 0 & 0 \\ 0 & 5-\lambda & 5 \\ 0 & 5 & 5-\lambda \end{vmatrix} = 0, \quad \text{or}$$

$$(6-\lambda)(3-\lambda)[(5-\lambda)^2 - 25] - 2\cdot 2[(5-\lambda)^2 - 25] = 0, \quad \text{or}$$

$$(\lambda^2 - 9\lambda + 14)(\lambda^2 - 10\lambda) = 0, \quad \text{or}$$

$$(\lambda - 7)(\lambda - 2)\lambda(\lambda - 10) = 0.$$

The roots are 7, 2, 0, 10, so these are the eigenvalues. Let us find an eigenvector with eigenvalue 7, that is, find $\mathbf{X} = (x_1, \ldots, x_4)$ with $T\mathbf{X} = 7\mathbf{X}$. Then,

$$\begin{pmatrix} 6 & 2 & 0 & 0 \\ 2 & 3 & 0 & 0 \\ 0 & 0 & 5 & 5 \\ 0 & 0 & 5 & 5 \end{pmatrix} \begin{pmatrix} x_1 \\ x_2 \\ x_3 \\ x_4 \end{pmatrix} = 7 \begin{pmatrix} x_1 \\ x_2 \\ x_3 \\ x_4 \end{pmatrix},$$

$$6x_1 + 2x_2 = 7x_1$$
$$2x_1 + 3x_2 = 7x_2$$
$$5x_3 + 5x_4 = 7x_3$$
$$5x_3 + 5x_4 = 7x_4$$

so

$$-x_1 + 2x_2 = 0$$
$$2x_1 - 4x_2 = 0$$
$$-2x_3 + 5x_4 = 0$$
$$x_3 = x_4.$$

Hence,

$$x_1 = 2x_2, \; 3x_3 = 0, \; x_3 = x_4.$$

So $\mathbf{X} = (2, 1, 0, 0)$ is an eigenvector with eigenvalue 7. The other eigenvectors with eigenvector 7 are scalar multiples of $\mathbf{X}$. (Why?)

Exercise 1. Find the eigenvectors of the transformation T of example 1 corresponding to eigenvalues 2, 0 and 10.

Exercise 2. Find all the eigenvalues and eigenvectors for the linear transformation of $\mathbb{R}$ with matrix

$$m = \begin{pmatrix} -3 & 1 & 1 & 1 \\ 1 & -3 & 1 & 1 \\ 1 & 1 & -3 & 1 \\ 1 & 1 & 1 & -3 \end{pmatrix}.$$

Hint: Recall the transformation Q in Chapter 4.1, and reason geometrically.

Quadratic Forms in n Variables

In Chapter 3.8 we considered a quadric surface in $\mathbb{R}^3$ with the equation

$$ax^2 + 2bxy + 2cxz + dy^2 + 2eyz + fz^2 = 1, \tag{1}$$

where a, b, c, d, e, f are constants.

If we denote by $Q(x, y, z)$ the polynomial $ax^2 + 2bxy + 2cxz + dy^2 + 2eyz + fz^2$, in the variables x, y, z, then equation (1) becomes: $Q(x, y, z) = 1$.

We proved Theorem 3.12, which told us that we may choose new coordinates u, v, w in $\mathbb{R}^3$ in such a way that if we express $Q(x, y, z)$ in terms of these new coordinates, we have the formula

$$Q(x, y, z) = t_1 u^2 + t_2 v^2 + t_3 w^2, \tag{2}$$

where t_1, t_2, t_3 are constants. It follows that when expressed in the new coordinates, our surface has the simple equation:

$$t_1 u^2 + t_2 v^2 + t_3 w^2 = 1. \tag{3}$$

From equation (3), if $t_1, t_2,$ and t_3 are not equal to zero, it is easy to tell whether the surface is an ellipsoid (possibly a sphere) or a hyperboloid.

We shall now study the corresponding situation in n variables. We consider a homogeneous second-degree polynomial in the n variables $x_1, \ldots, x_n$:

$$Q(x_1, \ldots, x_n) = \sum_{i,j=1}^{n} a_{ij} x_i x_j,$$

where the a_{ij} are constants and $a_{ij} = a_{ji}$ for each i, j. Writing this out, and combining terms $a_{ij} x_i x_j$ and $a_{ji} x_j x_i$, we get

$$Q(x_1, \ldots, x_n) = a_{11} x_1^2 + 2a_{12} x_1 x_2 + \cdots + 2a_{1n} x_1 x_n$$
$$+ a_{22} x_2^2 + 2a_{23} x_2 x_3 + \cdots + 2a_{2n} x_2 x_n + \cdots + a_{nn} x_n^2.$$

Our problem is to choose new coordinates $u_1, \ldots, u_n$ in $\mathbb{R}^n$ such that expressed in the new coordinates

$$Q(x_1, \ldots, x_n) = t_1 u_1^2 + t_2 u_2^2 + \cdots + t_n u_n^2. \tag{4}$$

Such polynomials $Q(x_1, \ldots, x_n)$ often occur in problems in geometry, statistics, mechanics, advanced calculus, etc., and formula (4) greatly simplifies dealing with them.

To obtain formula (4), we shall use the Spectral Theorem we proved in Chapter 7.0. Let m denote the symmetric matrix with entries a_{ij}, and let A be the linear transformation of $\mathbb{R}^n$ whose matrix is m. If $\mathbf{X}$ is the column vector $(x_1, \ldots, x_n)$, then

$$AX = \begin{pmatrix} a_{11} & a_{12} & \cdots & a_{1n} \\ a_{21} & a_{22} & \cdots & a_{2n} \\ \vdots & & & \\ a_{n1} & a_{n2} & \cdots & a_{nn} \end{pmatrix} \begin{pmatrix} x_1 \\ x_2 \\ \vdots \\ x_n \end{pmatrix},$$

and so, $\mathbf{X} \cdot A\mathbf{X} = \sum_{i=1}^n x_i (\sum_{j=1}^n a_{ij} x_j) = \sum_{i,j} a_{ij} x_i x_j = Q(x_1, \ldots, x_n)$.

Since A has a symmetric matrix, and so is self-adjoint, the Spectral Theorem applies to A and there exists an orthonormal basis $\mathbf{G}_1, \mathbf{G}_2, \ldots, \mathbf{G}_n$ of $\mathbb{R}^n$ with $A\mathbf{G}_j = t_j \mathbf{G}_j, j = 1, \ldots, n$, where the t_j are the eigenvalues of A. Then if $\mathbf{X}$ is a vector in $\mathbb{R}^n$, we have $\mathbf{X} = \sum_{i=1}^n u_i \mathbf{G}_i$, where $u_1, \ldots, u_n$ are the coordinates of the vector $\mathbf{X}$ in the coordinate system whose axes lie along the basis vectors $\mathbf{G}_1, \ldots, \mathbf{G}_n$. Then

$$\mathbf{X} \cdot A\mathbf{X} = \left(\sum_{i=1}^n u_i \mathbf{G}_i \right) \cdot \left(\sum_{i=1}^n u_i t_i \mathbf{G}_i \right) = \sum_{i=1}^n u_i^2 t_i,$$

so $Q(x_1, \ldots, x_n) = \sum_{i=1}^n t_i u_i^2$. Thus, we have the following:

Theorem 7.2. Let Q be a quadratic polynomial in n variables given by $Q(x_1, \ldots, x_n) = \sum_{i,j=1}^n a_{ij} x_i x_j$, where the a_{ij} are constants such that $a_{ij} = a_{ji}$. Then there exists a coordinate system with coordinates denoted $u_1, \ldots, u_n$ such that for every vector $(x_1, \ldots, x_n)$ we have

$$Q(x_1, \ldots, x_n) = \sum_{i=1}^n t_i u_i^2, \tag{5}$$

where $(u_1, \ldots, u_n)$ are the new coordinates of $(x_1, \ldots, x_n)$ and $t_1, \ldots, t_n$ are fixed scalars depending upon Q. The new coordinate axes, $(u_i$-axes), lie along the eigenvectors of the linear transformation A whose matrix is $((a_{ij}))$.

Note: A homogeneous polynomial, each of whose terms is of the second degree, is called a *quadratic form*. Theorem 7.2 tells us that an arbitrary quadratic form in n variables can be written as a diagonal quadratic form: $\sum_{i=1}^n t_i u_i^2$ in suitable coordinates u_i.

EXAMPLE 1. Let $Q(x_1, x_2, x_3, x_4) = x_1 x_2 + 2x_3 x_4$. We wish to write Q as a diagonal quadratic form in new coordinates $u_1, \ldots, u_4$. First, we have to write Q in the form $\sum_{i,j=1}^{n} a_{ij} x_i x_j$ with $a_{ij} = a_{ji}$, and then we must form the matrix $m = ((a_{ij}))$.

$$
\begin{aligned}
x_1 x_2 + 2x_3 x_4 = 0x_1^2 &+ \tfrac{1}{2}x_1 x_2 + 0x_1 x_3 + 0x_1 x_4 + \\
\tfrac{1}{2}x_2 x_1 + 0x_2^2 &+ 0x_2 x_3 + 0x_2 x_4 + \\
0x_3 x_1 + 0x_3 x_2 &+ 0x_3^2 + 1x_3 x_4 + \\
0x_4 x_1 + 0x_4 x_2 &+ 1x_4 x_3 + 0x_4^2.
\end{aligned}
$$

So,

$$
m = \begin{pmatrix} 0 & \tfrac{1}{2} & 0 & 0 \\ \tfrac{1}{2} & 0 & 0 & 0 \\ 0 & 0 & 0 & 1 \\ 0 & 0 & 1 & 0 \end{pmatrix}.
$$

The eigenvalues of m are the roots of the characteristic equation

$$
\begin{vmatrix} -\lambda & \tfrac{1}{2} & 0 & 0 \\ \tfrac{1}{2} & -\lambda & 0 & 0 \\ 0 & 0 & -\lambda & 1 \\ 0 & 0 & 1 & -\lambda \end{vmatrix} = 0. \tag{6}
$$

Evaluating the determinant, as we did in Chapter 4, we see that equation (6) is equivalent to

$$
(\lambda^2 - \tfrac{1}{4})(\lambda^2 - 1) = 0.
$$

So the eigenvalues t_i are: $\tfrac{1}{2}$, $-\tfrac{1}{2}$, 1, -1. If we choose the u_i-axes along the corresponding eigenvectors, we obtain from Theorem 7.2 that,

$$
x_1 x_2 + 2x_3 x_4 = \tfrac{1}{2}u_1^2 - \tfrac{1}{2}u_2^2 + u_3^2 - u_4^2.
$$

Exercise 1. Find the corresponding eigenvectors in Example 1, and calculate the change of coordinate formulas that express u_1, u_2, u_3, u_4 in terms of x_1, x_2, x_3, x_4.

The diagonalization of a symmetric $n \times n$ matrix. If m is a given matrix and if we can find a diagonal matrix d and an invertible matrix r, such that

$$
m = r\, d\, r^{-1},
$$

then we say that we have *diagonalized* the matrix m. In Chapter 3.7, we showed how we may diagonalize a symmetric 3×3 matrix, and we shall now solve the corresponding problem for an arbitrary symmetric $n \times n$ matrix. As in the preceding result about quadratic forms, the Spectral Theorem plays a key role in our solution.

If $b = ((b_{ij}))$ is any $n \times n$ matrix and $\mathbf{X} = (x_1, \ldots, x_n)$ is a vector in $\mathbb{R}^n$, we write: $b\mathbf{X}$ for the vector $\mathbf{Y} = (y_1, \ldots, y_n)$, which is the result of b acting on $\mathbf{X}$. This means that

$$y_i = \sum_{j=1}^{n} b_{ij} x_j, \quad i = 1, \ldots, n.$$

If $\mathbf{E}_k$ denotes the vector in $\mathbb{R}^n$ whose kth entry is 1 and whose other entries are 0, then $b\mathbf{E}_k$ is the kth column vector of b.

An $n \times n$ matrix b is called an *orthogonal matrix*, just as in the case $n = 3$ we discussed in Chapter 3.6, if $b^{-1} = b^*$, that is, if the inverse of b equals the transpose of b.

Let m be a symmetric $n \times n$ matrix, and denote by A, the linear transformation of $\mathbb{R}^n$ whose matrix in m. By the Spectral Theorem, there exists an orthonormal basis $\mathbf{X}_1, \ldots, \mathbf{X}_n$ of $\mathbb{R}^n$ consisting of eigenvectors of A. Then $m\mathbf{X}_i = t_i \mathbf{X}_i$, where $t_1, \ldots, t_n$ are the eigenvalues of A. Let p be the matrix $(\mathbf{X}_1 | \mathbf{X}_2 | \cdots | \mathbf{X}_n)$, whose columns are the eigenvectors $\mathbf{X}_i$. Since $\mathbf{X}_1, \ldots, \mathbf{X}_n$ is a basis of $\mathbb{R}^n$, p has an inverse p^{-1}. We define by d the diagonal matrix with entries down its diagonal given by the eigenvalues t_i:

$$d = \begin{bmatrix} t_1 & 0 & \cdots & 0 \\ 0 & t_2 & \cdots & 0 \\ \vdots & & & \vdots \\ 0 & 0 & \cdots & t_n \end{bmatrix}.$$

We want to show that the matrices m and $p\, dp^{-1}$ are equal. Fix i. We have $p\mathbf{E}_i = \mathbf{X}_i$, so $\mathbf{E}_i = p^{-1}\mathbf{X}_i$. Also, $d\mathbf{E}_i = t_i \mathbf{E}_i$. So $(p\, dp^{-1})\mathbf{X}_i = p(dp^{-1}\mathbf{X}_i) = p(t_i\mathbf{E}_i) = t_i p\mathbf{E}_i = t_i\mathbf{X}_i$. Also, $m\mathbf{X}_i = t_i\mathbf{X}_i$. Thus, the result of applying $p\, dp^{-1}$ and m to $\mathbf{X}_i$ is the same for each i. Since the $\mathbf{X}_i$ forms a basis for $\mathbb{R}^n$, it follows that $(p\, dp^{-1})\mathbf{X} = m\mathbf{X}$ for every vector $\mathbf{X}$ in $\mathbb{R}^n$. We conclude that $m = p\, dp^{-1}$. Thus, we have proved the following.

Theorem 7.3. *Let m be an $n \times n$ symmetric matrix. If we define the matrices p and d as above, we obtain*
$$m = p\, dp^{-1}. \tag{7}$$

When $n = 3$, we had proved this result as Theorem 3.11 in Chapter 3.7. As in Corollary 1 of Theorem 3.11, we can show that the matrix p in (7) is an orthogonal matrix.

PROOF. Since the columns of p are the vectors $\mathbf{X}_i$, the rows of the transpose p^* are the vectors $\mathbf{X}_i$, so that

$$p^* = \begin{bmatrix} \mathbf{X}_1 \\ \mathbf{X}_2 \\ \vdots \\ \mathbf{X}_n \end{bmatrix}$$

and, hence, the product matrix

$$p^*p = \begin{pmatrix} \mathbf{X}_1 \\ \mathbf{X}_2 \\ \vdots \\ \mathbf{X}_n \end{pmatrix} (\mathbf{X}_1 | \cdots | \mathbf{X}_n) = \begin{pmatrix} \mathbf{X}_1 \cdot \mathbf{X}_1 & \mathbf{X}_1 \cdot \mathbf{X}_2 & \cdots & \mathbf{X}_1 \cdot \mathbf{X}_n \\ \mathbf{X}_2 \cdot \mathbf{X}_1 & \mathbf{X}_2 \cdot \mathbf{X}_2 & \cdots & \mathbf{X}_2 \cdot \mathbf{X}_n \\ \vdots & \vdots & & \vdots \\ \mathbf{X}_n \cdot \mathbf{X}_1 & \mathbf{X}_n \cdot \mathbf{X}_2 & \cdots & \mathbf{X}_n \cdot \mathbf{X}_n \end{pmatrix}$$

$$= \begin{pmatrix} 1 & 0 & \cdots & 0 \\ 0 & 1 & \cdots & 0 \\ \vdots & \vdots & & \vdots \\ 0 & 0 & \cdots & 1 \end{pmatrix}.$$

So p^*p is the identity matrix and, hence, $p^{-1} = p^*$, and so p is an orthogonal matrix, as we set out to prove.

Exercise 2. Express in the form (7) the matrix m that occurs in Example 1.

Exercise 3. Express in the form (7) the 4×4 matrix

$$\begin{pmatrix} 1 & 1 & 1 & 1 \\ 1 & 1 & 1 & 1 \\ 1 & 1 & 1 & 1 \\ 1 & 1 & 1 & 1 \end{pmatrix}.$$

Exercise 4. Write, as in (5), the quadratic form

$$x_1^2 + 2x_1 x_2 + 2x_1 x_3 + 2x_1 x_4 + x_2^2 + 2x_2 x_3 + 2x_2 x_4 + x_3^2 + 2x_3 x_4 + x_4^2.$$

Exercise 4. Express, as in (5), the quadratic form

$$6x_1 x_4 - 2x_2 x_3.$$

CHAPTER 8.0

Differential Systems

In this chapter we present examples from one of the most important applications of linear algebra, to systems of differential equations. We have seen how the use of matrices makes it possible for us to handle systems of k equations in n unknowns, and to interpret these as representing linear transformations between spaces. We now see how the use of linear algebra makes it possible to approach systems of equations involving derivatives of functions. For this chapter, a knowledge of calculus of one variable is assumed.

Our first examples lead to systems of two differential equations, and the mathematics involved in the analysis of such systems is already contained in Chapter 2. In §3 of this chapter, we show how the techniques we used in two dimensions can be combined with the linear algebra of three and more dimensions to give a general theory of differential systems of higher order.

EXAMPLE 1. We are given two tanks of capacity 100 gallons, each filled with a mixture of salt and water. The tanks are connected by pipes as shown in Fig. 8.1 and at all times the mixture in each tank is kept uniform by stirring.

The mixture from tank I flows into tank II through a pipe at 10 gal/min, and in the reverse direction, the mixture flows into tank I from tank II through a second pipe at 5 gal/min. Also, the mixture leaves tank II through a third pipe at 5 gal/min, while fresh water flows into tank I through another pipe at 5 gal/min.

Denote by $x(t)$ the amount of salt (in lbs) in tank I at time t, and by $y(t)$ the corresponding amount in tank II. Suppose, at time $t = 0$, there are x_0 lbs of salt in tank I, and 0 lbs of salt in tank II. Find expressions for $x(t)$ and $y(t)$ in terms of t.

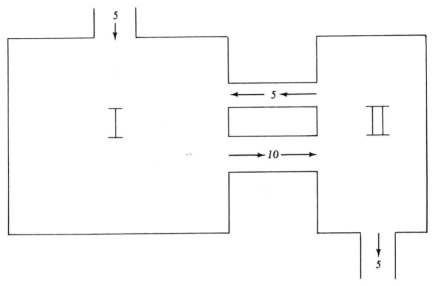

Figure 8.1

Consider the time interval from time t to time $t + \Delta t$. During that time interval, each gallon flowing into tank I from tank II contains $y(t)/100$ lbs of salt, while each gallon flowing from tank I to tank II contains $x(t)/100$ lbs of salt. Hence, the net change of the amount of salt in tank I during the time interval is

$$\Delta x = \frac{5y(t)\Delta t}{100} - \frac{10x(t)\Delta t}{100} \, ,$$

while the corresponding change for tank II is

$$\Delta y = \frac{10x(t)\Delta t}{100} - \frac{10y(t)\Delta t}{100} \, .$$

Dividing both equations by Δt and letting $\Delta t \to 0$, we get

$$\begin{cases} \dfrac{dx}{dt}(t) = \dfrac{5}{100}\, y(t) - \dfrac{10}{100}\, x(t), \\[2mm] \dfrac{dy}{dt}(t) = \dfrac{10}{100}\, x(t) - \dfrac{10}{100}\, y(t). \end{cases} \tag{1}$$

In addition, we know that

$$x(0) = x_0, \qquad y(0) = 0. \tag{2}$$

The functions $t \to x(t)$, $t \to y(t)$ must be determined from conditions (1) and (2).

A system of equations involving two unknown functions x and y which has the form

$$\begin{cases} \dfrac{dx}{dt} = ax + by, \\[2mm] \dfrac{dy}{dt} = cx + dy, \end{cases} \tag{3}$$

where a, b, c, d are given constants, is called a *differential system*.

Thus (1) is a differential system with $a = -\frac{10}{100}$, $b = \frac{5}{100}$, $c = \frac{10}{100}$, $d = -\frac{10}{100}$. The condition

$$x(0) = x_0, \qquad y(0) = y_0,$$

where x_0, y_0 are given constants, is called an *initial condition* for the system (3). Thus (2) is an initial condition.

We shall use the notion of a vector-valued function of t. A vector-valued function $\mathbf{X}(t)$ assigns to each number t a vector $\mathbf{X}(t) = \begin{pmatrix} x_1(t) \\ x_2(t) \end{pmatrix}$. Thus,

$$\mathbf{X}(t) = \begin{pmatrix} t^2 \\ t^3 + 1 \end{pmatrix} \quad \text{and} \quad \mathbf{X}(t) = \begin{pmatrix} \cos t \\ \sin t \end{pmatrix}$$

are vector-valued functions. If $\mathbf{X}(t) = \begin{pmatrix} x_1(t) \\ x_2(t) \end{pmatrix}$, then $t \rightarrow x_1(t)$ and $t \rightarrow x_2(t)$ are scalar-valued functions. We define the *derivative* of the function $t \rightarrow \mathbf{X}(t) = \begin{pmatrix} x_1(t) \\ x_2(t) \end{pmatrix}$ by

$$\frac{d\mathbf{X}}{dt} = \begin{pmatrix} dx_1/dt \\ dx_2/dt \end{pmatrix}.$$

Thus, if $\mathbf{X}(t) = \begin{pmatrix} t^2 \\ t^3 + 1 \end{pmatrix}$, then $d\mathbf{X}/dt = \begin{pmatrix} 2t \\ 3t^2 \end{pmatrix}$, while if $\mathbf{X}(t) = \begin{pmatrix} \cos t \\ \sin t \end{pmatrix}$, then $d\mathbf{X}/dt = \begin{pmatrix} -\sin t \\ \cos t \end{pmatrix}$. Note that $d\mathbf{X}/dt$ is again a vector-valued function.

Exercise 1. Fix a vector $\mathbf{Y}$. Define a vector-valued function $t \rightarrow \mathbf{Y}(t)$ by setting $\mathbf{Y}(t) = t^n \mathbf{Y}$. Show that

$$\frac{d\mathbf{Y}}{dt} = nt^{n-1}\mathbf{Y}.$$

Now let the scalar-valued functions $t \rightarrow x(t)$, $t \rightarrow y(t)$ be a solution of the differential system (3). In vector form, we can write

$$\begin{pmatrix} dx/dt \\ dy/dt \end{pmatrix} = \begin{pmatrix} ax + by \\ cx + dy \end{pmatrix}. \tag{4}$$

We define the vector-valued function $t \rightarrow \mathbf{X}(t)$ by $\mathbf{X}(t) = \begin{pmatrix} x(t) \\ y(t) \end{pmatrix}$. Then the left-hand side of (4) is $d\mathbf{X}/dt$, and the right-hand side of (4) is

$$\begin{pmatrix} ax + by \\ cx + dy \end{pmatrix} = \begin{pmatrix} a & b \\ c & d \end{pmatrix}\begin{pmatrix} x \\ y \end{pmatrix} = \begin{pmatrix} a & b \\ c & d \end{pmatrix}(\mathbf{X}(t)).$$

Thus (4) may be written in the form

$$\frac{d\mathbf{X}}{dt} = \begin{pmatrix} a & b \\ c & d \end{pmatrix}\mathbf{X}(t). \tag{5}$$

How shall we solve Eq. (5) for $\mathbf{X}(t)$? Recall that letting a matrix $\begin{pmatrix} a & b \\ c & d \end{pmatrix}$ act on a vector $\mathbf{X}$ to give the vector $\begin{pmatrix} a & b \\ c & d \end{pmatrix}\mathbf{X}$ is analogous to multiplying a number x by a scalar a to give the number ax. So Eq. (5) is analogous to the equation

$$\frac{dx}{dt} = ax, \tag{6}$$

where x is now a scalar-valued function of t and a is a given scalar. We know how to solve Eq. (6). The solutions have the form

$$x(t) = Ce^{at},$$

where C is a constant. Setting $t = 0$, we get $x(0) = C$, so

$$x(t) = x(0)e^{at} = e^{ta}(x(0)),$$

where we have changed the order of multiplication with malice afore-thought. Let us look for a solution to Eq. (5) by looking for an analogue of $e^{ta}(x(0))$. We take

$$\mathbf{X}(t) = e^{tm}(\mathbf{X}(0)), \qquad \text{with} \quad m = \begin{pmatrix} a & b \\ c & d \end{pmatrix}. \tag{7}$$

First we must define the *exponential of a matrix*. In §1 we shall define, given a matrix m, a matrix to be denoted e^m or $\exp(m)$ and to be called the *exponential of m*.

Applying the matrix e^{tm} to a fixed vector $\mathbf{X}(0)$, we then obtain a vector for each t, and thus we get the vector-valued function $t \to \mathbf{X}(t)$ defined in (7). We shall then show that $\mathbf{X}(t)$ solves (5).

In what follows we shall use the symbol I for *the matrix* $\begin{pmatrix} 1 & 0 \\ 0 & 1 \end{pmatrix}$, which is properly denoted $m(I)$. This simplifies the formulas, and should cause no confusion.

§1. The Exponential of a Matrix

Let $m = \begin{pmatrix} a & b \\ c & d \end{pmatrix}$ be a matrix. Since we have defined addition and multipli-cation of matrices, we can write expressions such as m^2 or $m^3 - 3m + I$. We interpret $m^3 - 3m + I$ as the result of applying the polynomial

$P(x) = x^3 - 3x + 1$ to the matrix m:

$$P(m) = m^3 - 3m + I.$$

More generally, if $Q(x)$ is the polynomial

$$Q(x) = c_n x^n + c_{n-1} x^{n-1} + \cdots + c_1 x + c_0,$$

where $c_n, c_{n-1}, \ldots, c_1, c_0$ are scalars, we set

$$Q(m) = c_n m^n + c_{n-1} m^{n-1} + \cdots + c_1 m + c_0 I,$$

and we regard $Q(m)$ as the matrix obtained by *applying the polynomial Q to the matrix m.*

We now replace the polynomial Q by the *exponential function* $\exp(x)$. We know that $\exp(x)$ is given by an infinite series

$$\exp(x) = 1 + x + \frac{x^2}{2!} + \cdots + \frac{x^n}{n!} + \cdots, \tag{8}$$

where the series converges for every number x. We wish to apply the function $\exp(x)$ to the matrix m. We *define*

$$\exp(m) = I + m + \frac{m^2}{2!} + \cdots + \frac{m^n}{n!} + \cdots. \tag{9}$$

An infinite series is understood as a *limit*. Thus, Eq. (8) means that the sequence of numbers

$$1, 1 + x, 1 + x + \frac{x^2}{2!}, \ldots, 1 + x + \frac{x^2}{2!} + \cdots + \frac{x^n}{n!}, \ldots$$

converges to the limit $\exp(x)$ as $n \to \infty$. Similarly, we interpret Eq. (9) to say that $\exp(m)$ is defined as the *limit of the sequence of matrices*

$$I, I + m, I + m + \frac{m^2}{2!}, \ldots, I + m + \frac{m^2}{2!} + \cdots + \frac{m^n}{n!}, \ldots.$$

Of course, $\exp(m)$ is then itself a matrix.

EXAMPLE 2. Let m be the diagonal matrix

$$m = \begin{pmatrix} s & 0 \\ 0 & t \end{pmatrix},$$

where s, t are scalar. What is $\exp(m)$? Recall the formula for $\begin{pmatrix} s & 0 \\ 0 & t \end{pmatrix}^n$ found in Chapter 2.7.

$$1 + m + \frac{m^2}{2!} + \cdots + \frac{m^n}{n!}$$

$$= \begin{pmatrix} 1 & 0 \\ 0 & 1 \end{pmatrix} + \begin{pmatrix} s & 0 \\ 0 & t \end{pmatrix} + \frac{1}{2!} \begin{pmatrix} s^2 & 0 \\ 0 & t^2 \end{pmatrix} + \frac{1}{3!} \begin{pmatrix} s^3 & 0 \\ 0 & t^3 \end{pmatrix} + \cdots + \frac{1}{n!} \begin{pmatrix} s^n & 0 \\ 0 & t^n \end{pmatrix}$$

$$= \begin{pmatrix} 1 & 0 \\ 0 & 1 \end{pmatrix} + \begin{pmatrix} s & 0 \\ 0 & t \end{pmatrix} + \begin{pmatrix} s^2/2! & 0 \\ 0 & t^2/2! \end{pmatrix}$$

$$+ \begin{pmatrix} s^3/3! & 0 \\ 0 & t^3/3! \end{pmatrix} + \cdots + \begin{pmatrix} s^n/n! & 0 \\ 0 & t^n/n! \end{pmatrix}$$

$$= \begin{pmatrix} 1 + s + s^2/2! + s^3/3! + \cdots + s^n/n! & 0 \\ 0 & 1 + t + t^2/2! + t^3/3! + \cdots + t^n/n! \end{pmatrix}.$$

Letting $n \to \infty$, we find that

$$\exp(m) = \lim_{n \to \infty} \left(I + m + \frac{m^2}{2!} + \cdots + \frac{m^n}{n!} \right)$$

$$= \begin{bmatrix} \lim_{n \to \infty} (1 + s + \cdots + s^n/n!) & 0 \\ 0 & \lim_{n \to \infty} (1 + t + \cdots + t^n/n!) \end{bmatrix} = \begin{bmatrix} e^s & 0 \\ 0 & e^t \end{bmatrix}.$$

Thus,

$$\exp \left[\begin{pmatrix} s & 0 \\ 0 & t \end{pmatrix} \right] = \begin{pmatrix} e^s & 0 \\ 0 & e^t \end{pmatrix}.$$

EXAMPLE 3. Find $\exp \left[\begin{pmatrix} 0 & 1 \\ 0 & 0 \end{pmatrix} \right]$.

$$\begin{pmatrix} 0 & 1 \\ 0 & 0 \end{pmatrix}^2 = \begin{pmatrix} 0 & 1 \\ 0 & 0 \end{pmatrix} \begin{pmatrix} 0 & 1 \\ 0 & 0 \end{pmatrix} = \begin{pmatrix} 0 & 0 \\ 0 & 0 \end{pmatrix}.$$

Hence, $\begin{pmatrix} 0 & 1 \\ 0 & 0 \end{pmatrix}^n = 0$ for $n = 2, 3, \ldots$. So

$$\exp \left[\begin{pmatrix} 0 & 1 \\ 0 & 0 \end{pmatrix} \right] = \begin{pmatrix} 1 & 0 \\ 0 & 1 \end{pmatrix} + \begin{pmatrix} 0 & 1 \\ 0 & 0 \end{pmatrix} = \begin{pmatrix} 1 & 1 \\ 0 & 1 \end{pmatrix}.$$

Let A be a linear transformation. We define $\exp(A)$ as the linear transformation whose matrix is $\exp[m(A)]$.

EXAMPLE 4. Let $R_{\pi/2}$ be rotation by $\pi/2$. Find $\exp(R_{\pi/2})$.

Set $m = m(R_{\pi/2}) = \begin{pmatrix} 0 & -1 \\ 1 & 0 \end{pmatrix}$.

$$m^2 = \begin{pmatrix} 0 & -1 \\ 1 & 0 \end{pmatrix} \begin{pmatrix} 0 & -1 \\ 1 & 0 \end{pmatrix} = \begin{pmatrix} -1 & 0 \\ 0 & -1 \end{pmatrix} = (-1)I.$$

Hence, for every positive integer k,

$$m^{2k} = ((-1)I)^k = (-1)^k I^k = (-1)^k I,$$

and so

$$m^{2k+1} = m^{2k} m = ((-1)^k I) m = (-1)^k m.$$

So

$$m^3 = (-1)m, \qquad m^4 = I, \qquad m^5 = m, \qquad m^6 = (-1)I, \qquad m^7 = (-1)m,$$

and so on. Hence,

$$\exp(m) = I + m + \frac{1}{2!}(-1)I + \frac{1}{3!}(-1)m + \frac{1}{4!}I$$

$$+ \frac{1}{5!}m + \frac{1}{6!}(-1)I + \frac{1}{7!}(-1)m + \cdots$$

$$= \left(1 - \frac{1}{2!} + \frac{1}{4!} - \frac{1}{6!} + \cdots\right)I + \left(1 - \frac{1}{3!} + \frac{1}{5!} - \frac{1}{7!} + \cdots\right)m.$$

We can simplify this formula by recalling that

$$\cos x = 1 - \frac{x^2}{2!} + \frac{x^4}{4!} - \frac{x^6}{6!} + \cdots,$$

and

$$\sin x = x - \frac{x^3}{3!} + \frac{x^5}{5!} - \frac{x^7}{7!} + \cdots,$$

so

$$\cos 1 = 1 - \frac{1}{2!} + \frac{1}{4!} - \frac{1}{6!} + \cdots,$$

and

$$\sin 1 = 1 - \frac{1}{3!} + \frac{1}{5!} - \frac{1}{7!} + \cdots.$$

So

$$\exp(m) = (\cos 1)I + (\sin 1)m = (\cos 1)\begin{pmatrix} 1 & 0 \\ 0 & 1 \end{pmatrix} + (\sin 1)\begin{pmatrix} 0 & -1 \\ 1 & 0 \end{pmatrix}$$

$$= \begin{pmatrix} \cos 1 & -\sin 1 \\ \sin 1 & \cos 1 \end{pmatrix}.$$

So $\exp(R_{\pi/2})$ is the linear transformation whose matrix is

$$\begin{pmatrix} \cos 1 & -\sin 1 \\ \sin 1 & \cos 1 \end{pmatrix}.$$

Exercise 2. Fix a scalar t and consider the matrix $\begin{pmatrix} 0 & -t \\ t & 0 \end{pmatrix}$. Show that

$$\exp\left[\begin{pmatrix} 0 & -t \\ t & 0 \end{pmatrix}\right] = \begin{pmatrix} \cos t & -\sin t \\ \sin t & \cos t \end{pmatrix}. \tag{10}$$

Exercise 3. Set $m = \begin{pmatrix} 1 & 1 \\ 1 & 1 \end{pmatrix}$.

(i) Calculate m^k for $k = 2, 3, 4, \ldots$.
(ii) Calculate $\exp(m)$ and simplify.

Exercise 4. Set $m = \begin{pmatrix} 3 & 1 \\ 0 & 3 \end{pmatrix}$.

(i) Calculate m^k for $k = 2, 3, 4, \ldots$.

(ii) Calculate $\exp(m)$ and simplify.

In Chapter 2.7, we considered a linear transformation A having eigenvalues t_1, t_2 with $t_1 \neq t_2$ and eigenvectors $\mathbf{X}_1 = \begin{pmatrix} x_1 \\ y_1 \end{pmatrix}$ and $\mathbf{X}_2 = \begin{pmatrix} x_2 \\ y_2 \end{pmatrix}$. We defined linear transformations P and D with

$$m(P) = \begin{pmatrix} x_1 & x_2 \\ y_1 & y_2 \end{pmatrix}, \qquad m(D) = \begin{pmatrix} t_1 & 0 \\ 0 & t_2 \end{pmatrix},$$

and we showed, in formula (5) of Chapter 2.7, that

$$(m(A))^n = m(P) \begin{pmatrix} t_1^n & 0 \\ 0 & t_2^n \end{pmatrix} m(P^{-1}), \qquad n = 1, 2, 3, \ldots .$$

It follows that

$$\exp(m(A)) = I + m(A) + \frac{1}{2!}(m(A))^2 + \cdots$$

$$= I + m(P) \begin{pmatrix} t_1 & 0 \\ 0 & t_2 \end{pmatrix} m(P^{-1}) + \frac{m(P)}{2!} \begin{pmatrix} t_1^2 & 0 \\ 0 & t_2^2 \end{pmatrix} m(P^{-1}) + \cdots$$

$$= m(P) \left[I + \begin{pmatrix} t_1 & 0 \\ 0 & t_2 \end{pmatrix} + \frac{1}{2!} \begin{pmatrix} t_1^2 & 0 \\ 0 & t_2^2 \end{pmatrix} + \cdots \right] m(P^{-1})$$

(where we have used that $m(P) \cdot m(P^{-1}) = I$)

$$= m(P) \begin{bmatrix} 1 + t_1 + (1/2!)t_1^2 + \cdots & 0 \\ 0 & 1 + t_2 + (1/2!)t_2^2 + \cdots \end{bmatrix} m(P^{-1})$$

$$= m(P) \begin{pmatrix} e^{t_1} & 0 \\ 0 & e^{t_2} \end{pmatrix} m(P^{-1}).$$

Thus, we have shown:

Theorem 8.1.

$$\exp(m(A)) = m(P) \begin{pmatrix} e^{t_1} & 0 \\ 0 & e^{t_2} \end{pmatrix} m(P^{-1}). \tag{11}$$

EXAMPLE 5. Calculate $\exp\left[\begin{pmatrix} 3 & 4 \\ 4 & -3 \end{pmatrix} \right]$.

Here

$$t_1 = 5, \qquad t_2 = -5, \qquad \mathbf{X}_1 = \begin{pmatrix} 2 \\ 1 \end{pmatrix}, \qquad \mathbf{X}_2 = \begin{pmatrix} -1 \\ 2 \end{pmatrix}.$$

So

$$m(P) = \begin{pmatrix} 2 & -1 \\ 1 & 2 \end{pmatrix}, \qquad m(P^{-1}) = \begin{pmatrix} 2/5 & 1/5 \\ -1/5 & 2/5 \end{pmatrix}.$$

By (11), we have

$$\exp\left[\begin{pmatrix} 3 & 4 \\ 4 & -3 \end{pmatrix}\right] = \begin{pmatrix} 2 & -1 \\ 1 & 2 \end{pmatrix} \begin{pmatrix} e^5 & 0 \\ 0 & e^{-5} \end{pmatrix} \begin{pmatrix} 2/5 & 1/5 \\ -1/5 & 2/5 \end{pmatrix}$$

$$= \begin{pmatrix} 2 & -1 \\ 1 & 2 \end{pmatrix} \begin{pmatrix} (2/5\ )e^5 & (1/5\ )e^5 \\ -(1/5\ )e^{-5} & (2/5\ )e^{-5} \end{pmatrix}$$

$$= \begin{pmatrix} (4/5\ )e^5 + (1/5\ )e^{-5} & (2/5\ )e^5 - (2/5\ )e^{-5} \\ (2/5\ )e^5 - (2/5\ )e^{-5} & (1/5\ )e^5 + (4/5\ )e^{-5} \end{pmatrix}.$$

EXAMPLE 6. Calculate $\exp\left[\begin{pmatrix} 1 & 0 \\ 0 & 0 \end{pmatrix}\right]$.

Since $\begin{vmatrix} 1-t & 0 \\ 0 & -t \end{vmatrix} = t^2 - t = t(t-1)$, the eigenvalues are $t_1 = 1$, $t_2 = 0$.
The corresponding eigenvectors are $\mathbf{X}_1 = \begin{pmatrix} 1 \\ 0 \end{pmatrix}$, $\mathbf{X}_2 = \begin{pmatrix} 0 \\ 1 \end{pmatrix}$. So $m(P) = \begin{pmatrix} 1 & 0 \\ 0 & 1 \end{pmatrix} = I$, and then $m(P^{-1}) = I$. Hence, by (11),

$$\exp\left[\begin{pmatrix} 1 & 0 \\ 0 & 0 \end{pmatrix}\right] = I\begin{pmatrix} e & 0 \\ 0 & 1 \end{pmatrix}I = \begin{pmatrix} e & 0 \\ 0 & 1 \end{pmatrix}.$$

Exercise 5.

(a) Compute $\begin{pmatrix} 1 & 0 \\ 0 & 0 \end{pmatrix}^n$ for $n = 1, 2, 3, \ldots$.

(b) Compute $\exp\left[\begin{pmatrix} 1 & 0 \\ 0 & 0 \end{pmatrix}\right]$ directly from the definition and compare your answer with the result of Example 6.

Exercise 6. Using Theorem 8.1 calculate $\exp\left[\begin{pmatrix} 1 & 3 \\ 3 & -1 \end{pmatrix}\right]$.

Exercise 7. Calculate $\exp\left[\begin{pmatrix} 3 & 0 \\ 4 & 2 \end{pmatrix}\right]$.

Exercise 8. Calculate $\exp\left[\begin{pmatrix} 1 & 2 \\ 1 & 1 \end{pmatrix}\right]$.

Recall Eq. (5): $d\mathbf{X}/dt = m(\mathbf{X}(t))$, where $m = \begin{pmatrix} a & b \\ c & d \end{pmatrix}$. We fix a vector $\mathbf{X}$ and define $\mathbf{X}(t) = \exp(tm)(\mathbf{X}_0)$. In §2, we shall show that $\mathbf{X}(t)$ solves (5) and satisfies the initial condition $\mathbf{X}(0) = \mathbf{X}_0$, and we shall study examples and applications.

§2. Solutions of Differential Systems

We fix a matrix m and a vector $\mathbf{X}_0$.

$$\exp(tm) = I + tm + \frac{t^2}{2!}m^2 + \frac{t^3 m^3}{3!} + \cdots,$$

so

$$(\exp(tm))(\mathbf{X}_0) = \mathbf{X}_0 + tm(\mathbf{X}_0) + \frac{t^2}{2!}m^2(\mathbf{X}_0) + \frac{t^3}{3!}m^3(\mathbf{X}_0) + \cdots.$$

Both sides of the last equation are vector-valued functions of t. It can be shown that the derivative of the sum of the infinite series is obtained by differentiating the series term by term. In other words,

$$\frac{d}{dt}\{(\exp(tm))(\mathbf{X}_0)\} = \frac{d}{dt}(tm(\mathbf{X}_0)) + \frac{d}{dt}\left(\frac{t^2}{2!}m^2(\mathbf{X}_0)\right) + \cdots. \quad (12)$$

The right-hand side of (12) is equal to

$$m(\mathbf{X}_0) + \frac{2t}{2!}m^2(\mathbf{X}_0) + \frac{3t^2}{3!}m^3(\mathbf{X}_0) + \frac{4t^3}{4!}m^4(\mathbf{X}_0) + \cdots$$

$$= m(\mathbf{X}_0) + tm^2(\mathbf{X}_0) + \frac{t^2}{2!}m^3(\mathbf{X}_0) + \frac{t^3}{3!}m^4(\mathbf{X}_0) + \cdots$$

$$= m(\mathbf{X}_0) + m(tm(\mathbf{X}_0)) + m\left(\frac{t^2}{2!}m^2(\mathbf{X}_0)\right) + m\left(\frac{t^3}{3!}m^3(\mathbf{X}_0)\right) + \cdots$$

$$= m\left\{\mathbf{X}_0 + tm(\mathbf{X}_0) + \frac{t^2}{2!}m^2(\mathbf{X}_0) + \frac{t^3}{3!}m^3(\mathbf{X}_0) + \cdots\right\}$$

$$= m\{(\exp(tm))(\mathbf{X}_0)\}.$$

So (12) gives us

$$\frac{d}{dt}\{(\exp(tm))(\mathbf{X}_0)\} = m\{(\exp(tm))(\mathbf{X}_0)\}. \quad (13)$$

We define $\mathbf{X}(t) = (\exp(tm))(\mathbf{X}_0))$. Then (13) states that

$$\frac{d\mathbf{X}}{dt}(t) = m(\mathbf{X}(t)). \quad (14)$$

In other words, we have shown that $\mathbf{X}(t)$ solves our original equation (5). Also, setting $t = 0$ in the definition of $\mathbf{X}(t)$, we find that

$$\mathbf{X}(0) = I(\mathbf{X}_0) = \mathbf{X}_0, \quad (15)$$

since $\exp(0) = I + 0 + 0 + \cdots = I$. So we have proved:

Theorem 8.2. *Let m be a matrix. Fix a vector $\mathbf{X}_0$. Set $\mathbf{X}(t) = (\exp(tm))(\mathbf{X}_0)$ for all t. Then,*

$$\frac{d\mathbf{X}}{dt} = m\mathbf{X}(t), \quad (16)$$

and

$$\mathbf{X}(0) = \mathbf{X}_0. \tag{17}$$

EXAMPLE 7. Solve the differential system

$$\begin{cases} \dfrac{dx}{dt} = -y, \\[2mm] \dfrac{dy}{dt} = x, \end{cases} \tag{18}$$

with the initial condition: $x(0) = 1$, $y(0) = 0$.

In vector form, with $\mathbf{X}(t) = \begin{pmatrix} x(t) \\ y(t) \end{pmatrix}$, we have

$$\frac{d\mathbf{X}}{dt} = \begin{pmatrix} 0 & -1 \\ 1 & 0 \end{pmatrix} (\mathbf{X})$$

with initial condition $\mathbf{X}(0) = \begin{pmatrix} 1 \\ 0 \end{pmatrix}$. Set $m = \begin{pmatrix} 0 & -1 \\ 1 & 0 \end{pmatrix}$, $\mathbf{X}_0 = \begin{pmatrix} 1 \\ 0 \end{pmatrix}$ and set

$$\mathbf{X}(t) = \exp(tm)(\mathbf{X}_0).$$

By Exercise 2 in this chapter,

$$\exp(tm) = \exp\left[\begin{pmatrix} 0 & -t \\ t & 0 \end{pmatrix} \right] = \begin{pmatrix} \cos t & -\sin t \\ \sin t & \cos t \end{pmatrix}.$$

So

$$\mathbf{X}(t) = \begin{pmatrix} \cos t & -\sin t \\ \sin t & \cos t \end{pmatrix} \begin{pmatrix} 1 \\ 0 \end{pmatrix} = \begin{pmatrix} \cos t \\ \sin t \end{pmatrix}.$$

Since $\mathbf{X}(t) = \begin{pmatrix} x(t) \\ y(t) \end{pmatrix}$, we obtain $x(t) = \cos t$, $y(t) = \sin t$. Inserting these functions in (18), we see that it checks. Also, $x(0) = 1$, $y(0) = 0$, so the initial condition checks also.

EXAMPLE 8. Solve the differential system (18) with initial condition $x(0) = x_0$, $y(0) = y_0$.

We take $\mathbf{X}_0 = \begin{pmatrix} x_0 \\ y_0 \end{pmatrix}$ and set

$$\mathbf{X}(t) = \left(\exp\begin{bmatrix} 0 & -t \\ t & 0 \end{bmatrix} \right)(\mathbf{X}_0) = \begin{pmatrix} \cos t & -\sin t \\ \sin t & \cos t \end{pmatrix} \begin{pmatrix} x_0 \\ y_0 \end{pmatrix},$$

so

$$\mathbf{X}(t) = \begin{pmatrix} (\cos t)x_0 - (\sin t)y_0 \\ (\sin t)x_0 + (\cos t)y_0 \end{pmatrix}.$$

So

$$x(t) = (\cos t)x_0 - (\sin t)y_0, \qquad y(t) = (\sin t)x_0 + (\cos t)y_0.$$

We check that these functions satisfy (18) and that $x(0) = x_0$, $y(0) = y_0$.

Exercise 9. Calculate $\exp\left[\begin{pmatrix} 3t & 4t \\ 4t & -3t \end{pmatrix}\right]$, where t is a given number.

Exercise 10. Using the result of Exercise 9, solve the system

$$\begin{cases} \dfrac{dx}{dt} = 3x + 4y \\[2mm] \dfrac{dy}{dt} = 4x - 3y \end{cases} \quad \text{with} \quad \begin{cases} x(0) = 1 \\[2mm] y(0) = 0 \end{cases}, \tag{19}$$

by using Theorem 8.2 with $m = \begin{pmatrix} 3 & 4 \\ 4 & -3 \end{pmatrix}$ and $\mathbf{X}_0 = \begin{pmatrix} 1 \\ 0 \end{pmatrix}$.

Exercise 11. Solve the system (19) with $x(0) = s_1$, $y(0) = s_2$.

Exercise 12. Solve the system

$$\begin{cases} \dfrac{dx}{dt} = 2x + 4y \\[2mm] \dfrac{dy}{dt} = 4x + 6y \end{cases} \quad \text{with} \quad \begin{cases} x(0) = 1 \\[2mm] y(0) = 0 \end{cases}. \tag{20}$$

Exercise 13. Solve the system (1) (at the beginning of this chapter) with initial condition (2).

EXAMPLE 9. Consider an electric circuit consisting of a condenser of capacitance C connected to a resistance of R ohms and an inductance of L henries. A switch is inserted in the circuit (see Fig. 8.2). The condenser is charged with a charge of Q_0 coulombs, with the switch open. At time $t = 0$, the switch is closed and the condenser begins to discharge, causing a current to flow in the circuit. Denote by $i(t)$ the current flowing at time t and by $Q(t)$ the charge on the condenser at time t. The laws of electricity tell us the following: the voltage drop at time t equals $(1/C)Q(t)$ across the condenser, while the voltage drop across the resistance is $Ri(t)$ and the voltage drop across the inductance is $L(di/dt)$. The sum of all the voltage drops equals 0 at every time $t > 0$, since the circuit is closed. Thus, we have

$$\frac{1}{C}Q(t) + Ri(t) + L\frac{di}{dt} = 0$$

or

$$\frac{di}{dt} = -\frac{1}{LC}Q(t) - \frac{R}{L}i(t).$$

Also, the current at time t equals the negative of dQ/dt or $i(t) = -dQ/dt$. So the two functions: $t \to i(t)$ and $t \to Q(t)$ satisfy

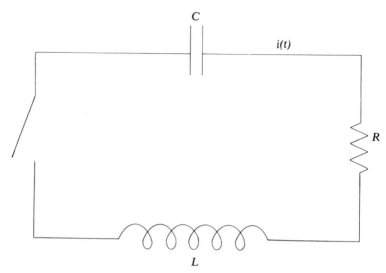

Figure 8.2

$$\begin{cases} \dfrac{di}{dt} = ai + bQ, \\ \dfrac{dQ}{dt} = -i, \end{cases} \tag{21}$$

where $a = -R/L$, $b = -1/LC$. So to calculate the current flowing in the circuit at any time t, we must solve the differential system (21) with initial condition $Q(0) = Q_0$, $i(0) = 0$.

EXAMPLE 10. Let c_1, c_2 be two scalars. We wish to solve the *second-order differential equation*

$$\frac{d^2x}{dt^2} + c_1 \frac{dx}{dt} + c_2 x = 0 \tag{22}$$

by a function $t \to x(t)$ defined for all t, and we want to satisfy the initial conditions

$$x(0) = x_0, \qquad \frac{dx}{dt}(0) = y_0. \tag{23}$$

We shall reduce the problem (22) to a *first-order differential system* of the form (3). To this end we define $y(t) = (dx/dt)(t)$. Then (22) can be written: $dy/dt + c_1 y + c_2 x = 0$ or

$$\frac{dy}{dt} = -c_2 x - c_1 y.$$

So x and y satisfy the differential system

$$\begin{cases} \dfrac{dx}{dt} = y, \\[2mm] \dfrac{dy}{dt} = -c_2 x - c_1 y. \end{cases} \tag{24}$$

EXAMPLE 11. We study the equation

$$\dfrac{d^2 x}{dt^2} + x = 0, \qquad x(0) = x_0, \qquad \dfrac{dx}{dt}(0) = y_0. \tag{25}$$

Setting $y = dx/dt$, (25) turns into

$$\begin{cases} \dfrac{dx}{dt} = y, \\[2mm] \dfrac{dy}{dt} = -x, \end{cases} \qquad x(0) = x_0, \quad y(0) = y_0. \tag{26}$$

Exercise 14. Fix a scalar t. Show that

$$\exp\!\left[t \begin{pmatrix} 0 & 1 \\ -1 & 0 \end{pmatrix} \right] = \begin{pmatrix} \cos t & \sin t \\ -\sin t & \cos t \end{pmatrix}.$$

Exercise 15. Using the result of Exercise 14, solve first the equations (26) and then Eq. (25).

In Theorem 8.2, we showed that the problem $d\mathbf{X}/dt = m\mathbf{X}(t)$, $\mathbf{X}(0) = \mathbf{X}_0$ has $\mathbf{X}(t) = (\exp(tm)(\mathbf{X}_0))$ as a solution for all t. We shall now show that this is the *only* solution, or, in other words, we shall prove uniqueness of solutions.

Suppose $\mathbf{X}$, $\mathbf{Y}$ are two solutions. Then $d\mathbf{X}/dt = m\mathbf{X}(t)$, $\mathbf{X}(0) = \mathbf{X}_0$ and $d\mathbf{Y}/dt = m\mathbf{Y}(t)$, $\mathbf{Y}(0) = \mathbf{X}_0$. Set $\mathbf{Z}(t) = \mathbf{X}(t) - \mathbf{Y}(t)$. Our aim is to prove that $\mathbf{Z}(t) = 0$ for all t. We have

$$\dfrac{d\mathbf{Z}}{dt} = \dfrac{d\mathbf{X}}{dt} - \dfrac{d\mathbf{Y}}{dt} = m\mathbf{X}(t) - m\mathbf{Y}(t)$$

$$= m(\mathbf{X}(t) - \mathbf{Y}(t)) = m\mathbf{Z}(t). \tag{27}$$

Also

$$\mathbf{Z}(0) = \mathbf{X}(0) - \mathbf{Y}(0) = \mathbf{X}_0 - \mathbf{X}_0 = 0. \tag{28}$$

We now shall use (27) and (28) to show that $\mathbf{Z}(t) = \mathbf{0}$ for all t. We denote by $f(t)$ the squared length of $\mathbf{Z}(t)$, i.e.,

$$f(t) = |\mathbf{Z}(t)|^2.$$

$t \to f(t)$ is a scalar-valued function. It satisfies

$$f(t) \geqslant 0 \qquad \text{for all } t \quad \text{and} \quad f(0) = 0.$$

Exercise 16. If $\mathbf{A}(t)$, $\mathbf{B}(t)$ are two vector-valued functions, then

$$\frac{d}{dt}(\mathbf{A}(t) \cdot \mathbf{B}(t)) = \mathbf{A}(t) \cdot \frac{d\mathbf{B}}{dt} + \mathbf{B}(t) \cdot \frac{d\mathbf{A}}{dt}.$$

It follows from Exercise 16 that

$$\frac{df}{dt} = \frac{d}{dt}(\mathbf{Z}(t) \cdot \mathbf{Z}(t)) = \mathbf{Z}(t) \cdot \frac{d\mathbf{Z}}{dt} + \mathbf{Z}(t) \cdot \frac{d\mathbf{Z}}{dt} = 2\mathbf{Z}(t) \cdot \frac{d\mathbf{Z}}{dt}.$$

Using (27), this gives

$$\frac{df}{dt}(t) = 2\mathbf{Z}(t) \cdot m\mathbf{Z}(t). \tag{29}$$

We set $m = \begin{pmatrix} a & b \\ c & d \end{pmatrix}$. Fix t and set $\mathbf{Z}(t) = \mathbf{Z} = \begin{pmatrix} z_1 \\ z_2 \end{pmatrix}$. Then

$$2\mathbf{Z}(t) \cdot m\mathbf{Z}(t) = 2\begin{pmatrix} z_1 \\ z_2 \end{pmatrix} \cdot \left\{ \begin{pmatrix} a & b \\ c & d \end{pmatrix}\begin{pmatrix} z_1 \\ z_2 \end{pmatrix} \right\}$$

$$= 2\begin{pmatrix} z_1 \\ z_2 \end{pmatrix} \cdot \begin{pmatrix} az_1 + bz_2 \\ cz_1 + dz_2 \end{pmatrix}$$

$$= 2(az_1^2 + bz_1z_2 + cz_2z_1 + dz_2^2).$$

Let K be a constant greater than $|a|$, $|b|$, $|c|$, $|d|$. Then

$$|2\mathbf{Z}(t) \cdot m\mathbf{Z}(t)| \leqslant 2(|a|z_1^2 + |b||z_1||z_2| + |c||z_2||z_1| + |d||z_2|^2)$$

$$\leqslant 2K(|z_1|^2 + 2|z_1||z_2| + |z_2|^2).$$

Also,

$$2|z_1||z_2| \leqslant |z_1|^2 + |z_2|^2.$$

So

$$|2\mathbf{Z}(t) \cdot m\mathbf{Z}(t)| \leqslant 2K(2|z_1|^2 + 2|z_2|^2) = 4K(|z_1|^2 + |z_2|^2)$$

$$= 4K|\mathbf{Z}|^2 = 4Kf(t).$$

By (29), setting $M = 4K$, this gives

$$\frac{df}{dt}(t) \leqslant Mf(t). \tag{30}$$

Consider the derivative

$$\frac{d}{dt}\left(\frac{f(t)}{e^{Mt}}\right) = \frac{e^{Mt}(df/dt) - f(t)Me^{M(t)}}{e^{2Mt}} = \frac{(df/dt) - Mf(t)}{e^{Mt}}.$$

By (30), the numerator of the right-hand term $\leqslant 0$ for all t. So

$$\frac{d}{dt}\left(\frac{f(t)}{e^{Mt}}\right) \leqslant 0,$$

so $f(t)/e^{Mt}$ is a decreasing function of t. Also, $f(t)/e^{Mt} \geqslant 0$ and $= 0$ at $t = 0$. But a decreasing function of t, defined on $t \geqslant 0$ which is $\geqslant 0$ for all t and $= 0$ at $t = 0$, is identically 0.

So $f(t)/e^{Mt} = 0$ for all t. Thus $|\mathbf{Z}(t)|^2 = f(t) = 0$, and so $\mathbf{Z}(t) = 0$, and so $\mathbf{X}(t) = \mathbf{Y}(t)$ for all t.

We have proved:

Uniqueness Property. *The only solution of the problem considered in Theorem 8.2 is* $\mathbf{X}(t) = (\exp(tm))(\mathbf{X}_0)$.

§3. Three-Dimensional Differential Systems

Just as in the case of two-dimensional matrices, we can define polynomials in a 3×3 (or an $n \times n$) matrix, and we can take a limit to form the exponential of a matrix.

EXAMPLE 12. If m is the 3×3 diagonal matrix with diagonal entries a, b, c, then m^n is the diagonal matrix with diagonal entries a^n, b^n, c^n and $\exp(m)$ is the diagonal matrix with diagonal entries $\exp(a)$, $\exp(b)$, and $\exp(c)$.

EXAMPLE 13. If m is a 3×3 upper triangular matrix, with 0 on the diagonal or below, then m^2 has 0 except in the upper-right-hand corner, and $m^3 = 0$. Thus, $\exp(m) = I + m + m^2/2$.

As in the two-dimensional case, the calculation of polynomials and of exponentials of a matrix is greatly simplified if the matrix is diagonalized. If A is a linear transformation from $\mathbb{R}^3$ to $\mathbb{R}^3$ with three distinct eigenvalues t_1, t_2, t_3, then $A = PDP^{-1}$ for some invertible matrix P with columns given by the eigenvectors corresponding to the eigenvalues which are the diagonal entries of the diagonal matrix D. Then, $[m(A)]^n = m(P)[m(D)]^n m(P^{-1})$ for all positive integers n, so $\exp(m(A)) = m(P)m(\exp(D))m(P^{-1})$.

The same method enables us to calculate the exponential of any matrix representing a transformation A, such that $\mathbb{R}^3$ has a basis consisting of eigenvectors of A.

In the case of a vector function $\mathbf{X}(t)$ in $\mathbb{R}^3$, we may solve the differential system $d\mathbf{X}/dt = m\mathbf{X}(t)$ just as we did in the two-dimensional case. If the initial condition is $\mathbf{X}(0) = \mathbf{X}_0$, then the solution of the system is $\mathbf{X}(t) = (\exp(tm))(\mathbf{X}_0)$. The method of proof used in the two-dimensional case can be used to show that this solution is unique.

EXAMPLE 14. The third-order ordinary differential equation $x'''(t) + ax''(t) + bx'(t) + cx(t) = 0$ can be expressed as a system of three first-order equations by setting $y(t) = x'(t)$ and $z(t) = x''(t)$. The single equation is equivalent to the system

$$x'(t) = y(t)$$
$$y'(t) = z(t)$$
$$z'(t) = -cx(t) - by(t) - az(t).$$

This allows us to solve a third-order differential equation by solving a differential system.

Least Squares Approximation

We consider two variable quantities x and y. We make n simultaneous measurements of both quantities and obtain n pairs of values: x_1, y_1, x_2, y_2, ..., x_n, y_n, and we call the corresponding points (x_1, y_1), ..., (x_n, y_n) in the xy plane the *data points*.

In certain situations, when we plot the data points on a piece of graph paper, we discover that they almost lie on a straight line.

This leads us to expect that there is a relation

$$y = ax + b \tag{1}$$

between the two quantities, where a and b are certain constants. This relation may be exact in theory, but the data points (x_i, y_i) fail to lie exactly on one straight line because of small errors of measurement. We now ask, what is the best choice of a and b to give the correct relation (1)?

We should choose the line L with equation $y = ax + b$ in such a way that the total "deviation" of the points (x_i, y_i) from L is as small as possible. Let us fix a line $y = mx + b$ and let (x_j, z_j) be the point on this line with x coordinate x_j, so that $z_j = mx_j + b, j = 1, ..., n$.

Two possible measures of the deviation of the data points from L are the sums

$$\sum_{j=1}^{n} |y_j - z_j| \quad \text{and} \quad \sum_{j=1}^{n} |y_j - z_j|^2.$$

It turns out that the second expression is easier to deal with. So we define the deviation of the data points from the line L by

$$D(m, b) = \sum_{j=1}^{n} |y_j - z_j|^2 = \sum_{j=1}^{n} (y_j - (mx_j + b))^2. \tag{2}$$

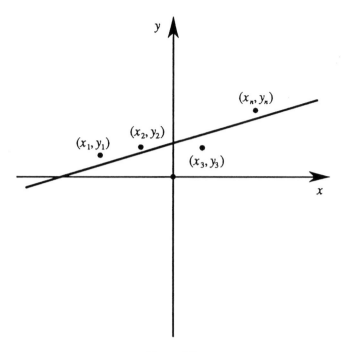

Figure 8.3

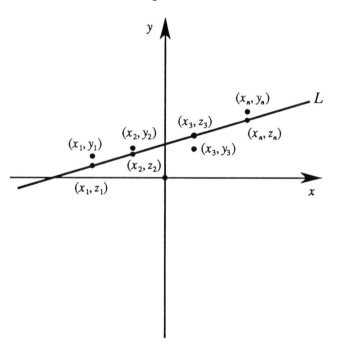

Figure 8.4

We wish to minimize $D(m, b)$, i.e., to find scalars $\bar{m}, \bar{b}$ such that $D(\bar{m}, \bar{b}) \leqslant D(m, b)$ for all m, b. The line

$$\bar{L} : y = \bar{m}x + \bar{b}$$

then is the line of least deviation for our data points. We may use the line $\bar{L}$ to predict the results of future measurements of the quantities x and y.

We may interpret the expression $D(m, b)$ geometrically. Let

$$\mathbf{X} = \begin{bmatrix} x_1 \\ x_2 \\ \vdots \\ x_n \end{bmatrix}, \quad \mathbf{Y} = \begin{bmatrix} y_1 \\ y_2 \\ \vdots \\ y_n \end{bmatrix}, \quad \mathbf{1} = \begin{bmatrix} 1 \\ 1 \\ \vdots \\ 1 \end{bmatrix}.$$

Then $\mathbf{X}, \mathbf{Y}, \mathbf{1}$ are vectors in $\mathbb{R}^n$ and the squared length

$$|\mathbf{Y} - (m\mathbf{X} + b\mathbf{1})|^2 = \sum_{i=1}^{n} (y_i - (mx_i + b))^2.$$

So

$$|\mathbf{Y} - (m\mathbf{X} + b\mathbf{1})|^2 = D(m, b).$$

The totality of all vectors $m\mathbf{X} + b\mathbf{1}$ in $\mathbb{R}^n$ with m, b in $\mathbb{R}$ is a 2-dimensional subspace of $\mathbb{R}^n$, which we denote by Π. The pair of vectors $\mathbf{X}, \mathbf{1}$ is a basis of Π.

If $\bar{m}, \bar{b}$ are the scalars that minimize the deviation D, then the distance from $\mathbf{Y}$ to $\bar{m}\mathbf{X} + \bar{b}\mathbf{1}$ is smaller than the distance from $\mathbf{Y}$ to any other point $m\mathbf{X} + b\mathbf{1}$ in Π. So $\bar{m}\mathbf{X} + \bar{b}\mathbf{1}$ is the nearest point to $\mathbf{Y}$ in Π. It follows that $\mathbf{Y} - (\bar{m}\mathbf{X} + \bar{b}\mathbf{1})$ is perpendicular to Π. In particular, then, we have

$$(\mathbf{Y} - (\bar{m}\mathbf{X} + \bar{b}\mathbf{1}), \mathbf{X}) = 0,$$

and

$$(\mathbf{Y} - (\bar{m}\mathbf{X} + \bar{b}\mathbf{1}), \mathbf{1}) = 0.$$

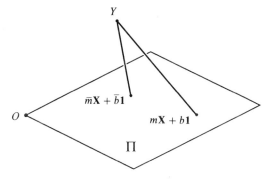

Figure 8.5

So we have

$$\begin{cases} (\mathbf{Y}, \mathbf{X}) = \bar{m}(\mathbf{X}, \mathbf{X}) + \bar{b}(\mathbf{1}, \mathbf{X}), \\ (\mathbf{Y}, \mathbf{1}) = \bar{m}(\mathbf{X}, \mathbf{1}) + \bar{b}(\mathbf{1}, \mathbf{1}). \end{cases} \tag{3}$$

We now solve the system (3).

$$(\mathbf{X}, \mathbf{1})(\mathbf{Y}, \mathbf{X}) = \bar{m}(\mathbf{X}, \mathbf{1})(\mathbf{X}, \mathbf{X}) + \bar{b}(\mathbf{X}, \mathbf{1})(\mathbf{1}, \mathbf{X}),$$

$$(\mathbf{X}, \mathbf{X})(\mathbf{Y}, \mathbf{1}) = \bar{m}(\mathbf{X}, \mathbf{X})(\mathbf{X}, \mathbf{1}) + \bar{b}(\mathbf{X}, \mathbf{X})(\mathbf{1}, \mathbf{1}).$$

So

$$\frac{(\mathbf{X}, \mathbf{1})(\mathbf{Y}, \mathbf{X}) - (\mathbf{X}, \mathbf{X})(\mathbf{Y}, \mathbf{1})}{(\mathbf{X}, \mathbf{1})^2 - (\mathbf{X}, \mathbf{X})(\mathbf{1}, \mathbf{1})} = \bar{b}. \tag{4a}$$

Also,

$$(\mathbf{1}, \mathbf{1})(\mathbf{Y}, \mathbf{X}) = \bar{m}(\mathbf{1}, \mathbf{1})(\mathbf{X}, \mathbf{X}) + \bar{b}(\mathbf{1}, \mathbf{1})(\mathbf{1}, \mathbf{X}),$$

$$(\mathbf{1}, \mathbf{X})(\mathbf{Y}, \mathbf{1}) = \bar{m}(\mathbf{1}, \mathbf{X})(\mathbf{X}, \mathbf{1}) + \bar{b}(\mathbf{1}, \mathbf{X})(\mathbf{1}, \mathbf{1}).$$

So

$$\frac{(\mathbf{1}, \mathbf{1})(\mathbf{Y}, \mathbf{X}) - (\mathbf{1}, \mathbf{X})(\mathbf{Y}, \mathbf{1})}{(\mathbf{1}, \mathbf{1})(\mathbf{X}, \mathbf{X}) - (\mathbf{1}, \mathbf{X})(\mathbf{X}, \mathbf{1})} = \bar{m}. \tag{4b}$$

Expressed in terms of x_i, y_i, this gives

$$\bar{b} = \frac{(\sum x_i)(\sum x_i y_i) - (\sum x_i^2)(\sum y_i)}{(\sum x_i)^2 - n(\sum x_i^2)}, \tag{5a}$$

$$\bar{m} = \frac{n \sum x_i y_i - (\sum x_i)(\sum y_i)}{n(\sum x_i^2) - (\sum x_i)^2}. \tag{5b}$$

Here, each sum is taken from $i = 1$ to $i = n$. Finally, the line $\bar{L}$, which gives the least deviation from our data points, is given by:

$$\bar{L} : y = \bar{m}x + \bar{b},$$

where $\bar{m}$, $\bar{b}$ are given by (5a) and (5b).

Exercise 1. Find an equation for the line of least deviation for the data points: $(x_1, y_1) = (1, 1)$, $(x_2, y_2) = (2, 2)$ $(x_3, y_3) = (3, 4)$.

Exercise 2. The weight (in ounces) w and the age (in months) t of lobsters from a certain area are believed to be approximately related by the formula: $w = mt + b$, where m and b are certain constants. Lobsters are caught and t and w are determined, giving

$$t = 4, \quad w = 7 \qquad t = 16, \quad w = 11$$

$$t = 8, \quad w = 9 \qquad t = 18, \quad w = 11.$$

(a) Find the straight line of least deviation for these data points.
(b) Predict the weight of a two-year-old lobster.

Exercise 3. Let $(x_1, y_1), \ldots, (x_n y_n)$ be n data points, such that

$$\sum_{i=1}^{n} x_i = 0 \quad \text{and} \quad \sum_{i=1}^{n} y_i = 0.$$

Show that the line $\bar{L}$ of least deviation passes through the origin, and find its slope.

Exercise 4. Let $(x_1, y_1), \ldots, (x_n, y_n)$ be n arbitrary data points. Put

$$\bar{x} = (1/n) \sum_{i=1}^{n} x_i, \; \bar{y} = (1/n) \sum_{i=1}^{n} y_i.$$

Show that the line $\bar{L}$ of least deviation for these data points passes through the point $(\bar{x}, \bar{y})$.

CHAPTER 8.2

Curvature of Function Graphs

One of the nicest applications of linear algebra to geometry comes in the study of total curvature of graphs of functions of two variables. We shall see how a very natural construction leads to a self-adjoint linear transformation, the eigenvalues and eigenvectors of which have particular geometric significance.

Consider a function $f(x, y)$ of two variables, and consider its graph in three-dimensional space. This graph is a surface with equation: $z = f(x, y)$. We will assume that $f(0, 0) = 0$, so the graph goes through the origin. We can always arrange this by translating the graph in a vertical direction. Furthermore, we will assume that the tangent plane to the graph at the origin is the horizontal plane, so the partial derivatives $f_x(0, 0)$ and $f_y(0, 0)$ are both zero. In order to investigate the way in which the graph is curved in a neighborhood of the origin, we may consider slices of the graph by vertical planes through the origin.

One such plane is the x–z coordinate plane with equation: $y = 0$. The portion of the graph lying in this plane is the curve of points $(x, 0, f(x, 0))$. We may determine whether this curve is concave up or concave down by computing the second derivative $f_{xx}(x, 0)$. If this value is positive for $x = 0$, the curve is concave upwards at the origin, and if it is negative, concave downwards. Similarly, the portion of the surface lying in the y–z coordinate plane is given by putting $x = 0$, so we get a curve of points $(0, y, f(0, y))$, which is concave up or down depending on whether the second derivative $f_{yy}(0, y)$ is positive or negative at 0.

For example, if $f(x, y) = ax^2 + by^2$, then $f_{xx}(0, 0) = 2a$ and $f_{yy}(0, 0) = 2b$, so these slice curves are concave up or down in a neighborhood of the origin depending on whether a and b are positive or negative.

It may happen that this way of determining the shape of the surface is

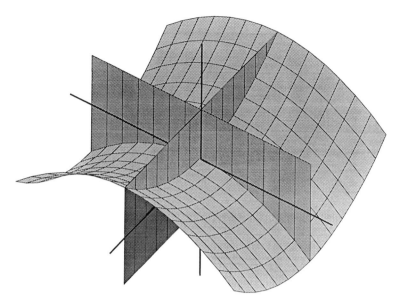

Figure 8.6

inconclusive, for example, if $f(x, y) = 2xy$, then $f(x, 0) = 0$ and $f(0, y) = 0$, so $f_{xx}(x, 0) = 0$ and $f_{yy}(0, y) = 0$, for all x and y. However, in this case, the slice over the line $x = y$ is a parabola pointing up, since $f(x, x) = x^2$, while over the line $x = -y$, the slice curve is a parabola pointing down, since $f(x, -x) = -x^2$. This surface is shown in Figure 8.7.

How much information do we need in order to determine the shape of the surface? To answer this question, we first recall the formula for the curvature of the curve: $z = z(t)$ in the plane. The curvature is defined to be $z''(t)/(1 + z'(t)^2)^{3/2}$ so at $t = 0$, this curvature is $z''(0)/(1 + z'(0)^2)^{3/2} = z''(0)$, provided that $z'(0) = 0$.

Returning to our graph $z = f(x, y)$, we now consider the slice curve above the line $x = t \cos \phi$, $y = t \sin \phi$, with slope $\tan \phi$. Here ϕ is a fixed angle, and t is the variable parameter. The z coordinate $z(t) = f(t \cos \phi, t \sin \phi)$, and so, by the chain rule, $z'(t) = f_x(t \cos \phi, t \sin \phi) \cos \phi + f_y(t \cos \phi, t \sin \phi) \sin \phi$. Note that $z'(0) = 0$, since $f_x(0, 0) = 0 = f_y(0, 0)$. Then $z''(t) = f_{xx}(t \cos \phi, t \sin \phi) \cos^2 \phi + 2f_{xy}(t \cos \phi, t \sin \phi) \cos \phi \sin \phi + f_{yy}(t \cos \phi, t \sin \phi) \sin^2 \phi$. So we have

$$z''(0) = f_{xx}(0, 0) \cos^2 \phi + 2f_{xy}(0, 0) \cos \phi \sin \phi + f_{yy}(0, 0) \sin^2 \phi. \quad (1)$$

Thus it is that the curvature of the slice above the line with slope $\tan \phi$ is a quadratic expression in $\cos \phi$ and $\sin \phi$, with coefficients that are partial derivatives of the function $f(x, y)$ evaluated at the origin.

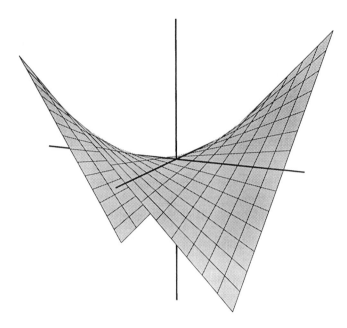

Figure 8.7

Definition. The *normal curvature of the graph* $z = f(x, y)$ *in the direction* ϕ is the curvature of the slice curve over the line with slope $\tan \phi$ at the origin. It is denoted $k(\phi)$.

The normal curvature $k(\phi) = z''(0)$ and, hence, is given by formula (1). We can get a geometrically more enlightening formula for $k(\phi)$ as follows:

We denote by Π the xy plane, $z = 0$, and on this plane we consider the linear transformation A with matrix

$$\begin{pmatrix} f_{xx}(0, 0) & f_{xy}(0, 0) \\ f_{xy}(0, 0) & f_{yy}(0, 0) \end{pmatrix},$$

called the *Hessian* of f at $(0, 0)$. Fix an angle ϕ and denote by $\mathbf{X}$ the vector $\begin{pmatrix} \cos \phi \\ \sin \phi \end{pmatrix}$ in Π. Then

$$\mathbf{AX} \cdot \mathbf{X} = \begin{pmatrix} f_{xx}(0, 0) & f_{xy}(0, 0) \\ f_{xy}(0, 0) & f_{yy}(0, 0) \end{pmatrix} \begin{pmatrix} \cos \phi \\ \sin \phi \end{pmatrix} \cdot \begin{pmatrix} \cos \phi \\ \sin \phi \end{pmatrix}$$

$$= f_{xx}(0, 0) \cos^2 \phi + 2f_{xy}(0, 0) \cos \phi \sin \phi + f_{yy}(0, 0) \sin^2 \phi.$$

By (1) it follows that

$$\mathbf{AX} \cdot \mathbf{X} = k(\phi). \qquad (2)$$

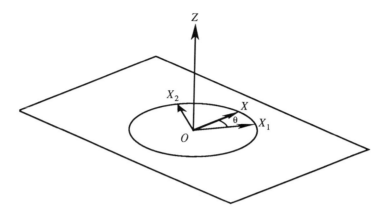

Figure 8.8

Now the matrix of A is symmetric, so by the results in Chapter 2.6, A has unit eigenvectors $\mathbf{X}_1$, $\mathbf{X}_2$ in Π with $\mathbf{X}_1 \cdot \mathbf{X}_2 = 0$. Denoting the corresponding eigenvalues by k_1, k_2, then,

$$AX_1 = k_1 X_1 \quad \text{and} \quad AX_2 = k_2 X_2.$$

We choose the indices so that $k_1 \geqslant k_2$.

Now let $\mathbf{X} = (\cos \phi, \sin \phi)$ be a unit vector in Π and let θ denote the angle measured counterclockwise in Π, from $\mathbf{X}_1$ to $\mathbf{X}$.

Then $\mathbf{X} = (\cos \theta)\mathbf{X}_1 + (\sin \theta)\mathbf{X}_2$, and so, $AX \cdot X = ((\cos \theta)k_1 X_1 + (\sin \theta)k_2 X_2) \cdot ((\cos \theta)X_1 + (\sin \theta)X_2)$, or $AX \cdot X = \cos^2 \theta k_1 + \sin^2 \theta k_2$. Formula (2) leads to *Euler's formula*:

$$k(\phi) = (\cos^2 \theta)k_1 + (\sin^2 \theta)k_2. \tag{3}$$

We choose the angle θ_0 so that $\mathbf{X}_1 = (\cos \theta_0, \sin \theta_0)$. Then $\phi = \theta_0 + \theta$.

When $\theta = 0$, and so $\phi = \theta_0$, then $k(\phi) = k_1$ and when $\theta = \dfrac{\pi}{2}$ and so $\phi = \theta_0 + \dfrac{\pi}{2}$, then $k(\phi) = k_2$ (see Figure 8.9).

Exercise 1. Show that the normal curvature always lies between the values k_2 and k_1, that is, that

$$k_2 \leqslant k(\phi) \leqslant k_1 \quad \text{for all } \phi.$$

The two extreme normal curvatures k_1 and k_2 are called the *principal curvatures* and the corresponding directions: $\phi = \theta_0$ and $\phi = \theta_0 + \dfrac{\pi}{2}$ are called the *principal directions* of the surface $z = f(x, y)$ at the origin. The principal directions, thus, lie along the eigenvectors $\mathbf{X}_1$ and $\mathbf{X}_2$.

Note: If $k_1 = k_2$, then all vectors $\mathbf{X}$ in the tangent plane are eigenvectors.

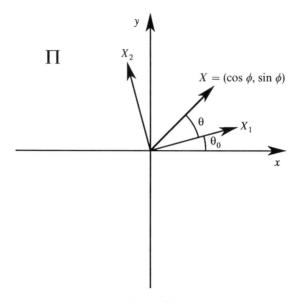

Figure 8.9

EXAMPLE 1. Let $\sum$ denote the cylinder: $x^2 + (z - 1)^2 = 1$ with axis along the y axis and radius 1. $\sum$ passes through the origin and its tangent plane at the origin is the horizontal plane. We shall calculate the principal directions and the principal curvatures for $\sum$ at $(0, 0, 0)$. We write $\sum$, near the origin, as the graph

$$z = 1 - \sqrt{1 - x^2} = f(x, y).$$

Then $f_{yy} = f_{xy} = 0$, and $f_x = \dfrac{x}{\sqrt{1 - x^2}}$. So $f_x = f_y = 0$ at $(0, 0)$. Also

$$f_{xx} = \frac{1}{(1 - x^2)^{3/2}},$$

so $f_{xx}(0, 0) = 1$. So

$$\begin{pmatrix} f_{xx}(0, 0) & f_{xy}(0, 0) \\ f_{xy}(0, 0) & f_{yy}(0, 0) \end{pmatrix} = \begin{pmatrix} 1 & 0 \\ 0 & 0 \end{pmatrix}.$$

The eigenvalues of this matrix are 0 and 1, so $k_1 = 1$ and $k_2 = 0$. The corresponding eigenvectors are $(0, 1)$ for k_2 and $(1, 0)$, for k_1. Thus the principal directions in this case are along the coordinate axes.

Exercise 2. Find the normal curvature $k(\phi)$ in Example 1 when $\phi = \dfrac{\pi}{4}$ and sketch the corresponding slice curve.

Exercise 3. Let Σ be the sphere: $x^2 + y^2 + (z - a)^2 = a^2$.

(a) Find the principal curvatures of Σ at the origin.
(b) Calculate $k(\phi)$ for an arbitrary angle ϕ.

Exercise 4. Let Σ be the surface: $z = (\sin x)(\sin y)$. Find the principal curvatures and the principal directions for Σ at the origin.

.

Index

Undergraduate Texts in Mathematics